Data-Rate-Constrained State Estimation and Control of Complex Networked Systems

This book presents research developments and novel methodologies on data-rate-constrained control and state estimation for complex networked systems with different kinds of encoding-decoding mechanisms. It describes framework of state estimator and controller design, stability and performance analysis for data-rate constrained complex systems with various kinds of encoding-decoding schemes and so forth. Simulations given in this book are constructed by applying MATLAB® software package.

Features:

- Gives a systematic investigation of the control and state estimation for complex networked systems subject to the data rate constraint.
- Develops control/filtering algorithms in a unified framework.
- Includes comparisons for different coding-decoding techniques proposed.
- Discusses theoretical value and practical application for the resource-constrained communication environment.
- Provides performance analysis as well as the parameterizations of filters and FD units.

This book is aimed at researchers and graduate students in electrical engineering, signal processing, control systems and complex networks.

Data-Rate-Constrained State Estimation and Control of Complex Networked Systems

Licheng Wang, Zidong Wang, and Guoliang Wei

CRC Press
Taylor & Francis Group
Boca Raton London New York

CRC Press is an imprint of the
Taylor & Francis Group, an **informa** business

Designed cover image: Shutterstock

First edition published 2025
by CRC Press
2385 NW Executive Center Drive, Suite 320, Boca Raton FL 33431

and by CRC Press
4 Park Square, Milton Park, Abingdon, Oxon, OX14 4RN

CRC Press is an imprint of Taylor & Francis Group, LLC

© 2025 Licheng Wang, Zidong Wang, and Guoliang Wei

ISBN: 978-1-032-64542-1 (hbk)
ISBN: 978-1-032-87862-1 (pbk)
ISBN: 978-1-003-53485-3 (ebk)

DOI: 10.1201/9781003534853

Typeset in Palatino
by SPi Technologies India Pvt Ltd (Straive)

To my supervisor and my family.

Contents

List of Figures..x
List of Tables...xiii
Preface..xiv
Foreword..xvi
Acknowledgment...xvii
Contributors...xviii
Symbols..xix

1. Introduction...1
    1.1   Theoretical Outlines..2
    1.2   Main Encoding-Decoding Schemes in NSs 5
    1.3   Encoding-Decoding-Based Control and Filtering Problems......... 13
    1.4   Distributed Control and State Estimation Problems
          with Limited Communication Capacity..21
    1.5   Outline...25

**2. Gain-Scheduled State Estimation for Discrete-Time Complex
Networks under Bit-Rate Constraints** 29
    2.1   Problem Formulation...30
          2.1.1   System Model..30
          2.1.2   Measurement Model Subject to Bit-Rate Constraint........... 31
          2.1.3   Bit-Rate-Constrained Encoding-Decoding Scheme........... 31
    2.2   Main Results...35
    2.3   An Illustrative Example..41
    2.4   Summary...45

**3. Partial-Neurons-Based State Estimation for Artificial Neural
Networks under Constrained Bit Rate: The Finite-Time Case**.................46
    3.1   Problem Formulation...47
          3.1.1   Encoding-Decoding Mechanism under Bit
                 Rate Constraint...49
          3.1.2   Estimator Structure 51
    3.2   Main Results...52
    3.3   An Illustrative Example..59
    3.4   Summary...64

4. Synchronization Control for a Class of Discrete-Time Dynamical Networks with Packet Dropouts: A Coding-Decoding-Based Approach 65

   4.1   Problem Formulation 66

   4.2   Main Results 70

       4.2.1   Preliminaries 70

       4.2.2   State Estimator Design 72

       4.2.3   Detectability Analysis 74

       4.2.4   Synchronization Control 80

   4.3   An Illustrative Example 83

   4.4   Summary 87

5. Observer-Based Consensus Control for Discrete-Time Multi-Agent Systems with Coding-Decoding Communication Protocol 88

   5.1   Problem Formulation 89

       5.1.1   Traditional Observer Structure 90

       5.1.2   CDCP-Based Observer Structure 90

   5.2   Design of CDCP and Performance Analysis of MASs 96

       5.2.1   Coding-Decoding Communication Protocol 96

       5.2.2   Consensus Analysis 101

   5.3   An Illustrative Example 107

   5.4   Summary 109

6. Recursive Filtering with Measurement Fading: A Multiple Description Coding Scheme 110

   6.1   Problem Formulation 111

       6.1.1   Channel Fading 112

       6.1.2   Multiple Description Coding Scheme 113

       6.1.3   The Filter Structure 115

   6.2   Design of the MDC and Analysis of the Decoding Error 116

       6.2.1   Design of the MDC 116

       6.2.2   Analysis of the Decoding Error 118

       6.2.3   A Unified Measurement Model 121

   6.3   The Recursive Filtering Strategy 122

   6.4   An Illustrative Example 131

   6.5   Summary 140

7. Stabilization of Linear Discrete-Time Systems over Resource-Constrained Networks under Dynamical Multiple Description Coding Scheme 141

   7.1   Problem Formulation 142

   7.2   Main Results 145

       7.2.1   Decoding Scheme 148

       7.2.2   Design of the Encoder-Decoder Pair 149

7.3   Illustrative Examples 161
7.4   Summary 164

**8. An Event-Triggered Encoding Approach to Control of Linear
      Systems under Bit Rate Conditions** 166
8.1   Problem Formulation 168
8.2   Circumventing the "Zeno" Phenomenon 170
8.3   Convergence Analysis of $\tilde{x}_a(t)$ 174
8.4   A Sufficient Condition of the Bit Rate to Guarantee the
       Convergence of the Decoding Error 174
8.5   A Necessary Condition of the Bit Rate 175
8.6   Analysis and Synthesis of the Addressed Event-Based
       Control Issue 176
8.7   Illustrative Examples 180
8.8   Summary 184

**9. Event-Based State Estimation under Constrained Bit Rate:
      An Encoding-Decoding Approach** 185
9.1   Problem Formulation 186
       9.1.1   System Model 186
       9.1.2   Event-Based Encoder-Decoder Design 187
       9.1.3   State Estimator Structure 190
9.2   Main Results 191
       9.2.1   Elimination of the Zeno Phenomenon 191
       9.2.2   The Boundedness Analysis of the Decoding Error 196
       9.2.3   The Design of the State Estimator 201
9.3   An Illustrative Example 204
9.4   Summary 210

10. Conclusions and Future Topics 211

Bibliography 213

Index 226

List of Figures

Figure 1.1 Structure of the general communication system. 3

Figure 1.2 Framework of data-rate-constrained control and state estimation issues. 5

Figure 1.3 The summarization of several typical encoding-decoding schemes in NCSs. 6

Figure 1.4 Multiple description encoding-based data communication in NSs. 11

Figure 1.5 NCSs with the encoding-decoding-based communication protocol. 14

Figure 1.6 Filtering system model with encoding-decoding-based data transmission scheme. 16

Figure 1.7 The architecture of distributed systems with encoding-decoding communication mechanism. 22

Figure 2.1 State trajectory $u_{i1}(t)$ $(i = 1, 2, 3)$ and its estimation. 42

Figure 2.2 State trajectory $u_{i2}(t)$ $(= 1, 2, 3)$ and its estimation. 43

Figure 2.3 Measurement output $s_i(t)$ $(i = 1, 2, 3)$ and its decoded value. 43

Figure 2.4 MSE of node i $(i = 1, 2, 3)$. 44

Figure 2.5 Decoding error $\tilde{s}_i(t)$ $(i = 1, 2, 3)$. 44

Figure 2.6 The time-varying probability $p(t)$. 45

Figure 3.1 State trajectory $u_1(t)$ and its estimation. 60

Figure 3.2 State trajectory $u_2(t)$ and its estimation. 61

Figure 3.3 State trajectory $u_3(t)$ and its estimation. 61

Figure 3.4 Estimation error of neuron i $(i = 1, 2, 3)$. 62

Figure 3.5 The measurement output $s_1(t)$ of the first neuron and its decoded value. 62

Figure 3.6 The measurement output $s_2(t)$ of the second neuron and its decoded value. 63

Figure 3.7 Decoding error $\tilde{s}_i(t)$ $(i = 1, 2)$. 63

Figure 4.1 State trajectories $x_{i1}(k)$ and their decoded values $\hat{x}_{i1}(k)$ $(i = 1, 2, 3)$. 84

Figure 4.2 State trajectories $x_{i2}(k)$ and their decoded values $\hat{x}_{i2}(k)$ $(i = 1, 2, 3)$. 85

Figure 4.3 Decoding errors $w_{ij}(k)$ $(i = 1, 2, 3; j = 1, 2)$. 85

Figure 4.4 State trajectories $x_{ij}(k)$ of the uncontrolled nodes i $(i = 1, 2, 3; j = 1, 2)$. 86

Figure 4.5 State trajectories $x_{ij}(k)$ of the controlled nodes i $(i = 1, 2, 3; j = 1, 2)$. 87

Figure 5.1 State trajectories $x_{ij}(k)$ $(i = 1, 2 \ldots, 5; j = 1, 2)$ with control input. 108

Figure 5.2 State trajectories $x_{ij}(k)$ $(i = 1, 2 \ldots, 5; j = 1, 2)$ without control input. 108

Figure 5.3 The decoding errors $\vec{\xi}_{ij}(k)$ $(i = 1, 2 \ldots, 5; j = 1, 2)$. 109

Figure 6.1 Structure of two-description coding-decoding for NSs with unreliable channels. 113

Figure 6.2 Nested index assignment for $d = 8$ and $p = 1$ in [156]. 117

Figure 6.3 The diagram of three-tank system. 131

Figure 6.4 The fading measurement components y_{ik} $(i = 1, 2)$ and their decoded values $\bar{y}_{ik}$. 135

Figure 6.5 The actual state components x_{ik} $(i = 1, 2, 3)$ and their estimates $\hat{x}_{ik}$. 135

Figure 6.6 The FE of x_{1k} and its upper bound. 136

Figure 6.7 The FE of x_{2k} and its upper bound. 136

Figure 6.8 The FE of x_{3k} and its upper bound. 137

Figure 6.9 The upper bounds of filtering errors of x_{ik} $(i = 1, 2, 3)$ with single description and two descriptions. 137

Figure 6.10 The effect of fading channel parameter M on filtering performance. 138

Figure 6.11 The filtering errors of x_{ik} $(i = 1, 2, 3)$ and their upper bounds with different noise intensities. 138

Figure 6.12 The filtering errors of x_{ik} $(i = 1, 2, 3)$ and their upper bounds with different packet-arrival probabilities. 139

Figure 6.13 The effect of the model uncertainty on the filtering performance. 139

Figure 7.1 The structure of the closed-loop control system with the MDC scheme. 145

Figure 7.2 Nested index assignment for $l = 8, d = 1$. 147

Figure 7.3 State trajectories of the open-loop system. 162

Figure 7.4 State trajectories of the closed-loop system. 163

Figure 7.5 State trajectories and their decoded values. 163

Figure 7.6 The trajectories of decoding errors. 164

Figure 7.7 The Bernoulli sequences $\gamma_i(\tau h)$ $(i = 1, 2)$. 165

Figure 8.1 The schematic of the event-based control with encoding-decoding scheme. 168

Figure 8.2 Dynamical evolutions of auxiliary system and decoder. 181

Figure 8.3 The decoding errors between the actual system states and their decoded values. 182

Figure 8.4 State response without control signal $u(t)$. 182

Figure 8.5 State response with control signal $u(t)$. 183

Figure 8.6 The threshold of the event generation function and the norm of $\tilde{x}_a(t)$. 183

Figure 9.1 Schematic of state estimation problem under constrained bit rate...191

Figure 9.2 Measurement $y(t)$ and its decoded value $y_d(t)$..........................205

Figure 9.3 The norm of the decoding error $\tilde{y}_d(t)$.............................206

Figure 9.4 The state trajectories $x_i(t)$ and their estimates $\hat{x}_i(t)$ $(i = 1, 2)$..206

Figure 9.5 The estimation errors $\tilde{x}_i(t)$ $(i = 1, 2)$.........................207

Figure 9.6 Triggering threshold $\sigma(t)$ and the norm of the decoding error $\|\tilde{y}_d(t)\|$. 207

Figure 9.7 The state trajectories $x_i(t)$ $(i = 1, 2, 3)$ and their estimates. 209

Figure 9.8 The estimation errors $\tilde{x}_i(t)$ $(i = 1, 2, 3)$. 209

List of Tables

Table 6.1 System Parameters 133
Table 9.1 Bit Rate $\mathcal{R}_0$ Under Different Threshold Values ϱ 210
Table 9.2 Bit Rate $\mathcal{R}_0$ Under Different Disturbance Intensities 210

Preface

With ever-rapid developments in sensing, computing, communication and integration technologies, the modern industrial systems possess high scalability and integrated architecture. In particular, compared with traditional industrial control systems where information interactions are implemented by a wire-based point-to-point way, modern networked control systems upgrade in such a data exchange manner to the shared-network-based one (e.g. field bus, IP/Ethernet, Bluetooth), which greatly enhances the reliability, efficiency and flexibility of the systems. On the other hand, the introduction of the communication network in industrial systems also brings in some issues, a representative one of which is the data rate-constraint. In practical applications, most of the systems suffer from limited date rate of communication network due either to physical constraint or to cost consideration, which gives rise to various undesirable network-induced phenomena. The data encoding-decoding scheme is one of the efficient approaches to deal with the scarcity issue of the communication resource and thus might avoid the performance deteriorations of data communication. The existing control/state estimation methods are designed without consideration of the date rate issue and this necessitates re-designing or developing new methodologies to accommodate the imbedding of different encoding-decoding mechanisms.

The objective of this book is to present the state-of-the-art research developments and novel methodologies on data-rate-constrained control and state estimation for complex networked systems with different kinds of encoding-decoding mechanisms. The contents of this book can be divided into two parts, where the first part (Chapters 2–5) addresses the data-constrained control/state estimation problems under uniform-quantization-based encoding-decoding scheme for complicated systems (e.g. neural networks, complex networks and multi-agent systems) and the second part (Chapters 6–9) introduces both the multiple description encoding mechanism and the event-based encoding protocol into the networked systems to cope with the data constraint issue. The work provides a framework of state estimator and controller design, stability and performance analysis for data-rate constrained complex systems with various kinds of encoding-decoding schemes, including the static/dynamic uniform-quantization-based scheme, the multiple description coding scheme, the sign-based coding scheme, etc. Some related techniques and theories are applied, which include the recursive Riccati-like equations, hybrid system theory, matrix theory, and mathematical optimization methods, to fulfill the specific analysis and synthesis issues. In addition, this book provides up-to-date reference materials for researchers who wish to explore the control and state estimation problems with constrained data rates.

The concise frame and description of the book are given as follows. Chapter 1 introduces the recent advances on data-rate-constrained control and state estimation problems for networked systems and the outline of the book. Chapter 2 investigates the state estimation for discrete-time complex networks subject to date-rate constraints. Chapter 3 is concerned with the finite-time state estimation problem for artificial neural networks under constrained bit rate, where only partial neurons measurement outputs are available for the state estimation task. Chapter 4 considers the synchronization control for a class of discrete-time dynamical networks with a uniform-based encoding-decoding method. Chapter 5 studies the observer-based consensus control for discrete-time multi-agent systems with coding-decoding communication protocol. Chapter 6 deals with the recursive filtering problem for time-varying systems where a static multiple description coding scheme is introduced to facilitate the data transmission. Chapter 7 addresses the stabilization issue of linear discrete-time systems over resource-constrained networks under dynamical multiple description coding scheme. In Chapter 8, an event-triggered encoding approach is employed to the control issue of linear systems under bit rate conditions. Chapter 9 discusses a sign-based coding approach to the state estimation problem under constrained bit rate. Chapter 10 gives the conclusion and some possible future research directions. Simulations given in this book are constructed by applying The MathWorks MATLAB software package.

This book is a research monograph whose intended audience is graduate and postgraduate students as well as researchers.

Licheng Wang
Shanghai, China

Foreword

I am delighted to introduce the book *Data-Rate-Constrained State Estimation and Control of Complex Networked Systems*. When I came to know about this book project undertaken by three of the most active researchers in the field, I was pleased that this book is coming in the early stage of a field that will need it more than most fields do. In most emerging research fields, a book can play a significant role in bringing some maturity to the field. Research fields advance through research papers. In research papers, however, only a limited perspective could be provided about the field, its application potential, and the techniques required and already developed in the field. A book gives such a chance. I liked the idea that there will be a book that will try to unify the field by bringing in disparate topics already available in several papers that are not easy to find and understand. I was supportive of this book project even before I had seen any material on it. The project was a brilliant and bold idea by three active researchers. Now that I have it on my screen, it appears to be even a better idea.

Data-rate-constrained state estimation and control started gaining recognition in the last century as a field. Signal processing, tracking, output regulation, communication and optimization technologies had advanced enough that researchers and technologists started developing different data encoding approaches such that the resulting control and state estimation performance is less influenced by the distortion of the transmitted data. By properly selecting encoding-decoding criterion to restrain the signal distortion, the data encoding-decoding approaches aspire to create a holistic picture for complex networked systems subject to limited data rate.

Since the data distortion caused by the finite bit rate of the communication media has a major impact on the overall system performance, how to well attenuate or even eliminate the negative effects is of significant importance. In addition, taking into account various encoding-decoding strategies, different encoding-decoding approaches result in different degrees of data distortion. Thus, the necessity (of researching state estimation and control theories/methodologies with various encoding-decoding strategies under a data-rate-constrained communication manner) arises, paving a way for people to hypothesize relationships among data sources and explore support for system monitoring/tracking.

Once successfully developed, such theories/methodologies will find vast applications in many diverse domains and keep getting more applications. In fact, many new filtering/control approaches are direct outgrowth of state estimation and control subject to constrained bit rate and it is likely to become a powerful system estimation/control tool.

Acknowledgment

The authors would like to express their deep appreciation to those who have been directly involved in various aspects of the research leading to this book. Special thanks go to Professor Bo Shen from Donghua University, Shanghai, China, Professor Fuad E. Alsaadi from King Abdulaziz University, Jeddah, Saudi Arabia, Professor Xiaohui Liu from Brunel University London, London, UK, Professor Derui Ding from University of Shanghai for Science and Technology, Shanghai, China, Professor Tingwen Huang from Texas A&M University at Qatar, Doha, Qatar, Professor Hongli Dong from Northeast Petroleum University, Daqing, China, and Professor Yurong Liu from Yangzhou University, Yangzhou, China.

The writing of this book was supported in part by the National Natural Science Foundation of China under Grants 62473247, 62473261, 62003213, 61903253, 61873148, 61933007, 61873169, 61873059 and 61973219, the China Postdoctoral Science Foundation under Grants 2019TQ0202 and 2020M671172, the National Postdoctoral Program for Innovative Talents under Grant BX20180202, the Shanghai Pujiang Program under Grant 19PJ1408100, the Royal Society of the UK and the Alexander von Humboldt Foundation of Germany.

Contributors

Bo Shen
Donghua University
Shanghai, China

Fuad E. Alsaadi
King Abdulaziz University
Jeddah, Saudi Arabia

Xiaohui Liu
Brunel University London
London, U.K.

Tingwen Huang
Texas A&M University
at Qatar
Doha, Qatar

Derui Ding
University of Shanghai for Science
and Technology
Shanghai, China

Hongli Dong
Northeast Petroleum University
Daqing, China

Shuai Liu
University of Shanghai for Science
and Technology
Shanghai, China

Yurong Liu
Yangzhou University
Yangzhou, China

Symbols

$\otimes$	The Kronecker product
$\mathbb{R}^n$	The n-dimensional Euclidean space
$\mathbb{R}^{n \times m}$	The set of all $n \times m$ real matrices
$\mathbb{R}^+$	The set of all positive real numbers
$\mathbb{N}$	The set of natural numbers
$\mathbb{S}^n_+$	The set of $n \times m$ positive definite matrices
A^T or A'	The transpose of matrix A
$A^\dagger$	The Moore-Penrose pseudo inverse of A
$A > 0$	The matrix A is positive definite
$A \geq 0$	The matrix A is positive semidefinite
$A < 0$	The matrix A is negative definite
$A \leq 0$	The matrix A is negative semidefinite
$\| \cdot \|$	The Euclidian norm of real vectors or the spectral norm of real matrices
$\| \cdot \|_{\max}$	The maximum singular value of a matrix
$\rho(A)$	The spectral radius of matrix A
$\delta(\cdot)$	The Kronecker delta function
$\mathrm{tr}(A)$	The trace of matrix A
$\|x\|_P^2$	Equals to $x^T P x$ when x is a vector
$\mathbb{P}\{\cdot\}$	The occurrence probability of the event "$\cdot$"
$\mathbb{E}\{x\}$	The expectation of stochastic variable x
$\mathrm{Var}\{x\}$	The variance of stochastic variable x
$\mathbb{E}\{x|y\}$	The conditional expectation of x given y
I	The identity matrix of compatible dimension

0 The zero matrix of compatible dimension

$\mathbf{0}_n$ The $n \times n$ zero matrix

$\mathbf{1}_n$ The $n \times 1$ column vector with all elements equal to 1

$vec\{x_1, x_2\}$ The column vector $\begin{bmatrix} x_1^T & x_2^T \end{bmatrix}^T$

$vec_n\{x_i\}$ The column vector $vec\{x_1^T, x_2^T, \cdots x_n^T\}$

$\text{diag}\{x_i\}$ The block diagonal matrix with ith block being x_i and all other entries being zero

$\text{diag}_n\{A_i\}$ The block diagonal matrix $\text{diag}\{A_1, A_2, \cdots, A_n\}$

$\{M_{ij}\}_{n \times n}$ The partitioned matrix with M_{ij} being (i, j)-th block submatrix

1

Introduction

It has been well recognized that the publication titled *A Mathematical Theory of Communication* [142,143] by C. E. Shannon in 1948 marked the establishment of the information theory. From then on, constant research effort from the researchers of different fields has been put on its developments and considerable progress has been made in the aspects of theory and application. Coincidentally, in the same year, another celebrated mathematician N. Wiener published his famous treatise *Cybernetics: Or Control and Communication in the Animal and the Machine* [169], which laid down the theoretical foundation for the servomechanisms (whether electrical, mechanical or hydraulic), automatic navigation, analog computing, artificial intelligence, neuroscience and reliable communications, and made the birth of control theory. Nevertheless, during the past few decades, the information theory and control theory have almost been regarded as two independent disciplines with few common features and connections due to their different concerns.

Broadly speaking, the information theory mainly focuses on how to obtain, store, process and transmit the data but seldom pays attention to the specific usage of the data itself. On the contrary, the control theory concerns how to make use of the obtained data to achieve the prescribed tasks through certain strategies. For general systems, the most basic but important idea to realize desirable system performances is the *feedback*, which means the obtained signals (usually the system outputs) should be fed back to the systems' inputs in certain manners by designing proper feedback mechanisms. However, in most cases, such an idea is usually implemented upon the assumption that the signal transmission between the components is with infinite precision, namely, the signals can be perfectly transmitted and utilized to contribute to the feedback schemes.

In recent years, owing to the rapid development of the networked communication technology, networked systems (NSs) have emerged as a new yet popular system architecture with an extensive utilization in the industry for its low power cost, simple installation, easy maintenance and high reliability. Different from the non-networked systems where the data flows through the components in a point-to-point way, the NSs mean that all the data communications between the components are executed

DOI: 10.1201/9781003534853-1

via a commonly shared communication network (channel). Although the NSs possess various merits mentioned above, the introduction of the communication network also brings about some new and noteworthy problems including 1) how the network-based data exchange affects the system performance, and 2) how to maximize the utilization of limited network bandwidth while ensuring the required system performance.

If the network environment is ideal, that is, the bandwidth of the network is sufficiently large, the assumption of the perfect data transmission would make sense and it is reasonable to ignore the influence from the network communication on the system performance. However, in situations that the network resources are subject to certain constraints (e.g. limited bandwidth of the channel), it is no longer proper to entirely neglect the distortion of the transmitted data caused by the limited communication capacity of the network. Especially, when it comes to large-scale NSs such as wireless sensor networks (WSNs), multi-agent systems (MASs) and smart grids, the limited network resources impose severe restrictions on the data communication quality and even deteriorate the overall system performances. On the other hand, from the standpoint of the network security, the NSs are usually fragile and vulnerable to be attacked, for example, the data flowing through the system components are easily stolen by the adversaries for some hostile purposes, see [20,50,96,183]. Therefore, in order to make the data transmission of the NSs more efficient and secure via a resource-constrained communication network, the encoding-decoding technique stands out as an ideal candidate for its advantage in data compression and encryption.

1.1 Theoretical Outlines

Normally, as shown in Figure 1.1, a classic communication system always contains three basic parts: the information source, the information channel and the information sink, based on which, the encoding-decoding technique can be accordingly divided into two main categories: the source encoding-decoding and the channel encoding-decoding. In this paper, our attention will be specifically focused on the source encoding-decoding technique, which aims to transform the original data into certain codewords. Generally speaking, the source information encoding technique possesses the following three prominent features: 1) realizing the data compression, 2) facilitating the digital data transmission and 3) enhancing the data security. From the aspect of mathematics, the process of encoding can be regarded as a mapping that maps the original data to certain codewords. Correspondingly, the decoding process can be viewed as the "inverse" operation which intends to recover the original data as exactly as possible by excavating the information

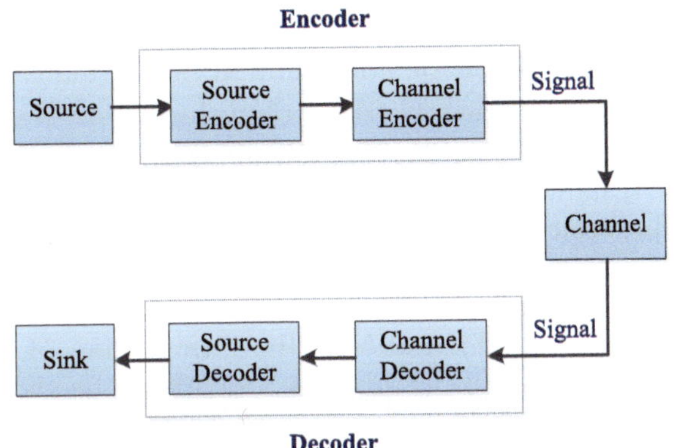

FIGURE 1.1
Structure of the general communication system.

contained in the codewords. To be more specific, in the networked communication systems, let $x(k)$, $\mathcal{F}(\cdot)$, $\mathcal{H}$, $\mathcal{G}(\cdot)$ and $x_d(k)$ be the original signal before being encoded, the encoding mapping, the encoding alphabet, the decoding mapping, and the decoded signal, respectively, where k is the discrete time step and the $\mathcal{H} \triangleq \{s_1, s_2, \ldots, s_m\}$ is a set whose elements s_i ($i = 1$, $2, \ldots, m$) are certain deterministic codewords. In most existing literature, the authors' attempts have always been made to design a pair of encoder-decoder appropriately such that at each time instant k, the decoding error $e_d(k) \triangleq x_d(k) - x(k)$ is possibly small at the minimum cost of the network bandwidth.

In view of the aforementioned discussion, it is obvious that there are two important performance indices to evaluate the designed encoding-decoding procedure, namely, the decoding error and the bandwidth condition. Intuitively, the large network bandwidth would result in a small decoding error. For the NSs, if the network resource is not the concern (e.g. the bandwidth can be sufficiently large), the decoding error would probably be very small and hence has little impact on the system performance. In this case, the coding-decoding-based NSs reduce to the traditional NSs for the analysis and synthesis issues. However, in most practical situations, due to the limited network resources, the network bandwidth serving as a major concern should be incorporated into the system design. In this sense, for the bandwidth-constrained systems, the concept of "minimum bit rate" is proposed to characterize the smallest needed channel capacity for the codeword transmission, above which the performance of the controlled (or filtering error) systems is explored including asymptotical/exponential stability, ultimately bounded stability, input-to-state stability (ISS), etc.

In the past few years, a rich body of research results with respect to the minimum bit-rate-based control/filtering problem for both linear and nonlinear NSs have appeared. Among them, it has been shown in [155] and [115] that for linear systems, the minimum bit rate R (bits/sample) should satisfy $R > \frac{1}{\log(2)} \sum_{\mathrm{Re}(\lambda_i(A))>0} (\lambda_i(A))$, where log is the base-$e$ logarithm, Re($\cdot$) denotes the real part of a complex-valued scalar, A is the system matrix, and $\lambda_i\{A\}$ stands for the ith eigenvalue of matrix A. Furthermore, the literature [123] revealed that in order to globally (or semiglobally) stabilize the $n-$dimensional nonlinear systems with a feedforward structure, the necessary bit rate should be $n + 1$ (or n).

Compared with the traditional NSs, the introduction of the encoding-decoding communication mechanism would make the dynamics of the NSs more complicated and diverse, which accordingly imposes certain challenges on the analysis and synthesis issues. Moreover, because the inevitable data distortion resulted from the encoding-decoding process, the traditional approaches cannot be directly generalized to cope with the performance analysis issues of data-rate-constrained NSs. In this case, the influence of the encoding-decoding mechanism on the system performance should be well taken into consideration. In other words, how to choose suitable theoretical tools to quantitatively evaluate the relationship between the system dynamics and the encoding-decoding communication protocol appears to be particularly important. To date, there are generally three theoretical methodologies to deal with the encoding-decoding-based analysis and synthesis problems for the NSs subject to data-rate constraint, namely, the information-theory-based methodology, the ISS-based methodology and the hybrid-system-based methodology.

The theoretical basis of the information theory methodology is the celebrated Shannon information theory, based on which, all information including the original data, codewords and decoded data are described as random variables [28,29,117]. Within such a framework, the main objective is to find general necessary conditions for the channel capacity measured by the concept of "entropy", under which the original data can be reconstructed from the decoded data with the acceptable distortion. Then, by resorting to the decoded data, the desired control or filtering performance can be guaranteed. On the other hand, for most of the effective encoding schemes, a basic requirement is that the decoding error $e_d(k)$ should be bounded or even convergent. Thus, it is natural to view the decoded error as the bounded disturbance of the dynamical system, and the ISS-based approach [87,124] has been proposed as a theoretical tool to handle the analysis and synthesis issues for these systems. For the continuous-time systems with a digital transmission manner, the received codewords by the decoder at a certain discrete instant can be seen as an impulse sequence that affects the decoder dynamics. Consequently, the hybrid-system-based approach [86,89] is valid for analyzing such a kind of systems' dynamical behaviors.

In this paper, our purpose is to offer an up-to-date review of existing research results on encoding-decoding-based control and filtering problems

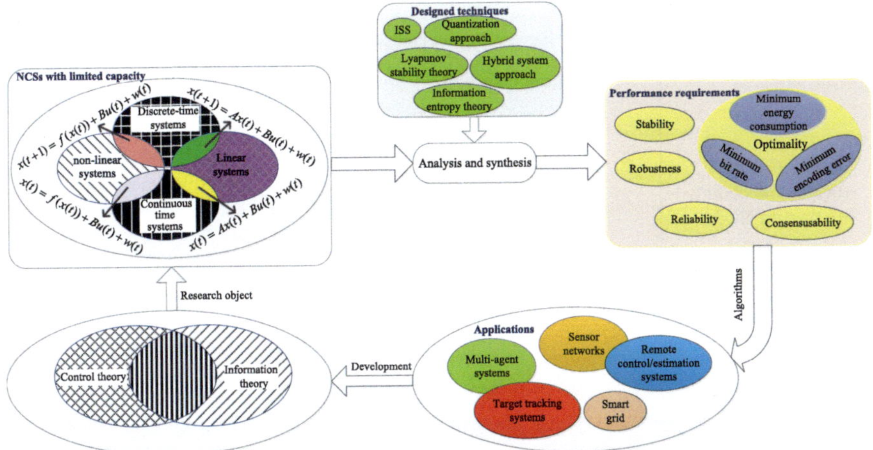

FIGURE 1.2

Framework of data-rate-constrained control and state estimation issues.

for the NSs. Various usually adopted design schemes of the encoding-decoding procedure in the NSs and their individual characteristics are reviewed. Subsequently, recent advances on the encoding-decoding-based control and filtering problems for the NSs are, respectively, surveyed. Then, the control and filtering issues for distributed systems with encoding-decoding communication scheme are discussed. Finally, conclusions and future research directions are presented. The main contents that are reviewed in this paper and the framework are shown in Figure 1.2.

1.2 Main Encoding-Decoding Schemes in NSs

In the framework of the coding-decoding-based data communications for NSs, it is of crucial importance to construct a suitable pair of encoder and decoder which could greatly restrain the impacts from the resulted data distortion on the system performance. To this end, we are going to give a clear look on some commonly utilized encoding-decoding schemes with the following system model

$$x(t+1) = Ax(t) + Bu(t) + Dw(t) \tag{1.1}$$

or

$$\dot{x}(t) = Ax(t) + Bu(t) + Dw(t) \tag{1.2}$$

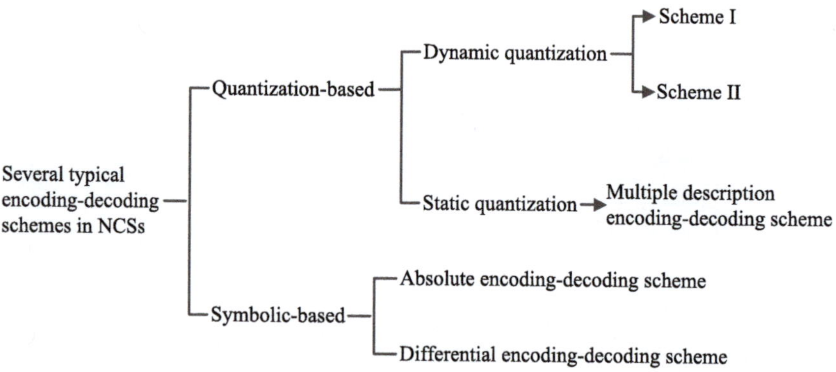

FIGURE 1.3

The summarization of several typical encoding-decoding schemes in NCSs.

with the measurement output

$$y(t) = Cx(t) + Ev(t) \tag{1.3}$$

where $x(t) \in \mathbb{R}^{n_x}$, $u(t) \in \mathbb{R}^{n_u}$, and $y(t) \in \mathbb{R}^{n_y}$ are, respectively, the system state, the system input, and the measurement output. $w(t) \in \mathbb{R}^{n_w}$ and $v(t) \in \mathbb{R}^{n_v}$ are the process and the measurement noises. A, B, C, D, and E are the known matrices with compatible dimensions. For analysis convenience, in this paper, it is assumed that the system state $x(t)$ is accessible for the stabilization problem, while for the filtering issue, it is supposed that only the measurement output $y(t)$ is observable. Till now, in the existing literature, two representative design solutions of the encoding-decoding procedure have been proposed to deal with the control and filtering problems in the NSs, namely, the quantization-based encoding-decoding technique and the symbolic-based encoding-decoding technique. Several typical encoding-decoding schemes in the NCSs are exhibited in Figure 1.3. In the following, we are in a position to successively illustrate these encoding-decoding design approaches mentioned above.

- **Quantization-Based Encoding-Decoding Mechanism**
  The quantization-based encoding-decoding method is one of the most common methods for the NSs with the limited communication capacity, based on which, a number of coder-decoder structures have been presented in the existing literature, see [59,69,79,140,190]. In addition, according to the case whether the quantizer parameters can be dynamically adjusted, the quantizer can be classified into the static quantizer and the dynamic one. In the following, in terms of the type of the used quantizer, three typical coder-decoder structures are listed and their corresponding characteristics are summarized.

Dynamic-Quantization-Based Encoding-Decoding Scheme I

Such a method was proposed in [140] and a similar idea appeared in the earlier literature [85]. For the sake of making this approach more understandable, first, let us briefly recall some basic concepts of the uniform quantization, and then we will show how the uniform quantization approach is applied to the encoder-decoder procedure.

Let the scaling parameter (the length of the quantization interval) $a > 0$ and integer q be given, and the $\mathcal{B}_a$ denotes the hyperrectangle in $\mathbb{R}^n$ centered at 0 with the edge length a, that is, $\mathcal{B}_a = \{\delta \in \mathbb{R}^n : |\delta|_\infty \leq a\}$, where $|\cdot|_\infty$ is the infinite vector norm. Then, we uniformly partition the hyperrectangles $\mathcal{B}_a$ into q^n sub-hyperrectangles $I^1_{s_1}(a) \times I^2_{s_2}(a) \times \cdots \times I^n_{s_n}(a)$, where $s_1, s_2, \ldots, s_n \in \{1, 2, \ldots, q\}$ and

$$I^i_1(a) \triangleq \left\{ \delta_i \Big| -a \leq \delta_i < -a + \frac{2a}{q} \right\}$$

$$I^i_2(a) \triangleq \left\{ \delta_i \Big| -a + \frac{2a}{q} \leq \delta_i < -a + \frac{4a}{q} \right\} \tag{1.4}$$

$$\vdots$$

$$I^i_q(a) \triangleq \left\{ \delta_i \Big| a - \frac{2a}{q} \leq \delta_i \leq a \right\}$$

where δ_i is the ith element of the vector δ. Here, each sub-hyperrectangle $I^1_{s_1}(a) \times I^2_{s_2}(a) \times \cdots \times I^n_{s_n}(a)$ is labeled with a sequence $\{s_1, s_2, \ldots, s_n\}$, which represents the specific region of the sub-hyperrectangle and will be sent to the decoder for the reconstruction of the original data. Moreover, the center of the sub-hyperrectangle $I^1_{s_1}(a) \times I^2_{s_2}(a) \times \cdots \times I^n_{s_n}(a)$ can be defined as

$$\eta_a(s_1, s_2, \ldots, s_n) \triangleq [b_1 \quad b_2 \quad \cdots \quad b_n]^T \tag{1.5}$$

where $b_i = -a + \frac{(2s_i - 1)a}{q}$, $i = 1, 2, \ldots, n$.

In what follows, we employ the proposed encoding-decoding protocol to the continuous-time system (1.2) for the stabilization problem. To be more specific, our goal is to design a pair of encoder-decoder and certain decoded-signal-based control strategy such that the closed-loop system is asymptotically stable. For simplicity of illustration, we assume that the disturbance $w(t)$ in (1.2) satisfies $w(t) \equiv 0$ and the main idea for the design of the encoding-decoding-based control strategy can be outlined in the following steps. To start with, at each encoding time jT ($j = 1, 2, \ldots$ and "T" is the encoding period), by calculating the error $e(jT)$ between the system state $x(jT)$ and the decoded state $\hat{x}(jT)$ which will be defined later, where $e(jT) = x(jT) - \hat{x}(jT^-)$, we are able to determine the uniform quantization parameter "a" and thus partition the hyperrectangle B_a into q^n equal sub-hyperrectangles. Here, $\hat{x}(jT^-)$ denotes the limit of $\hat{x}(jT)$ at the point of jT from the left, that is, $\hat{x}(jT^-) \triangleq \lim_{\epsilon>0,\epsilon\to 0} \hat{x}(jT - \epsilon)$. Subsequently, generate the corresponding codeword sequence $\{s_1, s_2, \ldots, s_n\}$ according to the region where the error "$e(jT)$" belongs to and then transmit the codeword sequence $\{s_1, s_2, \ldots, s_n\}$ to the decoder to obtain the decoded state $\hat{x}(jT)$. Finally, design the controller "$u(jT)$" in terms of the decoded state $\hat{x}(jT)$ to stabilize the original system.

Based on the above analysis, the detailed form of the encoder and decoder is given as follows:

Coder: For $e(jT) \in I_{s_1}^1 (a(jT)) \times I_{s_2}^2 (a(jT)) \times \cdots \times I_{s_n}^n (a(jT)) \subset B_{a(jT)}$, we have

$$g(e(jT)) = \{s_1,\ s_2, \ldots, s_n\} \tag{1.6}$$

where $a((j+1)T) = \varphi(T, q, x(0))a(jT)$ for $j = 1, 2, \ldots$ with $\varphi(T, q, x(0))$ being the function of certain form with respect to the encoding period T, the quantization level q and the initial condition $x(0)$.

Decoder:

$$\begin{cases} \hat{x}(0) = 0 \\[6pt] \dot{\hat{x}}(t) = A\hat{x}(t) + Bu(t), \forall t \in [jT,\ (j+1)T) \\[6pt] \hat{x}(jT) = \hat{x}(jT^-) + \eta_a(s_1, s_2, \ldots, s_n) \\[6pt] u(t) = K\hat{x}(t). \end{cases} \tag{1.7}$$

From structures of the coder (1.6) and the decoder (1.7), we have the following observations. 1) The proposed encoding approach is a kind of differential encoding technique. It is obvious that, in (1.6), only the decode error $e(jT)$ is encoded at the encoding time jT. In fact, by using the differential quantization encoding technique to encode the signal difference, the amount of the information to be sent is greatly reduced, see [27]. 2) It can be seen from the iteration relationship between $a((j+1)T)$ and $a(jT)$ that the quantization region dynamically varies at each encoding instant, and such a quantizer is referred to as the dynamic quantizer. 3) In order to enforce the decoding error be convergent, it is required that $\varphi(T, q, x(0)) < 1$, from which, we can see that the convergence of the encoding-decoding algorithm is closely associated with the encoding period, the quantization level and the initial condition of the system.

Apart from the continuous-time case, for the discrete-time counterpart, the encoding-decoding communication strategy for the NSs has been considered in [190], and the corresponding extended control/filtering problems subject to the network-induced phenomena have been investigated in [161,173].

Dynamic-Quantization-Based Encoding-Decoding Scheme II

This kind of encoding-decoding scheme is based on the so-called "zooming in/zooming out" quantization technique which was proposed in [171]. For the discrete-time system (1.1) with $w(t) \equiv 0$, the following encoder is given by

$$
\begin{cases}
\xi(0) = 0 \\
\xi(t) = g(t-1)s(t) + \xi(t-1) \\
s(t) = \mathcal{Q}\left(\dfrac{1}{g(t-1)}(x(t) - \xi(t-1))\right)
\end{cases}
\tag{1.8}
$$

where $\xi(t)$ is the encoder's internal state, $s(t)$ is the generated codeword which needs to be sent to the decoder, $\mathcal{Q}(\cdot)$ is a finite-level uniform quantizer, and $g(t)$ is a scaling function which is used to adjust the amplitude of the signal to be quantized. In most of the literature, the quantizer $\mathcal{Q}(\cdot)$ is chosen as the symmetric one with the following form

$$
\mathcal{Q}(z) =
\begin{cases}
0, & 1/2 \leq z < 1/2 \\
i, & (2i-1)/2 \leq z < (2i+1)/2 \\
& i = 1, 2, \ldots, L-1 \\
L, & z \geq (2L-1)/2 \\
-\mathcal{Q}(-z), & z \leq -1/2
\end{cases}
\tag{1.9}
$$

where L is the quantizer's saturation value and the associated quantization level is $2L + 1$.

By using the generated codeword $s(t)$, the decoder Ψ is designed as

$$
\begin{cases}
\hat{x}(0) = 0 \\
\hat{x}(t) = g(t-1)s(t) + \hat{x}(t-1)
\end{cases}
\tag{1.10}
$$

where $\hat{x}(t)$ is the output of the decoder, namely, the decoded value.

It can be concluded from the structure of the above encoder-decoder that 1) although the saturation value L is fixed *a priori*, due to the introduction of the scaling function $g(t)$, the signal to be quantized is enforced to be pre-processed first by the "zooming out" or "zooming in" technique, and hence this type of encoding approach is still of the dynamic-quantization-based encoding, 2) as discussed above, encoding the difference between $x(t)$ and $\xi(t-1)$ with scaling rather than the system state $x(t)$ itself possibly reduces the bits which are utilized to represent the codeword, 3) the saturation value L is closely related to the network bandwidth, that is, a small L results in the requirement of a small network bandwidth. Especially, when $L = 1$, it means that only one bit is necessary to send the codeword, and 4) the encoding period is consistent with the system sampling period, which is different from the former encoding approach that the encoding period can be the multiple of the system sampling period, see [140,173,190].

Such a "zooming in/zooming out" quantization-based encoding-decoding scheme was first proposed in [7], under which, a consensus algorithm has been designed for a class of discrete-time MASs that the information exchange among the agents was implemented through a digital communication network. Recently, some extensions have emerged, see [66,70,78,79,179]; however, most of which are concerned with the discrete-time systems and the results for the continuous-time counterpart are relatively few.

Static-Quantization-Based Encoding-Decoding Scheme

Having presented the above two different kinds of encoding-decoding schemes, now we would like to give a simple discussion about another encoding-decoding scheme, namely, the multiple description (MD) encoding-decoding scheme. The original idea of the MD encoding-decoding scheme comes from the investigation of the audio encoding issue in the telephone network by Bell Laboratories in the 1970s. As the rapid development of the network-based (e.g. internet) communication techniques, in recent decades, the MD encoding-decoding technology has gained considerable research attention from many researchers in different disciplines.

The idea of the MD encoding-decoding scheme is based on the diversity principle which could effectively enhance the transmission capacity and reliability of the channels, see [23,37,59,156,157]. From the point of the information theory view, the diversity principle can be understood as follows: encode the information source into multiple descriptions with the identical importance and then sent to the decoder via the parallel individual channel. The framework of the MD encoding-decoding scheme is built on the following two assumptions: 1) there are multi-channels between the encoder and the decoder and all the channels failing to work simultaneously is a small probability event, and 2) the original signal can be reconstructed with an acceptable quality as long as at least one of the descriptions is successfully delivered to the decoder, and when more descriptions are received by the decoder, the signal is reconstructed with a higher quality. Therefore, such an approach is able to improve the transmission reliability by rejecting the channel fluctuations resulted from the fading [47], shadowing, crosstalk, etc.

So far, according to the adopted method in the encoding-decoding process, the multiple description encoding-decoding can be classified into various classes including multiple description quantization coding (MDQC), the multiple description transformation coding (MDTC), the forward-error-correction-based multiple description coding (FECMDC), etc. In this paper, our attention is focused on the MDQC where the static uniform quantization method is applied to the addressed encoding-decoding issue.

Loosely speaking, the static uniform quantizer-based multiple description coding procedure can be divided into two steps. The first step is the index generation process which mainly employs the scalar quantizer to transform the information source into certain indices. Then, represent the indices by multiple descriptions in terms of a proper index assignment strategy. For convenience, a brief view on the encoding scheme with two descriptions is provided where the measurement output (1.3) is treated as the information source to be encoded. The diagram of the MDC for the NSs with two descriptions is shown in Figure 1.4.

Encoder:

$$\begin{cases} \theta_l(t) = f_{1l}(y_l(t)), & l = 1, \cdots, n_y \\ \vartheta_l(t) = f_{2l}(y_l(t)), & l = 1, \cdots, n_y \end{cases} \tag{1.11}$$

Decoder:

$$\bar{y}_l(t) = \begin{cases} g_{1l}(\theta_l(t)), & \text{when } \lambda_1(t) = 1, \lambda_2(t) = 0 \\ g_{2l}(\vartheta_l(t)), & \text{when } \lambda_1(t) = 0, \lambda_2(t) = 1 \\ g_{cl}(\theta_l(t), \vartheta_l(t)), & \text{when } \lambda_1(t) = 1, \lambda_2(t) = 1 \end{cases} \tag{1.12}$$

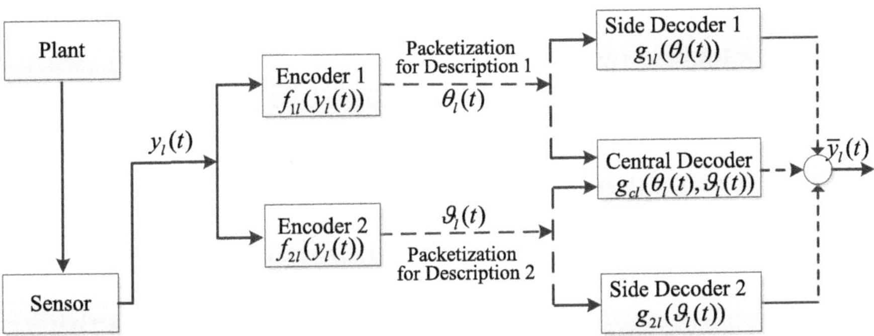

FIGURE 1.4
Multiple description encoding-based data communication in NSs.

where $f_{1l}(\,\cdot\,)$ and $f_{2l}(\,\cdot\,)$ are two coding functions whose outputs $\theta_l(t)$ and $\vartheta_l(t)$ can be regarded as the individual descriptions of the source $y_l(t)$ with $y_l(t)$ being the lth component of $y(t)$. $g_{1l}(\,\cdot\,)$ and $g_{2l}(\,\cdot\,)$ are two side decoding functions and $g_{cl}(\cdot,\cdot)$ is the central decoding function. $\lambda_i(t)$ $(i=1,2)$ are two independent random variables characterizing the packet dropout phenomena during the transmissions of the individual description, which satisfy the Bernoulli binary distribution taking values on either 1 or 0 with mathematical expectations $\bar{\lambda}_i$ and variances σ_i^2, respectively.

The early research effort on the MDC was mainly devoted to its theoretical research. Since the scalar quantization-based MDC method was first proposed in [156], the research direction has been moved from the pure theoretical investigation to its practical applications. From then on, the MDC has experienced a huge development in the theory and applications, and myriad coding schemes have appeared. It should be pointed out that most results of the MDC have been centered on the applications in distributed storage systems [60], diversity communication systems (antenna diversity) [133], and image/audio/video encoding [3,5,13,176]. However, to the best of the authors' knowledge, the relevant results of the MDC with respect to the control and filtering problems of NSs are very few.

- **Symbolic-Based Encoding-Decoding Mechanism**
  This kind of encoding schemes, also called as the "binary encoding", mainly takes the sign of the source into account, and thus only one bit is utilized to encode the signal. For convenience, in the sequel, we call it as the binary encoding instead of the symbolic-based encoding. Usually, the binary encoding technique including two versions, that is, the absolute (static) encoding technique [32,33] and the differential (dynamic) encoding technique [63,89]. Since the dynamic binary encoding technique is more complicated than the static one, in the following, we will concentrate our attention on the dynamic binary encoding-decoding technique.

  As we know, such a kind of dynamic encoding scheme was first proposed in [63], which is highly dependent on the change of certain system parameters. The core of the encoding rule can be expressed as follows with the system model (1.1) and (1.3).

$$\textbf{Coder}:\ s(t_k) = \text{sign}(y(t_k) - y(t_{k-1}))$$

$$\textbf{Decoder}:\ \bar{y}(t_k) = g(s(t_k)) \tag{1.13}$$

where sign($\cdot$) is the sign function, $g(\cdot)$ is the decoding function, $s(t_k)$ is the codeword generated at t_k, and $\bar{y}(t_k)$ is the decoded value of the measurement output $y(t_k)$. It follows from (1.13) that at each encoding instant t_k, only the sign of the change between the two successive measurement outputs is transmitted to the decoder, which implies that only one bit is needed for the codeword delivery during each encoding period. For the decoder design, the expected decoding function $g(\cdot)$ generally consists of two parts, namely, $g(s(t_k)) = \bar{y}(t_{k-1}) + g_0(s(t_k))$. The first term is the decoded value at the last encoding instant and the second one is a compensation function which aims to predict the difference between $y(t_k)$ and $y(t_{k-1})$. It is obvious that such an encoding scheme greatly reduces the communication burden and hence becomes a good candidate for the network-based control/filtering systems subjected to communication constraints. On the other hand, since only one-bit information could be obtained at each decoding instant, how to design an effective prediction function $g_0(s(t_k))$ so as to ensure the convergence of the decoding error appears to be the main challenge in the design of the encoder-decoder procedure.

1.3 Encoding-Decoding-Based Control and Filtering Problems

In this section, we will go over the recent theoretical developments of both the encoding-decoding-based control and filtering problems from various aspects including controller/filter design and the system performance evaluation.

- **Encoding-Decoding-Based Stabilization Problems**
  Linear Systems
  Among the various system performance indices for the networked control systems (NCSs), stability is the most basic yet crucial one, and thus it needs to be first considered in the system design. However, it should be noted that the introduction of the encoding-decoding-based communication scheme inevitably leads to the occurrence of the signal distortion, which would pose considerable additional complexities on the stability analysis and controller synthesis issues. A schematic structure of the NCSs with encoding-decoding-based data transmission protocol can be described in Figure 1.5.

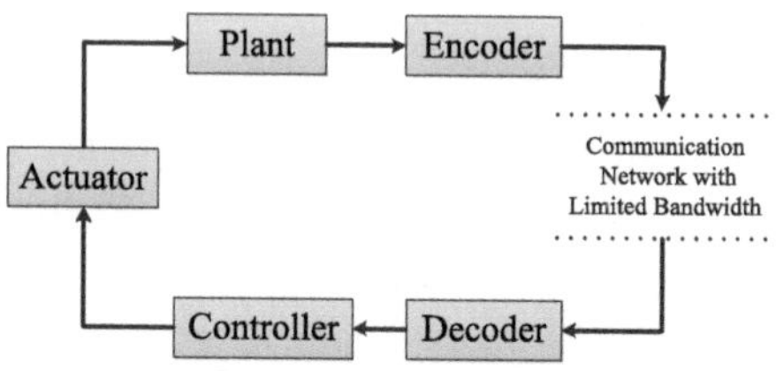

FIGURE 1.5

NCSs with the encoding-decoding-based communication protocol.

In [16], the pioneering work has been reported on the NCSs with quantization-based state feedback control strategy, which plays an important role in the subsequent fruitful results on the encoding-decoding schemes. In [171], the stabilization issue for a class of linear time-invariant system subject to the bandwidth constraint has been considered, where the encoding-decoding-based communication protocol has been introduced. For the quantization-based encoding scheme, a sufficient condition for the existence of the bit rate which enables to stabilize the system has been derived. Moreover, when the system reduces to a scalar one, a tight sufficient and necessary condition has been obtained. Nevertheless, the concrete forms of both the encoder and decoder are not explicitly given in their paper. The quadratic stabilization problem for sampled-data systems has been discussed in [53] under a bandwidth-constrained situation, where the memoryless uniform and logarithmic quantizers have been designed to overcome the circumstances of the limited bit rate of the communication channel and an upper bound on the bit rate has been acquired to achieve the stabilization purpose. Furthermore, both the state encoding and output encoding schemes have been developed in [85] for the continuous-time linear systems, and the encoding procedure has been well described in terms of the uniform quantization method. Meanwhile, the quantitative relationship between the number of the values taken from the encoder and the open-loop system parameters has been derived, under which the globally asymptotic stability of the closed-loop system has been guaranteed.

Nonlinear Systems

Subsequently, a number of research results have been extended to the encoding-decoding issues for the nonlinear NCSs, see [55,80, 85,87,119,123–125,140,190]. In [87], by utilizing the decoded system state, a feedback control strategy has been proposed to stabilize a class of continuous-time Lipschitz-type nonlinear systems. The ISS theory has been employed to characterize the impact of the decoding error on the closed-loop system state, based on which, it has been demonstrated that the closed-loop system can achieve globally asymptotical stability as long as the encoding frequency, the amount of the information provided by the encoder and the system parameters satisfy certain inequality conditions. In [124], the integral ISS condition, which is a weaker assumption than the ISS, has been obtained to ensure the stability of the closed-loop system under the same requirement on the channel capacity with [87]. Then, [122] extended the result to any smooth nonlinear systems, in which, it has been revealed that if the original nonlinear systems can be stabilized by certain control strategy without encoding process, it can also be stabilized by the state feedback scheme with the designed encoding communication protocol provided the channel capacity is larger than a given lower bound. Both the detectability and stabilizability problems have been investigated in [140] for the continuous-time systems with Lipschitz nonlinearities and bounded noises.

With the help of "zooming" technique, an observer-based encoding-decoding procedure has been put forward to globally stabilize the closed-loop system without the knowledge of the bound of the initial condition. Motivated by the approach proposed in [140], the subsequent work has been done in a large body of literature. For example, [190] has applied the approach to the discrete-time descriptor systems, and the nonlinearities existing in both the system state and the measurement equations have been considered in [10]. Recently, the encoding-scheme-based stabilization issue for the nonlinear uncertain systems has been studied in [80], and the traditional ISS condition has been relaxed to the differentiable ISS one. Moreover, by proposing a new uncertain nonlinear system model, the results obtained in [87] have been further generalized.

- **Encoding-Decoding-Based State Estimation Problems**
  In order to eliminate the confusion, it should be stated that the state estimation problem considered here is different from the channel state estimation issue [104,186] which belongs to the scope of information theory and aims to recover the state of the channel to facilitate the information transmission. In the category of control and signal processing, the state estimation (also called filtering) problem is to provide a possible accurate reconstruction of the state of the target plant and has long been a focus of research due to its engineering insights in many branches, see [46,49,52,67,76, 77,100,103,106,107,145,147,159,167,184,185,191]. For the data-rate-constrained NCSs, a typical filter structure with the encoder-decoder pair is shown in Figure 1.6. In this subsection, an overview will be given on the progress made in the encoding-decoding-based filtering/state estimation problems in the past few years.

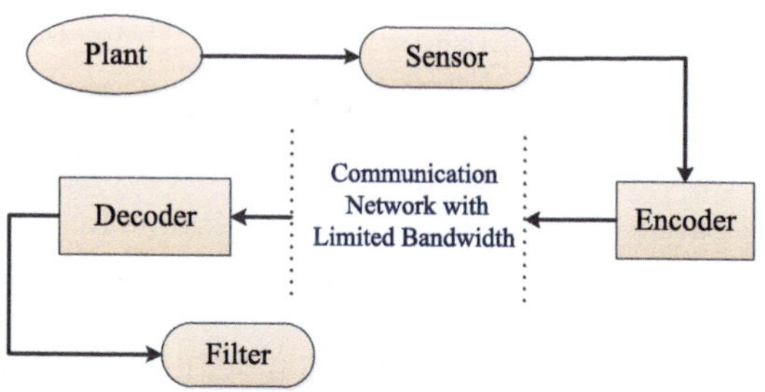

FIGURE 1.6

Filtering system model with encoding-decoding-based data transmission scheme.

Some initial attempts devoted to the state estimation problems with the compressed measurement data can be found in [4,15], and then a rich body of related results have appeared due to the wide utilization of the prevalent network communication techniques in a large range of the practical applications, see [12,31,111,112, 144]. For instance, the state estimation problem has been studied in [111] for the linear discrete-time unstable system with a noisy memoryless channel of communication constraints. The estimator has been explicitly designed and an equivalent relationship between estimation performance and the communication capacity of the channel has been revealed. For the same system model, its almost sure observability/stabilizability has been discussed in [112], where the communication channel was noisy and subject to the limited transmission capacity. It has been proved that the capability of the channel to guarantee the almost sure stabilizability/observability of the plant corresponds to its capability to deliver information with zero probability of error. The communication-constrained remote state estimation problem has been addressed in [12], where both the source coding and channel coding have been concerned. Without the assumption of the channel feedback existing in the communication loop, two encoding approaches have been proposed, and then the analysis results on the performance trade-off between the convergence rate of the decoding error and computational complexity have been derived.

It is noted that the results on the encoding-decoding-based state estimation problems recalled above are generally based on the information theoretic aspects, while, in recent years, there also has a surge of research attention on this issue from the point of non-information-theoretic view, see [41,110,117,150,177,178]. In [110], the robust set-valued state estimation problem has been addressed for the continuous-time uncertain systems via a digital communication with limited bit rate, where two encoding-decoding-based state estimation schemes have been presented from the aspects of the estimation precision and the algorithm implementation, respectively. The concept of "maximin information functional" has been established in [117] under a nonstochastic information theory framework, by which, it has been verified that the largest maximin information rate through a memoryless and error-prone channel reflects the zero-error capacity of the channel and the result has been applied to the state estimation issue for a linear time-invariant system. By applying the linear temporal coding technique, the optimal state estimation problem has been investigated in [41] for a class of discrete-time system with an unreliable channel. Both the orthogonal projection principle and the innovation sequence method have been employed to derive the minimum mean square estimation algorithm. The extended work has been done in [150], where the state estimation problem for the linear discrete-time stochastic systems subject to the packet dropout has been considered. In order to compensate the negative effect of the packet loss on the estimation performance, a linear coding temporal scheme has been employed to the raw measurements, by which some necessary and sufficient conditions for the estimation error system have been established in the mean square sense.

- **Encoding-Decoding-Based Control/State Estimation Subject to Network-Enhanced Complexities**
  In the remote networked control/estimation systems, because of the power restriction, network congestion or error, it is unavoidable that the data transmission via the digital channel suffers from the communication imperfections, for example, transmission delay, message loss, channel noise, etc. In this sense, for the encoding-decoding-based control/filtering problem in NCSs, the inherent complexities of the analysis and design issues have been typically enhanced and a number of relevant results have been reported in [24, 26, 38, 54, 56, 101, 126, 158, 165, 168].

Time-Delays

In [126], the minimal data-rate stabilization problem has been solved for the nonlinear NSs with transmission delays, which can be arbitrarily large. By applying the tool of the hybrid system theory, the original system has been converted into a couple of impulsive delayed nonlinear systems, and the semi-global and local stability of the closed-system has been guaranteed. The stabilization problem has been concerned in [26] for a class of distributed linear systems by a remote centralized control scheme, in which the random packet dropout and communication delay caused by an erasure network have been simultaneously considered. Three scenarios of the availability of the feedback channel have been, respectively, discussed and different encoding strategies have been correspondingly provided to ensure the closed-loop stability in the mean absolute sense. In [38], the channel coding technique has been adopted to reduce the occurrence of packet dropouts for the NCSs. However, the redundancy of the channel coding would lead to time delays. In order to balance such a trade-off, the ON-OFF channel coding technique has been utilized where the coding operation was adaptively implemented according to the expectation value of the tracking error. The data-rate-constrained stabilization problem has been investigated in [158] for a class of continuous-time linear systems. Due to the limited communication capacity of the channel between the sensor and controller, the sampled data has been encoded to certain codewords by the spherical polar coordinates encoding scheme. Dependent on whether the control inputs were needed to be sent to the decoder, two coding methods have been developed to achieve the stability of the controlled system.

Packet-Dropouts

For the randomly occurring message loss during the data transmission, a subband coding technique has been employed in [54] to deal with the controller and encoder-decoder pair design issues. To facilitate the analysis of the encoding-decoding procedure, the addressed control problem has been tackled in the frequency domain, and the stochastic stability and the $\mathcal{H}_\infty$ performance have been ensured in terms of the solution to certain linear matrix inequalities (LMIs). In [56], the attention has been focused on the stabilization problem for the discrete-time linear NCSs, where the data transmission has been implemented via a lossy channel with finite data rate. For the aim of the data compression, the control signal has been encoded to finite values by the uniform quantization method and a buffer has been designed in order to compensate the packet-dropout effect.

For the sake of improving the data transmission reliability via the erasure channel and obtaining a relatively satisfactory control/estimation performance, the multiple description encoding technique has been adopted in the NCSs in recent years. The MD encoding approach has been first introduced in the NCSs in [59], in which the Kalman filtering problem has been investigated for linear stochastic systems. Due to the packet dropout existing in the data packet transmission via the communication network, the MD encoding scheme has counteracted the effects of the data loss on the filtering performance. It has been demonstrated that compared with the single description encoding scheme, the MD one could significantly enhance the filtering performance. In [120], the state estimation problem has been addressed in the sensor networks together with the power control mechanism. The sensors were required to transmit the measurements to the gateway via a lossy channel subject to random packet dropouts, and then a novel MD quantizer has been applied to improve the estimation robustness. For more details with respect to the encoding-decoding-based filtering issues with lossy/noisy channels, we refer the readers to [92, 121, 130, 151] and the references therein.

Event-Triggering Mechanism

It is recognized that the signal delivery through the communication channel, especially for the wireless network, would account for a major ratio in the total power consumption of the NCSs. Therefore, for the purpose of keeping the use of the energy resource as small as possible and alleviating the network collision, the event-triggering data transmission mechanism has been successfully applied in the NCSs of digital communications [51, 71, 73, 89, 95, 102, 153, 164]. In [64], an early attempt has been made on the combination of the quantization-based control approach and the event-triggering approach to address the stabilization problem for linear continuous-time systems. Based on the quantization method utilized in [85], an encoder-decoder pair has been proposed to reduce the amount of the information to be sent via a bit-data-constrained channel; meanwhile, the event-triggered scheme has been employed to decrease the communication frequency. The boundedness of the closed-loop system state and a lower upper of the minimum time interval between two successive events have been explicitly studied. Very recently, for a linear scalar system with the communication delay, an event-triggered control problem combined with the quantization-based encoding-decoding scheme has been investigated in [61]. The additional information of the event-triggering time has been further excavated, and it has been shown that for the sufficiently small time delay, the timing information carried by the triggered events is large enough to stabilize the closed-loop system with any small information transmission rate. Subsequently, the related work has been extended to the vector systems in [62] and the binary differential coding approach in [89].

- **Optimal Encoding-Decoding Schemes**
  In the context of bit-rate-constrained control/estimation problems, the optimal coding scheme can be described from two aspects: 1) deriving the optimal control/filtering performance with a fixed bit rate of the network, and 2) determining the minimum information needed to transmit in order to achieve a given control/filtering task.

Optimal Control/Estimation Performance

In the first sense, there has been attracting considerable attention in recent years, see for example [2,17,30,32,57,139,166,188]. In [139], the optimal control problem for discrete-time nonlinear uncertain systems via digital communication channels with limited capacity has been taken into account. Based on the introduced concept "topological entropy", sufficient and necessary conditions, which are related to the bit rate of the channel and the topological entropy of the open-loop system, have been established for the solvability of the addressed optimal control problem. In[2], the control problem for a class of nonlinear systems subject to the communication constraints has been investigated. A performance index characterizing the difference between the quantized and unquantized systems has been defined. By optimizing such an index, the optimal dynamic quantizer in a closed form has been obtained. Zherlitsyn and Matveev [188] have considered the distributed estimation issue through a sensor network with the limited transmission capacity, in which an optimal-quantization-based encoding-decoding algorithm has been put forward to minimize the estimation error. The tracking problem has been addressed in [30] via a signal-noise-ratio-constrained communication network. Two practical network architectures have been illustrated to model the transmission channel. Moreover, the optimal encoder-decoder pair has been found, by which the tracking error variance has been minimized with minimum information transmissions.

Minimum Data-Rate

During the past 15 years, the minimum data-rate problem has gained a great deal of research attention in the network-based control and signal processing areas, see for example [44,114,115,131,138,155]. Some initial effort has been made in [114], in which the stabilization problem has been considered for linear discrete-time systems with the limited communication capacity. By resorting to the asymptotic quantization theory, a sufficient and necessary condition has been obtained for the existence of the minimum lower bound of the bit rate, above which the closed-loop system can be stabilized in the mth moment sense. The exponential stabilization problem with limited feedback data rates has been studied in [115], and moreover, the minimum data rate has been given which can achieve system stability. In [44], the minimum bit rate has been obtained to stabilize the linear time-invariant systems for both the continuous-time and the discrete-time cases. By applying differential pulse code modulation techniques, comparison results have been provided with respect to the derived minimum bit rate for the fix-step quantization and the variable-step quantization approaches. Very recently, a joint optimization problem has been discussed in [131]. For a NCS with the rate-limited communication channel, a general performance index including both the control cost and the communication cost has been constructed, and the corresponding optimal control and encoding problem has been solved.

1.4 Distributed Control and State Estimation Problems with Limited Communication Capacity

Compared with the simple NCS which consists of only a single plant and controller/estimator, distributed networked systems are more widespread in the real-world, such as the chemical process systems, the biological systems, the unmanned aerial vehicles, etc. For those large-scale distributed systems, each individual can be viewed as an agent which has the capacity of sensing, communication, computation, and implementation. In practical applications, the MASs and the WSNs are two typical examples of the distributed systems, and the corresponding distributed control and state estimation problems have attracted increasing research attention in recent years. Since the frequent information exchange among agents would certainly occupy considerable network resource and power energy, effective data-compression and energy-saving transmission mechanisms are vitally meaningful in real engineering systems. The typical framework of the distributed networked system with encoding-decoding communication is provided in Figure 1.7. In this section, we aim to provide some recent developments with respect to the encoding-decoding-based distributed control problem for the MASs and the state estimation problem for the WSNs, respectively.

- **Multi-Agent Systems with Encoding-Decoding-Based Information Transmission**

Without Network-Induced Phenomena

Consensus is one of the most important collective behaviors for the MASs, and the related control problems have received remarkable research interests in the past decade, see for example [39, 94, 109, 172, 179, 181]. It is known from [78] that the encoding-decoding-based consensus control protocol for MASs was first proposed in [7], in which, the average consensus performance has been investigated for a class of linear MASs by the bit-rate-constrained control protocol. According to different quantization strategies,

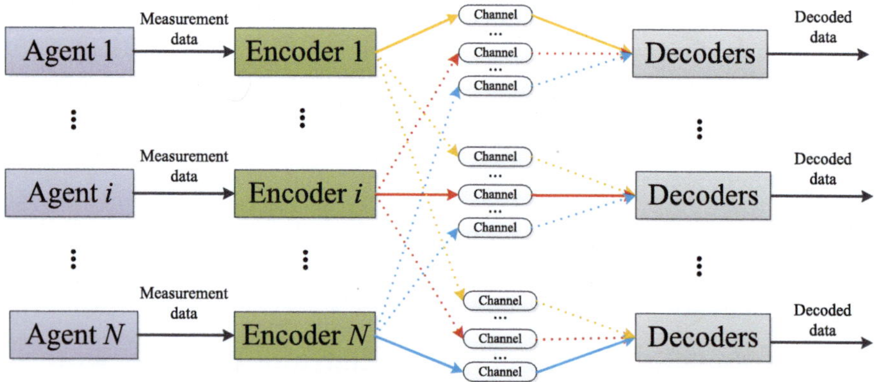

FIGURE 1.7

The architecture of distributed systems with encoding-decoding communication mechanism.

two encoding-decoding schemes have been provided which would guarantee the desired average consensus if the communication network is strongly connected. Under a connected communication network with limited data transmission capacity, [69] has explored the average consensus control problem for the first-order linear discrete-time MASs. By designing a dynamic encoding-decoding-based distributed control law, the exponential average consensus has been ensured for the requirement that only one bit data exchanged between two adjacent agents. Furthermore, an energy index has been defined to characterize the communication energy consumption for the consensus task, by optimizing which, a trade-off between the consensus rate and the quantization levels has been presented. Then, [78] has extended to the time-varying topologies case which is of jointly connected and the quantized observer-based encoding-decoding consensus control problem has been considered in [79] for the second-order discrete-time MASs. The related results with respect to high-order MASs can be found in [127].

Subject to Network-Induced Complexities

In the above results, a common assumption is that the communication channel is noiseless, that is, no time delays, dropouts, external disturbances, etc. during the data transmission process. However, this is not always true in the engineering practical owing to the limited network resources and the environmental complexity. In [93], the authors have discussed the average consensus problem of the MASs subject to bit-rate constraint, where the communication delays have been taken into consideration. By means of constructing a pair of dynamic encoder and decoder, a consensus control protocol has been designed to reach the average consensus performance with bounded time-delays. Because of the merits in the reduction of the energy and network resource, the event-triggering transmission mechanism has recently been adopted in the distributed control problem of the MASs with bandwidth restrictions, see for example [9,34,66,180]. In [66], in order to handle the consensus control problem for a class of discrete-time MASs via an imperfection communication network, both the event-triggering transmission mechanism and the dynamic encoding-decoding scheme have been employed. The proposed encoding scheme could adaptively select the number of the channel quantization level at each time step, and the relationship among the convergence rate, the number of the agents, the quantization interval, the network structure, and control gain has been well illustrated. Based on [66], very recently, the distributed consensus optimization problem for the MASs has been put forward in [70]. For each agent, a convex cost function has been defined and the objective optimization function was the sum of those individual cost functions. It also has been proved that as long as the system parameters satisfy certain conditions, the optimality of the consensus control scheme can be guaranteed with only one bit information exchange for each pair of adjacent agents. The similar problem with some extensions has been investigated in [72].

- **Wireless Sensor Networks with Encoding-Decoding-Based Communication Scheme**

  A wireless sensor network (WSN) is usually composed of numerous simple and low-cost sensor nodes geographically deployed in certain areas. With the collaboration such as integrated sensing, computation and communication, those sensor nodes can jointly perform various tasks in wide application fields including environmental monitoring, battlefield surveillance, civil infrastructure monitoring and control, intelligent solar farms, etc., [18,19,21,22,40,74,108,160, 162,168]. However, it should be stressed that due to the frequent communications among sensor nodes powered by irreplaceable batteries, the scarce network bandwidth and energy resources would give rise to a severe limitation on the practical applications of the WSNs. Therefore, developing certain useful communication techniques, which aims to improve the efficiency of the bandwidth utilization and reduce the energy consumption, appears to be an urgent task.

  So far, an alternative way to facilitate the information exchange in a bit-rate-constrained WSN is to design appropriate data encoding-decoding schemes so as to reduce the number of transmitted bits at each data transmission process, see [27,136,137]. In [137], the distributed filtering problem has been considered for a class of continuous-time stochastic systems with a low communication capacity. For each sensor, the observation value is required to be encoded into one bit information to be transmitted, based on which, a recursive algorithm of the Kalman filtering type has been designed. Also, it has been revealed that encoding-decoding-based filtering algorithm can achieve a close performance with the standard Kalman filtering algorithm. The distributed parameter estimation issue via the bandwidth-restricted WNSs has been studied in [25], in which the distributed adaptive quantization approach and a sequential transmission strategy have been employed. Based on the adaptive quantization approach, three quantization schemes have been proposed and the corresponding maximum likelihood estimators have been designed.

By using the observations from multiple sensors, [65] has researched the distributed state estimation problem for a discrete-time linear system. Because of the data-rate constraint, the individual sensors' innovations are required to be quantized before sending to the fusion center. The stability of the estimation scheme can be achieved if the data rate for each sensor is sufficiently large. In addition, an optimal rate allocation issue for the individual sensors has been discussed. [113] has studied the distributed parameter estimation problem in an energy-constrained WSN, where only one bit for each sensor has been allowed to be sent to the fusion center. On the basis of the variance of the measurement noises of individual sensors, the mean square estimation error can be minimized by designing an optimal bit assignment scheme. In [132], based on the Kalman filter structure, a distributed state estimation problem has been dealt with for the microgrid, and the systematic convolution encoding approach has been developed to facilitate smart grid communications. By applying the energy measurements from the bit-rate-constrained sensor networks, the target location estimation problem has been addressed in [91]. A position-based adaptive quantization scheme has been provided to improve the estimation performance, and the maximum likelihood estimator and the corresponding Cramér–Rao lower bound have been obtained. In [129], the distributed state estimation issue has been handled via a lossy WSN. By using the encoded measurements, the Kalman filter has been constructed to estimate the system state at the gateway, which also has controlled the transmission power of the radio power amplifier and the bit rate needed for each sensor.

1.5 Outline

Throughout this paper, we have summarized the existing results on the encoding-decoding-based control and state estimation/filtering issues for NSs. Several typical encoding-decoding schemes and the corresponding encoder-decoder structures have been systematically reviewed. Subsequently, for different kinds of NSs, a variety of encoding-decoding-based control and state estimation problems have been thoroughly surveyed. Finally, based on the literature review, possible relative directions for the further research work are listed as follows.

The framework of this book is shown as follows:

- In Chapter 1, the research background is first introduced, where the state estimation and control problems with constrained data rate are comprehensively reviewed. It is clearly pointed out that most existing literature has focused on the theoretical analysis for the data rate condition to certain control or state estimation issue. The corresponding results for providing concrete encoding-decoding procedures/algorithms have been much fewer, especially in the context of networked systems suffering multiple network induced phenomena. After presenting the motivation, the outline of this book is then listed.

- In Chapter 2, the gain-scheduled state estimation problem is studied for a class of complex networks subject to randomly occurring nonlinearities under bit-rate constraints. A set of random variables is employed to describe randomly occurring nonlinearities with a time-varying but bounded occurrence probability taking value in a known interval. Given a bit rate constraint, an encoding-decoding scheme is developed by which the boundedness of data distortion is guaranteed. The objective of this chapter is to propose a gain-scheduled state estimation algorithm by means of the decoded output measurements. With the help of the stochastic analysis technique as well as the Lyapunov stability theory, a sufficient condition is derived under which the prescribed estimation performance is ensured. Moreover, the estimator parameter is explicitly determined in terms of the solution to certain matrix inequalities.

- Chapter 3 studies the partial-neurons-based finite-time state estimation problem for a class of artificial neural networks with time-varying delays. Measurements information from only a small fractional of the artificial neurons are applied to the state estimation process. The data transmission from the sensor to the estimator is implemented via a bit-rate-constrained communication channel, and a data encoding-decoding scheme is developed to convert the original analog sensor measurements into certain digital codewords with fewer occupations of the network bandwidth. With the help of the Lyapunov stability theory, sufficient conditions are presented to guarantee the finite-time boundedness of the estimation error and the estimator gain matrix is parameterized in terms of the solution to certain matrix inequalities.

- Chapter 4 concerns the synchronization control problem for a class of discrete-time dynamical networks with packet dropouts via a coding-decoding-based approach. The data is transmitted through digital communication channels and only a sequence of finite coded signals is sent to the controller. A series of mutually independent Bernoulli distributed random variables is utilized to model the packet dropout phenomenon occurring in the transmissions of coded signals. The purpose of the addressed synchronization control problem is to design a suitable coding-decoding procedure for each node, based on which an efficient decoder-based control protocol is developed to guarantee that the closed-loop network achieves the desired synchronization performance.

- In Chapter 5, we deal with the consensus control problem for a class of discrete-time networked multi-agent systems with the coding-decoding communication protocol (CDCP). Under a directed communication topology, an observer-based control scheme is proposed for each agent by utilizing the relative measurement outputs between the agent itself and its neighboring ones. The signal delivery is in a digital manner, which means that only the sequence of finite coded signals is sent from the observer to the controller. To be specific, the observed data is encoded to certain codewords by a designed coder via the CDCP, and the received codewords are then decoded by the corresponding decoder at the controller side. The purpose of the addressed problem is to design an observer-based controller such that the close-loop multi-agent system achieves the expected consensus performance.

- In Chapter 6, the recursive filtering problem is investigated for a class of discrete-time stochastic nonlinear systems subject to fading channels where the multiple description coding scheme is employed during the data exchange between the devices. In order to facilitate the data transmission in a resource-constrained communication network, the multiple description coding scheme is adopted to encode the fading measurements into two descriptions with the identical importance. Two independent Bernoulli distributed random variables are introduced to govern the occurrences of the packet dropouts in two channels from the encoders to the decoders.

The channel fading phenomenon is characterized by the Mth-order Rice fading model whose coefficients are mutually independent random variables obeying certain probability distributions. The purpose of the problem addressed is to design a recursive filter such that, in the simultaneous presence of the stochastic noises, the channel fading and the data coding-decoding mechanism, an upper bound of the filtering error variance is obtained and then minimized at each time step. In virtue of the Riccati difference equation technique and the stochastic analysis approach, the explicit form of the desired filter parameters is derived by solving a sequence of coupled algebraic Riccati-like difference equations.

- Chapter 7 solves the stabilization problem is investigated for a class of linear discrete-time systems in a resource-constrained network environment where only the data with finite bits is permitted to transmit. The data transmissions from the sensors to the controller are subject to random packet dropouts. To improve the reliability of the data transmission, a novel dynamical multiple description coding scheme is developed with guaranteed convergence of the decoding error. By recurring to the matrix inequality theory and the stochastic analysis technique, sufficient conditions are established to ensure the stochastic stability of the closed-loop system. Furthermore, a parametric relationship is discovered that characterizes the influences from the packet-dropout probabilities and the encoder parameters on the stability of the closed-loop system.

- In Chapter 8, the control problem is taken into consideration for a class of continuous-time linear systems under the event-triggering encoding strategy. In order to accommodate the digital communication and save the network resource, a symbol-based difference coding scheme is proposed to convert the raw data into a symbolic one with only one-bit size. An event-triggering mechanism is put forward for the sake of reducing the frequency of both coder and controller updates. By deriving a lower bound of the inter-event execution interval, the Zeno behavior is successfully excluded. Then, a sufficient condition with respect to the required bit rate of the communication channel is derived to guarantee the convergence of the proposed encoding-decoding scheme. Moreover, a necessary condition of the bit rate is also given below which the encoding error is divergent. By utilizing the input-delay approach, the closed-loop system is transformed into a continuous time-delay one with a bounded time-varying delay. Sufficient conditions are established to guarantee the exponential stability of the closed-loop system. In addition, the explicit expression of the desired controller parameter is further obtained in terms of the solution to certain matrix inequality.

- In Chapter 9, we cope with the state estimation problem for a class of linear continuous-time systems under bit rate constraints. The data transmission between the sensors and the estimator is implemented through a digital communication network with limited bandwidth. By resorting to the singular value decomposition technique, an event-based encoding strategy is developed to encode the measurement signals into 1-bit codewords so as to reduce network resource consumption. The Zeno phenomenon is first proven to be excluded and then a sufficient condition is established under which the decoding error is ultimately bounded. Subsequently, a necessary condition is proposed to derive a lower bound of the bit rate below which the decoding error diverges. Moreover, the explicit expression of the desired state estimator is parameterized and the boundedness of the estimation errors is also analyzed.

- In Chapter 10, the conclusions and some potential topics for future work are given.

2

Gain-Scheduled State Estimation for Discrete-Time Complex Networks under Bit-Rate Constraints

Complex networks have been successfully applied in a broad scope of research areas such as social networks, transport networks, neural networks and power grids. By now, some popular models of complex networks (CNs) include, but are not limited to, switching CNs, time delay CNs and stochastic CNs. Built on such models, much effort has been focused on the analysis of the dynamical behaviors of various CNs such as stability, synchronization, control and state estimation.

It is noticeable that the state information is quite important for the dynamical analysis of CNs, which leads to the prosperity of the research on the filtering/state estimation problem for CNs [75,149]. On the other hand, due to the complexities caused by the strong couplings between nodes, the nonlinear function widely exists in the CN. However, due to unexpected broken links and sudden changes in the external environment, it is quite common that the nonlinear function appears in a random manner. Moreover, it is worth mentioning that the occurring probability of the randomly occurring phenomena may be time varying in engineering practice due to the unpredictably environmental change. In this context, an important question is how to reflect the effect of the time-varying probability on the control/filtering performance.

With the sustained and fast development of network technology, digitalized communication has displaced traditional analog communication as a new communication trend due to its distinct advantages in low-cost, easy wiring, and easy installation, operation and maintenance. This has also brought up the flourish of the networked systems with plenty of excellent results on the research of communication protocols and network-induced phenomena. Therefore, it is an interesting yet potential research topic to introduce digital communication mechanisms to the state estimation problem of CNs, which motivates our current investigation.

It is well recognized that the bit rate is an important indicator to evaluate the network communication capability, that is when the bit rate

DOI: 10.1201/9781003534853-2

is big, it means that the network has a strong communication capability. However, due to the limited network resources in real systems, the bit-rate-constrained problem has gained some attention and a variety of coding-decoding schemes have been developed, see [8,78,79,88,89,116,140,141, 190]. In particular, due to the large-scale and strong couplings, there is a great deal of data communication between nodes, which leads to a sizable demand in the bit rate. Therefore, it is of practical significance and theoretical importance to investigate the bit-rate-constrained state estimation problem for CNs with the following two crucial difficulties/challenges: 1) how to allocate the limited network resources to each network node and 2) how to design the coding-decoding scheme to take full advantage of the limited bit rate.

Concluding the above discussions, in this chapter, we strive to investigate the gain-scheduled state estimation problem for a kind of discrete-time complex networks (DTCNs) under the bit-rate constraints. An array of random variables is introduced to govern the nonlinearities whose occurring probability is a time-varying but bounded value with known upper and lower bounds. A bit-rate constraint model is established and an encoding-decoding mechanism is proposed, under which an upper bound of the decoding error is acquired. The primary purpose of the issue considered is to design a gain-scheduled state estimator to obtain an estimate of the network state with an acceptable accuracy according to available output measurements. By means of the stochastic analysis and Lyapunov stability theory, a sufficient condition is provided such that the estimation error dynamics achieve the exponentially mean square ultimate boundedness. The required estimator gain matrix is parameterized by solving a series of matrix inequalities. A numerical simulation is exploited to show the usefulness of the obtained gain-scheduled state estimator.

2.1 Problem Formulation

2.1.1 System Model

The dynamics of the considered DTCN with M nodes is described as follows:

$$u_i(t+1) = A_i u_i(t) + \gamma(t) f_i(u_i(t)) + \sum_{j=1}^{M} \sigma_{ij} \Lambda u_j(t) + D_i w_i(t) \qquad (2.1)$$

where $u_i(t) \in \mathbb{R}^{n_u}$ is the state vector of the i-th node. The process noise $w_i(t) \in \mathbb{R}$ is a zero-man Gaussian noise sequence with variance $\bar{w}_i$. $\sigma_{ij} > 0$ is the element located in the i-th row and the j-th column of the coupling configuration matrix $\Sigma = [\sigma_{ij}]_{M \times M}$. $\Lambda = \mathrm{diag}\{\lambda_1, \lambda_2, \ldots, \lambda_{n_u}\}$ is the inner-coupling matrix which determines the connection relationship between the elements of different nodes. A_i and D_i are constant matrices of appropriate dimensions. $\gamma(t)$ is a random variable satisfying the Bernoulli distribution with the following statistical properties:

$$\mathrm{Prob}\{\gamma(t) = 1\} = p(t), \quad \mathrm{Prob}\{\gamma(t) = 0\} = 1 - p(t) \tag{2.2}$$

where $p(t)$ is a time-varying value and satisfies $p(t) \in [p_1 \; p_2] \subseteq [0 \; 1]$ with $0 < p_1 < p_2$ being two positive scalars.

It is assumed that the nonlinear function $f_i(\cdot) : \mathbb{R}^{n_u} \to \mathbb{R}^{n_u}$ satisfies the following constraint:

$$[f_i(x) - f_i(y) - H_1(x - y)]^T [f_i(x) - f_i(y) - H_2(x - y)] \leq 0 \tag{2.3}$$

for any $x, y \in \mathbb{R}^{n_u}$, where H_1 and H_2 are known constant matrices.

2.1.2 Measurement Model Subject to Bit-Rate Constraint

The measurement output of the i-th node is given as follows:

$$s_i(t) = C_i u_i(t) + E_i v_i(t) \quad i \in \{1, 2, \ldots, M\} \tag{2.4}$$

where $s_i(t) \in \mathbb{R}^{n_s}$ is the measurement output of node i collected by sensor i, $v_i(t) \in \mathbb{R}$ is the zero-mean measurement noise with variance $\bar{v}_i$. C_i and E_i are given constant matrices with appropriate dimensions.

In this chapter, the data transmission from the sensor to the estimator is implemented by a wireless communication network of limited data rate, which would result in possible data collision or even packet dropout. Thus, it is necessary to allocate the available network resource to each node under certain protocols to alleviate those network-induced phenomena. As such, the model of the bit rate constraint is formulated as follows:

$$\sum_{i=1}^{M} \mathcal{R}_i \leq \mathcal{R} \tag{2.5}$$

where $\mathcal{R} \in \mathbb{R}^+$ is the total available bit rate in terms of the practical engineering specification and $\mathcal{R}_i \in \mathbb{R}^+$ is the assigned bit rate of the communication channel for node i.

2.1.3 Bit-Rate-Constrained Encoding-Decoding Scheme

For the digitized data transmission pattern, the raw data collected by the sensor should be converted into certain binary codes which are then sent to the destination via the digital channel. In this subsection, a uniform-quantization-based encoding-decoding scheme is presented for the data transmission under limited bit rate. In what follows, we introduce the quantization process for the measurement output $s_i(t)$. First, a hyperrectangle $\mathcal{S}_{m_i} = \{|s_{ij}(t)| \le l_i, j = 1, 2, \ldots, n_s\}$ is employed for node i in which the corresponding measurement output $s_i(t)$ is contained in, where $l_i > 0$ is a scaling parameter and $s_{ij}(t)$ is the j-th component of $s_i(t)$. Then, for node i, partitioning the interval $[-l_i \ l_i]$ into q_i subintervals uniformly which are described by

$$
\begin{aligned}
\mathcal{I}_{ij}^{(1)}(l_i) &\triangleq \left\{ s_{ij} \middle| -l_i \le s_{ij} < -l_i + \frac{2l_i}{q_i} \right\} \\[2mm]
\mathcal{I}_{ij}^{(2)}(l_i) &\triangleq \left\{ s_{ij} \middle| -l_i + \frac{2l_i}{q_i} \le s_{ij} < -l_i + \frac{4l_i}{q_i} \right\} \\
&\;\;\vdots \\
\mathcal{I}_{ij}^{(q_i)}(l_i) &\triangleq \left\{ s_{ij} \middle| l_i - \frac{2l_i}{q_i} \le s_{ij} \le l_i \right\}.
\end{aligned}
\tag{2.6}
$$

As such, the hyperrectangle $\mathcal{S}_{m_i} = \{|s_{ij}(t)| \le l_i, j = 1, 2, \ldots, n_s\}$ is uniformly divided into $q_i^{n_s}$ sub-hyperrectangles $\mathcal{I}_{ij}^{(\alpha_{i1})}(l_i) \times \mathcal{I}_{ij}^{(\alpha_{i2})}(l_i) \times \cdots \times \mathcal{I}_{ij}^{(\alpha_{in_s})}(l_i)$ where $\alpha_{i1}, \alpha_{i2}, \ldots, \alpha_{in_s} \in \{1, 2, \cdots, q_i\}$.

Taking the bit rate constraint condition (2.5) into account, for the uniform quantization scheme, the maximum quantization level for each sub-hyper-rectangle $\mathcal{I}_{ij}^{(\alpha_{i1})}(l_i) \times \mathcal{I}_{ij}^{(\alpha_{i2})}(l_i) \times \cdots \times \mathcal{I}_{ij}^{(\alpha_{in_s})}(l_i)$ is calculated as

$$
\bar{q}_i = \left\lfloor \sqrt[n_s]{2^{\mathcal{R}_i}} \right\rfloor
\tag{2.7}
$$

where $\lfloor \cdot \rfloor$ is the floor function. Thus, the center of the hyperrectangle $\mathcal{S}_{m_i}$ is determined by

$$
\eta_{l_i}^i(\alpha_{i1}, \alpha_{i2}, \ldots, \alpha_{in_s}) \triangleq \begin{bmatrix} b_{i1} & b_{i2} & \cdots & b_{in_s} \end{bmatrix}^T
\tag{2.8}
$$

where $b_{ij} = -l_i + \frac{(2\alpha_{ij}-1)l_i}{\bar{q}_i}$, $j = 1, 2, \ldots, n_s$. Thus, for any measurement output $s_i(t)$, there exists a unique sub-hyperrectangle $\mathcal{I}_{ij}^{(\alpha_{i1})}(l_i) \times \mathcal{I}_{ij}^{(\alpha_{i2})}(l_i) \times \cdots \times \mathcal{I}_{ij}^{(\alpha_{in_s})}(l_i)$ such that $s_i(t)$ is contained in. Moreover, the distance between any $s_i(t)$ in the sub-hyperrectangle $\mathcal{I}_{i1}^{(\alpha_{i1})}(l_i) \times \mathcal{I}_{i2}^{(\alpha_{i2})}(l_i) \times \cdots \times \mathcal{I}_{in_s}^{(\alpha_{in_s})}(l_i)$ and its center $\eta_{l_i}^i(\alpha_{i1}, \alpha_{i2}, \ldots, \alpha_{in_s})$ is bounded by

$$\left\| s_i(t) - \eta^i_{l_i}(\alpha_{i1}, \alpha_{i2}, \ldots, \alpha_{in_s}) \right\| \le \frac{\sqrt{n_s}l_i}{\sqrt[n_s]{2^{\mathcal{R}_i}}}. \tag{2.9}$$

Based on the above analysis, the following encoding-decoding scheme is proposed for node i subject to the constrained bit rate $\mathcal{R}_i$.

Encoding Scheme for Node i:

For the measurement output $s_i(t)$, if $s_i(t) \in \mathcal{I}^{(\alpha_{i1})}_{i1}(l_i) \times \mathcal{I}^{(\alpha_{i2})}_{i2}(l_i) \times \cdots \times \mathcal{I}^{(\alpha_{in_s})}_{in_s}(l_i)$, then the codewords corresponds to $s_i(t)$ is given as

$$c^{\mathcal{R}_i}_i = \begin{bmatrix} \alpha_{i1} & \alpha_{i2} & \cdots & \alpha_{in_s} \end{bmatrix}^T \quad \forall i \in \{1, 2, \ldots, M\} \tag{2.10}$$

Decoding Scheme for Node i:

For the decoder i, according to the received codewords $c^{\mathcal{R}_i}_i$ for node i, the corresponding decoded measurement output $\hat{s}_i(t)$ is obtained as

$$\hat{s}_i(t) = \eta^i_{l_i}(c^{\mathcal{R}_i}_i) = \begin{bmatrix} -l_i + \frac{(2\alpha_{i1}-1)l_i}{\lfloor \sqrt[n_s]{2^{\mathcal{R}_i}} \rfloor} \\ -l_i + \frac{(2\alpha_{i2}-1)l_i}{\lfloor \sqrt[n_s]{2^{\mathcal{R}_i}} \rfloor} \\ \cdots \\ -l_i + \frac{(2\alpha_{in_s}-1)l_i}{\lfloor \sqrt[n_s]{2^{\mathcal{R}_i}} \rfloor} \end{bmatrix}. \tag{2.11}$$

Define the decoding error for the measurement output of node i as $\tilde{s}_i(t) \triangleq s_i(t) - \hat{s}_i(t)$ and the augmented form as $\tilde{s}(t) \triangleq [\tilde{s}^T_1(t) \quad \tilde{s}^T_2(t) \quad \cdots \quad \tilde{s}^T_M(t)]^T$. It follows from (2.9)–(2.11) that the decoding error satisfies

$$\|\tilde{s}_i(t)\| \le \frac{\sqrt{n_s}l_i}{\lfloor \sqrt[n_s]{2^{\mathcal{R}_i}} \rfloor} \tag{2.12}$$

and

$$\|\tilde{s}(t)\| \le \sqrt{\sum_{i=1}^{n_s} \frac{n_s l^2_i}{\left(\lfloor \sqrt[n_s]{2^{\mathcal{R}_i}} \rfloor\right)^2}} \triangleq \varrho. \tag{2.13}$$

By utilizing the decoded measurement output $\hat{s}_i(t)$ ($i = 1, 2, \ldots, M$), the corresponding state estimator for $u_i(t)$ is designed as follows:

$$\hat{u}_i(t+1) = A_i\hat{u}_i(t) + p(t)f_i(\hat{u}_i(t)) + \sum_{j=1}^{M}\sigma_{ij}\Lambda\hat{u}_j(t)$$

$$+ K_i(t)(\hat{s}_i(t) - C_iu_i(t)) \tag{2.14}$$

where $p(t)$ is the time-varying probability given in (2.2), $K_i(t) = K_{i1} + p(t)K_{i2}$ with K_{i1} and K_{i2} being the estimator gain matrices to be determined. Denoting $e_i(t) \triangleq u_i(t) - \hat{u}_i(t)$ as the estimation error, it is inferred from (2.1)–(2.14) that the dynamics of the estimation error is described as

$$e_i(t+1) = (A_i - K_i(t)C_i)e_i(t) + \tilde{\gamma}(t)f_i(u_i(t)) + p(t)\tilde{f}_i(u_i(t))$$

$$+ \sum_{i=1}^{M}\sigma_{ij}\Gamma e_j(t) - K_i(t)E_iv_i(t) + K_i(t)\tilde{s}_i(t) \tag{2.15}$$

where $\tilde{\gamma}(t) \triangleq \gamma(t) - p(t)$ and $\tilde{f}_i(u_i(t)) \triangleq f_i(u_i(t)) - \hat{f}_i(u_i(t))$. Then, for node i, compacting the state vector $u_i(t)$ and the estimation error $e_i(t)$ as $\eta_i(t) \triangleq [e_i^T(t) \quad u_i^T(t)]^T$, we have the following error dynamics:

$$\eta_i(t+1) = \mathcal{A}_i(t)\eta_i(t) + p(t)\mathcal{F}_i(\eta_i(t)) + \tilde{\gamma}(t)\mathcal{W}\mathcal{F}_i(\eta_i(t))$$

$$+ \sum_{i=1}^{M}\sigma_{ij}\Lambda\eta_j(t) + \varepsilon_{1i}(t)\zeta_i(t) + \varepsilon_{2i}(t)\tilde{s}_i(t) \tag{2.16}$$

where

$$\mathcal{A}_i(t) \triangleq \begin{bmatrix} A_i - K_i(t)C_i & 0 \\ 0 & A_i \end{bmatrix}, \zeta_i(t) \triangleq \begin{bmatrix} w_i^T(t) & v_i^T(t) \end{bmatrix}^T,$$

$$\tilde{K}_{1i}(t) \triangleq \begin{bmatrix} 0 & -K_i(t)E_i \\ D_i & 0 \end{bmatrix}, \tilde{K}_{2i}(t) \triangleq \begin{bmatrix} K_i^T(t) & 0 \end{bmatrix}^T$$

$$\mathcal{W} \triangleq \begin{bmatrix} 0 & I \\ 0 & I \end{bmatrix}, \mathcal{F}_i(\eta_i(t)) \triangleq \begin{bmatrix} \tilde{f}_i^T(e_i(t)) & f_i^T(u_i(t)) \end{bmatrix}^T.$$

Setting $\eta(t) \triangleq [\eta_1^T(t) \quad \eta_2^T(t) \quad \cdots \quad \eta_M^T(t)]^T$ and stacking the error dynamics $\eta_i(t)$ of M nodes together, one has the following augmented system:

$$\eta(t+1) = \mathcal{A}(t)\eta(t) + p(t)\mathcal{F}(\eta(t)) + \tilde{\gamma}(t)\bar{\mathcal{W}}\mathcal{F}(\eta(t))$$

$$+ \tilde{K}_1(t)\zeta(t) + \tilde{K}_2(t)\tilde{s}(t) \tag{2.17}$$

where

$$\mathcal{A}(t) \triangleq \text{diag}_M\{\mathcal{A}_i(t)\} + \Sigma \otimes \Gamma, \tilde{K}_1(t) \triangleq \text{diag}_M\{\tilde{K}_{1i}(t)\},$$

$$\tilde{K}_2(t) \triangleq \mathrm{diag}_M\{\tilde{K}_{2i}(t)\}, \tilde{s}(t) \triangleq [\tilde{s}_1^T(t) \quad \tilde{s}_2^T(t) \quad \cdots \quad \tilde{s}_M^T(t)]^T,$$
$$\bar{W} \triangleq W \otimes I_M, \zeta(t) \triangleq [\zeta_1^T(t) \quad \zeta_2^T(t) \quad \cdots \quad \zeta_M(t)]^T$$
$$\mathcal{F}(\eta(t)) \triangleq [\mathcal{F}_1^T(\eta_1(t)) \quad \mathcal{F}_2^T(\eta_2(t)) \quad \cdots \quad \mathcal{F}_M^T(\eta_M(t))]^T.$$

The objective of our study is to develop a gain-scheduled state estimation strategy for DTCNs (2.1) where the signal transmission from the sensor to the estimator is executed by an encoding-decoding mechanism (2.10)–(2.11) subject to the bit rate constraint (2.5) such that the dynamics of the augmented system (2.17) is exponentially mean square ultimately bounded, that is,

$$\mathbb{E}\{\|\eta(k)\|^2\} \le e^{-vt}\beta_0\mathbb{E}\{\|\eta(0)\|^2\} + \delta \tag{2.18}$$

where $\beta_0 > 0$, $v > 0$ and δ are certain positive values.

2.2 Main Results

In this section, first, we shall first establish certain sufficient conditions to guarantee the exponentially mean square ultimate boundedness of the estimation error dynamics and then acquire the desired gain-scheduled state estimator gain by solving a series of matrix inequalities.

Theorem 2.1: *For a known positive scalar $0 < v < 1$ and the time-varying probability $p(t)$ belonging to $[p_1, p_2]$ as well as known gain matrices K_{i1} and K_{i2}, the augmented estimation error dynamics (2.17) is exponentially ultimately bounded in mean square sense if there exist two positive definite matrices $Q(p(t))$, $Q(p(t+1))$ and two positive scalars ϵ_1 and ϵ_2 such that the following matrix inequality*

$$\Pi(p(t)) \triangleq \Xi_0 + \Xi_1^T Q(p(t+1))\Xi_1 + \theta(p(t))\Xi_2^T Q(p(t+1))\Xi_2 < 0 \tag{2.19}$$

holds where

$$\Xi_0 \triangleq \begin{bmatrix} \Xi_0^{11} & \Xi_0^{12} & 0 \\ * & \Xi_0^{22} & 0 \\ * & * & -\epsilon_1 I_{M \times n_s} \end{bmatrix},$$
$$\Xi_1 \triangleq [\mathcal{A}(t) \quad p(t)I_{M \times n_u} \quad \tilde{K}_2(t)], \mathcal{I} \triangleq \mathrm{diag}_M\{I_{2n_u}\},$$
$$\Xi_2 \triangleq [0 \quad \bar{W} \quad 0], \theta(p(t)) \triangleq p(t)(1 - p(t)),$$
$$\Xi_0^{11} \triangleq -(1 - v)Q(p(t)) - \epsilon_2\phi_2, \Xi_0^{12} \triangleq -\epsilon_2\phi_1^T,$$
$$\Xi_0^{22} \triangleq -\epsilon_2\mathcal{I}, \hat{\phi}_1 \triangleq \mathrm{diag}\{\tilde{\phi}_1, \tilde{\phi}_1\}, \hat{\phi}_2 \triangleq \mathrm{diag}\{\tilde{\phi}_2, \tilde{\phi}_2\},$$

$$\mathcal{I} \triangleq \text{diag}_M\{I_{2n_u}\}, \phi_1 \triangleq \text{diag}_M\{\hat{\phi}_1\}, \phi_2 \triangleq \text{diag}_M\{\hat{\phi}_2\},$$

$$\hat{\phi}_1 \triangleq \text{diag}\{\tilde{\phi}_1, \tilde{\phi}_1\}, \hat{\phi}_2 \triangleq \text{diag}\{\tilde{\phi}_2, \tilde{\phi}_2\}, \phi_1 \triangleq \text{diag}_M\{\hat{\phi}_1\},$$

$$\phi_2 \triangleq \text{diag}_M\{\hat{\phi}_2\}, \tilde{\phi}_1 \triangleq -\frac{H_1^T + H_2^T}{2}, \tilde{\phi}_2 \triangleq \frac{H_1^T H_2 + H_2^T H_1}{2}.$$

Proof: *The Lyapunov function is chosen as*

$$V(t) = \eta^T(t)\mathcal{Q}(p(t))\eta(t) \tag{2.20}$$

and its difference along with the trajectory of (2.17) is defined as

$$\Delta V(t) = \eta^T(t+1)\mathcal{Q}(p(t+1))\eta(t+1) - \eta^T(t)\mathcal{Q}(p(t))\eta(t). \tag{2.21}$$

Next, the mathematical expectation of the term $\Delta V(t) + \nu V(t)$ is calculated as

$$\mathbb{E}\{\Delta V(t) + \nu V(t)\} = \mathbb{E}\{\eta^T(t+1)\mathcal{Q}(p(t+1))\eta(t+1) - (1-\nu)\eta^T(t)\mathcal{Q}(p(t))\eta(t)\}$$

$$= \mathbb{E}\left\{\left[\mathcal{A}(t)\eta(t) + p(t)\mathcal{F}(\eta(t)) + \tilde{K}_2(t)\tilde{s}(t)\right]^T \mathcal{Q}(p(t+1))\right.$$

$$\times [\mathcal{A}(t)\eta(t) + p(t)\mathcal{F}(\eta(t)) + \tilde{K}_2(t)\tilde{s}(t)]$$

$$+ \theta(p(t))\mathcal{F}^T(\eta(t))\bar{\mathcal{W}}^T \mathcal{Q}(p(t+1))\bar{\mathcal{W}}\mathcal{F}(\eta(t))$$

$$\zeta^T(t)\tilde{K}_1^T(t)\mathcal{Q}(p(t+1))\tilde{K}_1(t)\zeta(t)$$

$$\left. - (1-\nu)\eta^T(t)\mathcal{Q}(p(t))\eta(t)\right\}. \tag{2.22}$$

It is easy to verify that

$$\mathbb{E}\{\zeta^T(t)\tilde{K}_1^T(t)\mathcal{Q}(p(t+1))\tilde{K}_1(t)\zeta(t)\}$$

$$\leq \lambda_{\max}\{\mathcal{Q}(p(t+1))\}\mathbb{E}\left\{\tilde{K}_1^T(t)\tilde{K}_1(t)\right\} \sum_{i=1}^{M} (\bar{w}_i + \bar{v}_i) \tag{2.23}$$

$$\triangleq \varphi(t).$$

In addition, it is seen from (2.3) that

$$\begin{bmatrix} f_i(u_i(t)) \\ u_i(t) \end{bmatrix}^T \begin{bmatrix} I_{n_u} & \tilde{\phi}_1 \\ * & \tilde{\phi}_2 \end{bmatrix} \begin{bmatrix} f_i(u_i(t)) \\ u_i(t) \end{bmatrix} \leq 0 \tag{2.24}$$

and

$$\begin{bmatrix} f_i(e_i(t)) \\ e_i(t) \end{bmatrix}^T \begin{bmatrix} I_{n_u} & \tilde{\phi}_1 \\ * & \tilde{\phi}_2 \end{bmatrix} \begin{bmatrix} f_i(e_i(t)) \\ e_i(t) \end{bmatrix} \leq 0, \tag{2.25}$$

which further implies that

$$\begin{bmatrix} \mathcal{F}_i(\eta_i(t)) \\ \eta_i(t) \end{bmatrix}^T \begin{bmatrix} I_{2n_u} & \hat{\phi}_1 \\ * & \hat{\phi}_2 \end{bmatrix} \begin{bmatrix} \mathcal{F}_i(\eta_i(t)) \\ \eta_i(t) \end{bmatrix} \leq 0 \tag{2.26}$$

and

$$\begin{bmatrix} \mathcal{F}(\eta(t)) \\ \eta(t) \end{bmatrix}^T \begin{bmatrix} \mathcal{I} & \phi_1 \\ * & \phi_2 \end{bmatrix} \begin{bmatrix} \mathcal{F}(\eta(t)) \\ \eta(t) \end{bmatrix} \leq 0. \tag{2.27}$$

Considering the decoding error (2.13), *it follows from* (2.22)–(2.27) *that*

$$\begin{aligned}
\mathbb{E}&\{\Delta V(t) + v V(t)\} \\
&= \mathbb{E}\{\eta^T(t+1)\mathcal{Q}(p(t+1))\eta(t+1) - (1-v)\eta^T(t)\mathcal{Q}\eta(t)\} \\
&\leq \mathbb{E}\left\{ \left[\mathcal{A}(t)\eta(t) + p(t)\mathcal{F}(\eta(t)) + \tilde{K}_2(t)\tilde{s}(t)\right]^T \mathcal{Q}(p(t+1)) \right. \\
&\quad \times [\mathcal{A}(t)\eta(t) + p(t)\mathcal{F}(\eta(t)) + \tilde{K}_2(t)\tilde{s}(t)] \\
&\quad + \theta(p(t))\mathcal{F}^T(\eta(t))\bar{\mathcal{W}}^T \mathcal{Q}(p(t+1))\bar{\mathcal{W}}\mathcal{F}(\eta(t)) \\
&\quad - (1-v)\eta^T(t)\mathcal{Q}(p(t))\eta(t) - \epsilon_1(\tilde{s}^T(t)\tilde{s}(t) - \rho^2) + \varphi(t) \\
&\quad \left. - \epsilon_2 \begin{bmatrix} \mathcal{F}(\eta(t)) \\ \eta(t) \end{bmatrix}^T \begin{bmatrix} \mathcal{I} & \phi_1 \\ * & \phi_2 \end{bmatrix} \begin{bmatrix} \mathcal{F}(\eta(t)) \\ \eta(t) \end{bmatrix} \right\} \\
&= \xi^T(t)\Pi(t)\xi(t) + \bar{\varphi}(t)
\end{aligned} \tag{2.28}$$

where $\xi(t) \triangleq [\eta^T(t) \quad \mathcal{F}^T(\eta(t)) \quad \tilde{s}^T(t)]^T$ *and* $\bar{\varphi}(t) \triangleq \epsilon_1 \rho^2 + \varphi(t)$.

According to the condition (2.19) *in Theorem 2.1, one further calculates that*

$$\begin{aligned}
\lambda_{\min}&\{\mathcal{Q}(p(t+1))\}\mathbb{E}\{\|\eta(t)\|^2\} \\
&\leq \mathbb{E}\{V(t+1)\} \leq (1-v)\mathbb{E}\{V(t)\} + \varphi \\
&\leq (1-v)^2\mathbb{E}\{V(t-1)\} + (1-v)\bar{\varphi}(t) + \bar{\varphi}(t) \\
&\leq \cdots \\
&\leq (1-v)^{t+1}\mathbb{E}\{V(0)\} + (1-v)^t\bar{\varphi}(t) \\
&\quad + (1-v)^{t-1}\bar{\varphi}(t) + \cdots + (1-v)\bar{\varphi}(t) + \bar{\varphi}(t) \\
&\leq (1-v)^{t+1}\mathbb{E}\{V(0)\} + \frac{1-(1-v)^{t+1}}{v}\bar{\varphi}(t),
\end{aligned} \tag{2.29}$$

which indicates

$$\mathbb{E}\{\|\eta(t)\|^2\} \leq (1-v)^t \beta_0 \mathbb{E}\{\|\eta(0)\|^2\} + \frac{1}{v}\bar{\varphi}(t) \tag{2.30}$$

with $\beta_0 \triangleq (1-v)\frac{\lambda_{\max}\{\mathcal{Q}(p(0))\}}{\lambda_{\min}\{\mathcal{Q}(p(t+1))\}}$. *Hence, the proof of this theorem is ended.*

Up to now, the analysis issue of the addressed state estimation problem has been investigated for DTCNs with bit-rate-constrained encoding-decoding scheme. A sufficient condition has been provided in Theorem 2.1 to show the ultimate boundedness of the estimation error dynamics (2.17). In what follows, we shall focus our attention on the estimator design for the parameters K_{i1} and K_{i2}.

Theorem 2.2: *For a known positive scalar $0 < v < 1$ and the time-varying probability $p(t)$ belonging to $[p_1,\ p_2]$, if there exist positive definite matrices $\mathcal{Q}(p(t))$, $\mathcal{Q}(p(t+1))$, two positive scalars ϵ_1 and ϵ_2, real-valued matrices K_{i1} and K_{i2} and a nonsingular matrix $\mathcal{S}$ such that the following matrix inequality*

$$\Xi(p(t)) \triangleq \begin{bmatrix} \Xi_0 & \Xi_1^T & \theta(p(t))(\mathcal{S}\Xi_2)^T \\ * & \Xi_3 & 0 \\ * & * & \theta(p(t))\Xi_4^T \end{bmatrix} < 0 \qquad (2.31)$$

holds where $\Xi_3 = \Xi_4 \triangleq \mathcal{Q}(p(t+1)) - \mathcal{S}^T - \mathcal{S}$.

Proof: *Constructing the congruence transformation $\mathrm{diag}\{I, I, \mathcal{S}^{-1}\}$ to (2.31) leads to*

$$\begin{bmatrix} \Xi_0 & \Xi_1^T & \theta(p(t))\Xi_2^T \\ * & \Xi_3 & 0 \\ * & * & \theta(p(t))\bar{\Xi}_4^T \end{bmatrix} < 0 \qquad (2.32)$$

where $\bar{\Xi}_4 \triangleq \mathcal{S}^{-T}\mathcal{Q}(p(t+1))\mathcal{S}^{-1} - \mathcal{S}^{-T} - \mathcal{S}^{-1}$. Then, with the help of the Schur Complement Lemma, the inequality (2.19) holds if and only if the following inequality is true

$$\begin{bmatrix} \Xi_0 & \Xi_1^T & \theta(p(t))\Xi_2^T \\ * & -\mathcal{Q}(p(t+1))^{-1} & 0 \\ * & * & -\theta(p(t))\mathcal{Q}^{-1}(p(t+1)) \end{bmatrix} < 0. \qquad (2.33)$$

In addition, based on the fact $(\mathcal{S}^{-1} - \mathcal{Q}^{-1}(p(t+1)))^T \mathcal{Q}(p(t+1))(\mathcal{S}^{-1} - \mathcal{Q}^{-1}(p(t+1))) \geq 0$, one obtains the relationship that $-\mathcal{Q}(p(t+1))^{-1} \leq \mathcal{S}^{-T}\mathcal{Q}(p(t+1))\mathcal{S}^{-1} - \mathcal{S}^{-T} - \mathcal{S}^{-1}$. Then, it is not difficult to verify that (2.33) holds as long as (2.31) is true. As such, we can give the conclusion that the exponentially mean square ultimate boundedness of the augmented system (2.17) is attained for all $p(t) \in [p_1,\ p_2]$. Now, the proof of this theorem is ended.

It is worth mentioning that both the time-varying probabilities $p(t+1)$ and $p(t)$ are involved in the matrix inequality (2.31) simultaneously, which leads to that the matrix inequality (2.31) varies along with the time instant. Thus, the total number of the LMIs to be solved is accumulated and trends to be infinity when the time horizon reaches to infinite. Unfortunately, this is not acceptable in engineering practice because the computation burden caused by the implementation of the state estimation algorithm. In the next theorem, we will make our effort to overcome such a technical difficulty and propose an alternative method to transform the infinite number of LMIs to be solved into finite ones. To facilitate the technical development, we assume that the time-varying positive definite matrix is with the form $\mathcal{Q}(p(t)) = \mathcal{Q}_a + p(t)\mathcal{Q}_b$, where $\mathcal{Q}_a$ and $\mathcal{Q}_b$ are two positive definite matrices.

Theorem 2.3: *For a known positive scalar $0 < v < 1$ and the time-varying probability $p(t)$ belonging to $[p_1, p_2]$, if there exist two positive definite matrices $\mathcal{Q}_a$ and $\mathcal{Q}_b$, two positive scalars ϵ_1 and ϵ_2, real-valued matrices K_{i1} and K_{i2} and a nonsingular matrix $\mathcal{S}$ such that the following LMIs*

$$\Upsilon^{jkl} \triangleq \begin{bmatrix} \Phi_{0j} & \Phi_{1j}^T & \bar{\theta}(p_{jk})(\mathcal{S}\Phi_2)^T \\ * & \Phi_{3l} & 0 \\ * & * & \bar{\theta}(p_{jk})\Phi_{4l} \end{bmatrix} < 0, j, k, l = 1, 2 \tag{2.34}$$

are satisfied, where Φ_{0j} has the same form with Ξ_0 where only Ξ_0^{11} is replaced by $-(1-v)(\mathcal{Q}_a + p_j\mathcal{Q}_b) - \epsilon_2\phi_2$ and

$$\Phi_{1j} \triangleq [\bar{\mathcal{A}}_j \quad p_j I \quad \tilde{K}_{2j}], \quad \Phi_2 \triangleq \Xi_2, \tilde{K}_{2j} \triangleq \text{diag}_M\{\tilde{K}_{2ij}\}$$

$$\Phi_{3l} \triangleq \Phi_{4l} \triangleq \mathcal{Q}_a + p_l\mathcal{Q}_b - \mathcal{S}^T - \mathcal{S}, \quad \bar{\theta}(p_{jk}) \triangleq p_j(1 - p_k)$$

$$\bar{\mathcal{A}}_j \triangleq \text{diag}\{\bar{\mathcal{A}}_{1j}, \bar{\mathcal{A}}_{2j}, \cdots, \bar{\mathcal{A}}_{Mj}\},$$

$$\bar{\mathcal{A}}_{ij} \triangleq \begin{bmatrix} A_i - \bar{K}_{ip_j} & 0 \\ 0 & A_i \end{bmatrix}, \quad \tilde{K}_{2ij} \triangleq \begin{bmatrix} K_{ip_j} \\ 0 \end{bmatrix},$$

$$\bar{K}_{ip_j} \triangleq xK_{i1}C_i + p_jK_{i2}C_i.$$

Moreover, we derive an ultimate upper bound of dynamics (2.17) in mean square sense as $\gamma \triangleq \frac{1}{v}\bar{\rho}$, where $\bar{\rho} \triangleq \epsilon_1\rho^2 + 2\lambda_{\max}\{\mathcal{Q}_0 + p_2\mathcal{Q}_p\}\lambda_{\max}\{\Pi\}\sum_{i=1}^M (\bar{w}_i + \bar{v}_i)$ with $\Pi \triangleq \text{diag}_M\{\Pi_i\}$ and $\Pi_i = \text{diag}\{D_i^T D_i, 2E_i^T(K_{1i}^T K_{1i} + p_2^2 K_{2i}^T K_{2i})E_i\}$.

Proof: *Introduce four variables as follows*

$$\chi_1(t) \triangleq \frac{p_2 - p(t)}{p_2 - p_1}, \quad \chi_2(t) \triangleq \frac{p(t) - p_1}{p_2 - p_1},$$

$$\psi_1(t) \triangleq \frac{p_2 - p(t+1)}{p_2 - p_1}, \quad \psi_2(t) \triangleq \frac{p(t+1) - p_1}{p_2 - p_1}.$$

Then, we can rewrite $p(t)$ and $p(t+1)$ as

$$p(t) = \chi_1(t)p_1 + \alpha_2(t)p_2$$

and

$$p(t+1) = \psi_1(t)p_1 + \psi_2(t)p_2,$$

respectively, where $\chi_i(t) \geq 0$, $\psi_i(t) \geq 0$, $\sum_{i=1}^2 \chi_i(t) = 1$ and $\sum_{i=1}^2 \psi_i(t) = 1$. Moreover, one can further calculate that

$$\mathcal{Q}(p(t)) = \mathcal{Q}_a + \sum_{i=1}^2 \chi_i(t)p_i\mathcal{Q}_b$$

and

$$\mathcal{Q}(p(t+1)) = \mathcal{Q}_a + \sum_{i=1}^2 \psi_i(t)p_i\mathcal{Q}_b.$$

Based on the discussions made above, it is straightforward to turn out that $\Xi(p(t)) = \sum_{j,k,l=1}^2 \chi_j(t)\chi_k(t)\psi_l(t)\Upsilon^{jkl}$, which means that (2.31) is satisfied if (2.34) is met. On the other hand, by some algebraic manipulations, one has $\lambda_{\max}\{\mathcal{Q}(p(t+1))\} \leq \lambda_{\max}\{\mathcal{Q}_0 + p_2\mathcal{Q}_p\}$ and $\mathbb{E}\left\{\tilde{K}_1^T(t)\tilde{K}_1(t)\right\} \leq \lambda_{\max}\{\Pi\}I$. Thus, we can now conclude that $\lim_{k\to\infty}\mathbb{E}\{\|\xi_k\|^2\} \leq \frac{1-\mu}{\mu}\lim_{k\to\infty}\rho_{k+1} \leq \frac{1-\mu}{\mu}\bar{\rho} = \gamma$, which finishes the proof of this theorem.

Remark 2.1: *Till now, we have handled the state estimation issue for a kind of DTCNs subject to RONs under bit-rate constraints. Compared with the existing results on the state estimation of DTCNs, our main results have the following distinct characteristics: 1) a model of the bit-rate constraint is established for the DTCN according to the allowable bandwidth of the communication network under which an encoding-decoding mechanism is proposed; and 2) a general theoretical framework is provided for the estimation error dynamics, by which the exponentially ultimate boundedness is explored in the mean square sense. To be more specific, a sufficient condition is established on the performance analysis in Theorem 2.1 where the estimation error dynamics is exponentially mean square ultimately bounded. Then, the probability-dependent state estimator design problem is given in Theorem 2.2. To get rid of the difficulty resulted from the time-varying probability, Theorem 2.3 subsequently provides a finite number of matrix inequalities by replacing the time-varying probability by its known upper and lower bounds and the desired estimator gain matrix is further obtained.*

2.3 An Illustrative Example

A numerical example is given in this section to demonstrate the validness of the developed bit-rate-constrained state estimation strategy for DTCNs subject to RONs with a time-varying occurrence probability.

In this example, we consider that the network is composed of three nodes and the corresponding parameters (e.g. the system matrices, the coupling matrices, etc.) in the simulation are chosen as follows:

$$A_1 = \begin{bmatrix} 0.32 & 0.4 \\ 0.5 & 0.15 \end{bmatrix}, A_2 = \begin{bmatrix} 0.25 & 0.4 \\ 0.3 & 0.25 \end{bmatrix},$$

$$A_3 = \begin{bmatrix} 0.53 & 0.1 \\ 0.7 & 0.5 \end{bmatrix}, C_1 = \begin{bmatrix} 1.0 & 0.9 \end{bmatrix},$$

$$C_2 = \begin{bmatrix} 1.0 & 0.6 \end{bmatrix}, C_3 = \begin{bmatrix} 1.2 & 0.7 \end{bmatrix},$$

$$D_i = \begin{bmatrix} 0.5 & 0.5 \end{bmatrix}^T, \Lambda = I_2, E_i = 0.1,$$

$$\Sigma = \begin{bmatrix} -0.2 & 0.1 & 0.1 \\ 0.1 & -0.2 & 0.1 \\ 0.1 & 0.1 & -0.2 \end{bmatrix}.$$

Choose the nonlinear function $f_i(\,\cdot\,)$ as

$$f_i(x_i(t)) = \begin{bmatrix} 0.5x_{i1}(t) + \tanh(0.2x_{i1}(t)) \\ -\tanh(0.45x_{i2}(t)) + 0.65x_{i2}(t) \end{bmatrix}$$

and thus the matrices H_1 and H_2 are, respectively, given as

$$H_1 = \begin{bmatrix} -0.5 & 0 \\ 0 & 0.65 \end{bmatrix}, H_2 = \begin{bmatrix} -0.3 & 0 \\ 0 & 0.2 \end{bmatrix}.$$

Set the time-varying probability $p(t)$ of RONs as $p(t) = p_1 + (p_2 - p_1)|\cos(k)|$ with $p_1 = 0.3$ and $p_2 = 0.9$. Let the variances of the process noise $w_i(t)$ and the measurement noise $v_i(t)$ as $\bar{w}_i = 0.1$ and $\bar{v}_i = 0.1$. The total available bit rate of the DTCN is given as 18bps. According to the average allocation protocol, the individual bit rate for each node is $\mathcal{R}_1 = \mathcal{R}_2 = \mathcal{R}_3 = 6$bps and the quantization level is consequently calculated as $\bar{q}_i = 64$. The initial condition of the system state $u_i(t)$ and its estimation are set as $u_1(t) = [-0.10 \quad 0.14]^T$, $u_2(t) = [-0.28 \quad -0.10]^T$, $u_3(t) = [1.7 \quad 0.15]^T$ and $\hat{u}_i(t) = [0 \quad 0]^T$. By solving the LMIs (2.34) with the help of the Matlab LMI toolbox, we obtain the following estimator parameters:

$$K_{11} = \begin{bmatrix} 0.2250 & 0.3776 \end{bmatrix}^T, K_{12} = \begin{bmatrix} -0.1465 & 0.1798 \end{bmatrix}^T$$

$$K_{21} = \begin{bmatrix} 0.4399 & 0.4323 \end{bmatrix}^T, K_{22} = \begin{bmatrix} -0.3928 & 0.0480 \end{bmatrix}^T$$

$$K_{31} = \begin{bmatrix} 0.1941 & 0.4535 \end{bmatrix}^T, K_{32} = \begin{bmatrix} -0.2699 & 0.0605 \end{bmatrix}^T.$$

The simulation results are shown in Figures 2.1–2.6. Figures 2.1–2.2 describe the state strategies of node i and their estimates, and Figure 2.3 illustrates the measurement output of the DTCN and their decoded values. Figure 2.4 gives the root mean square error of the state estimation for each node. Figure 2.5 exhibits the decoding error for each component of the measurement output $s(t)$. Figure 2.6 shows the curve of the time-varying probability $p(t)$. From the provided simulation results, we can see that the developed gain-scheduled state estimation scheme achieves a desired estimation performance and the proposed encoding-decoding mechanism maintains an acceptable decoding accuracy under a bandwidth resource constraint. Consequently, the simulation results carried out above well confirm the applicability of the derived theoretical results.

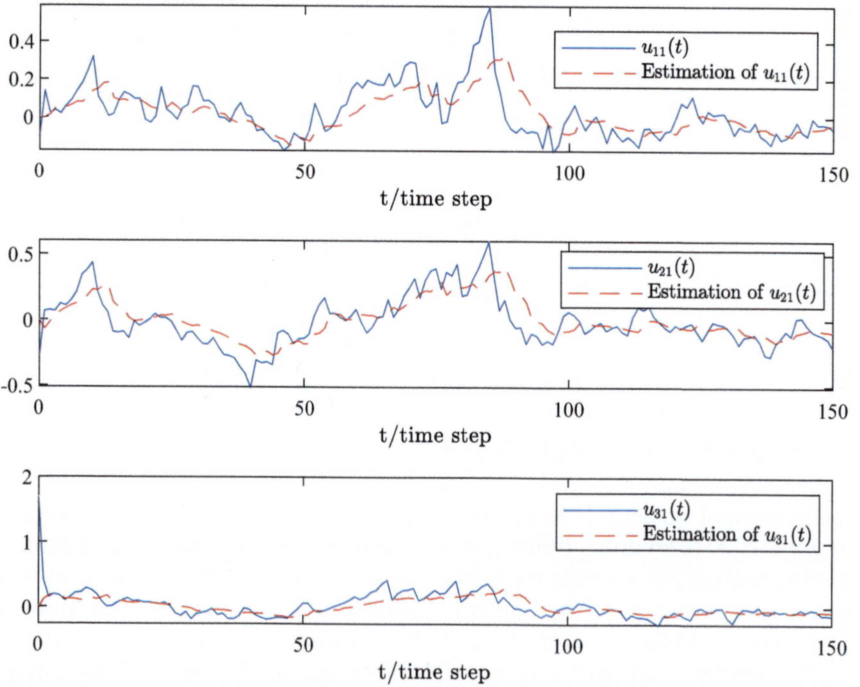

FIGURE 2.1

State trajectory $u_{i1}(t)$ ($i = 1, 2, 3$) and its estimation.

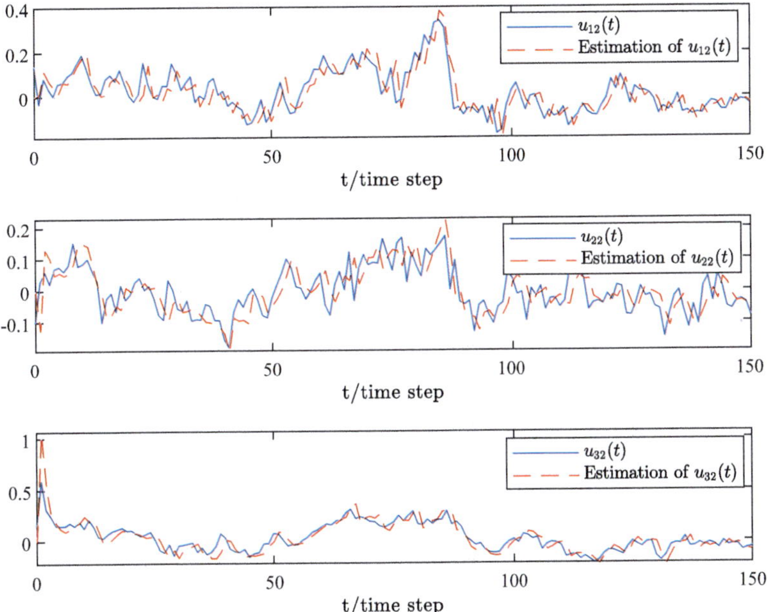

FIGURE 2.2

State trajectory $u_{i2}(t)$ ($= 1, 2, 3$) and its estimation.

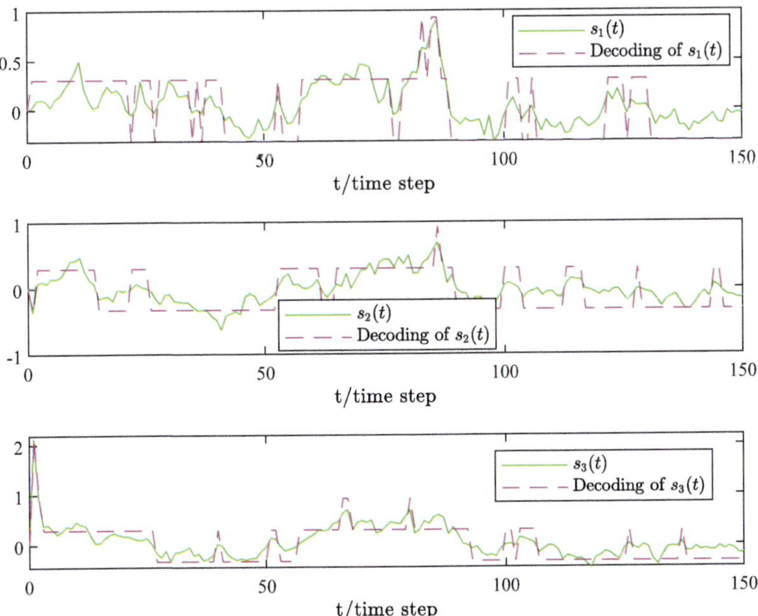

FIGURE 2.3

Measurement output $s_i(t)$ ($i = 1, 2, 3$) and its decoded value.

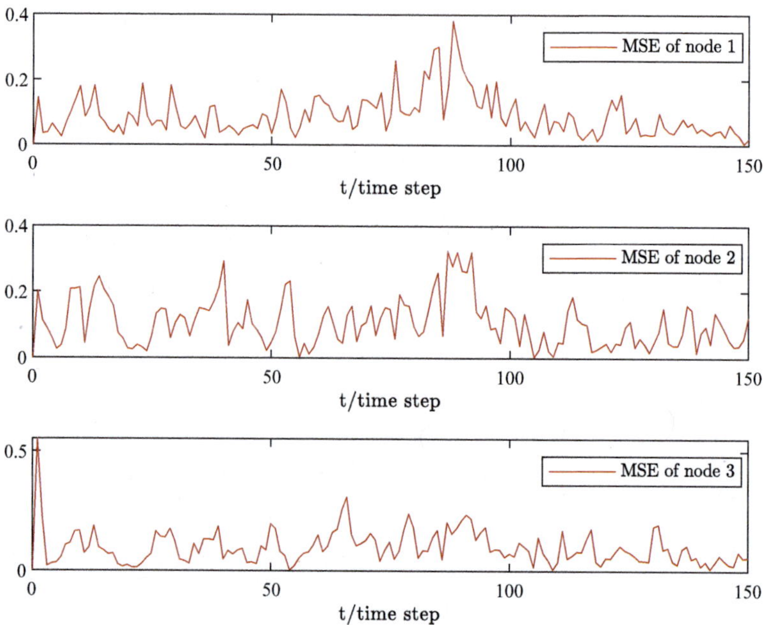

FIGURE 2.4
MSE of node i ($i = 1, 2, 3$).

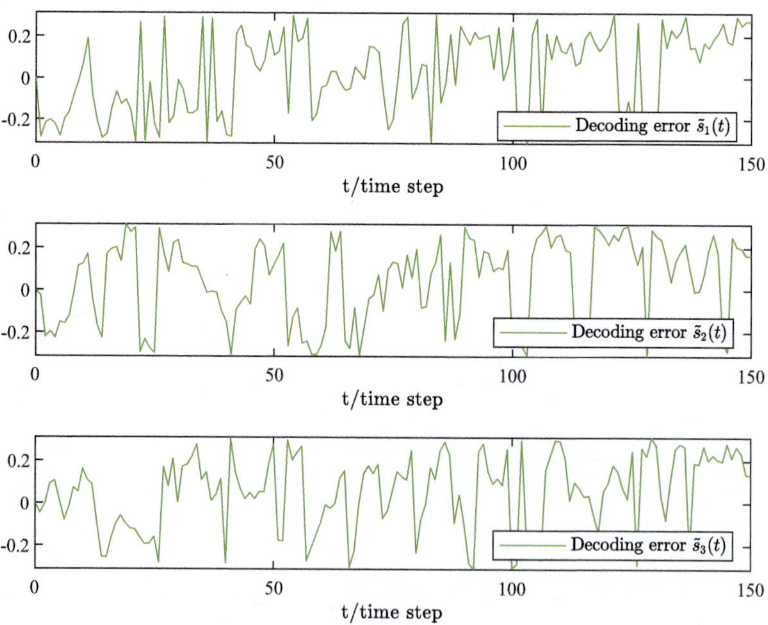

FIGURE 2.5
Decoding error $\tilde{s}_i(t)$ ($i = 1, 2, 3$).

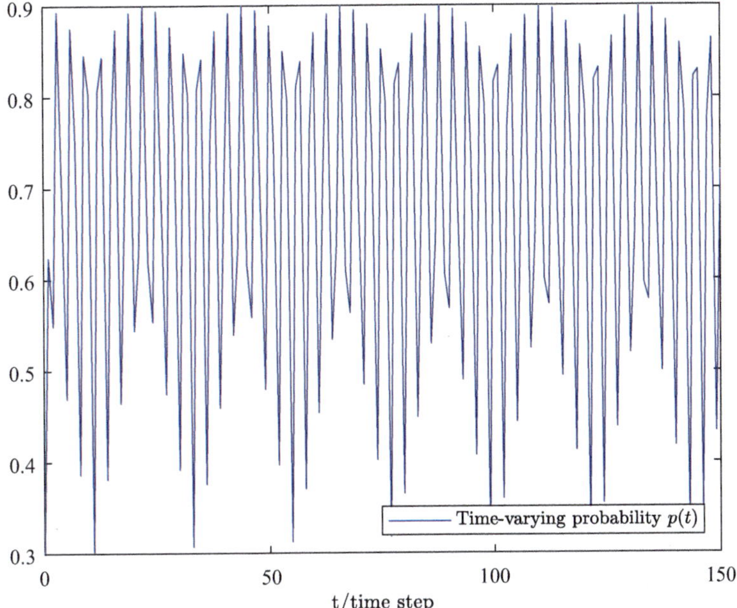

FIGURE 2.6
The time-varying probability $p(t)$.

2.4 Summary

This chapter has dealt with the gain-scheduled state estimation issue for a kind of DTCNs with RONs under bit-rate constraints. In a resource-constrained network circumstance, a bit-rate constraint model has been established under which an encoding-decoding scheme has been given and an upper bound of the decoding error has been acquired. The phenomenon of RONs has been characterized by an array of Bernoulli random variables. Sufficient conditions have been provided to ensure that the estimation error dynamics is exponentially mean square ultimate bounded. The required gain-scheduled state estimator gain matrix has been obtained in terms of the solution to certain matrix inequalities. In the end, the usefulness of the presented gain-scheduled state estimator design scheme has been validated via a numerical simulation example.

3

Partial-Neurons-Based State Estimation for Artificial Neural Networks under Constrained Bit Rate: The Finite-Time Case

With the blooming development of computer science as well as data science, artificial neural networks (ANNs) have been found extensive applications ranging from biomedicine, manufacturing and transportation, to smart home due to their excellent performance in simulating the human brain for decision-making and ratiocination. In general, the ANN is a kind of man-made complex network composed of a large number of interconnected neurons, which is able to perform various high-level tasks such as nonlinear approximation, associative memory, pattern recognition and optimization. Due to the strong coupling, highly nonlinearities and large scale of the network structure, the states of certain neurons are usually difficult to access directly. In order to have a clear understanding with respect to the relationship between the dynamical behaviors of ANNs (e.g. the stability, periodic oscillation, bifurcation, or even chaotic dynamics) and the performance in handling those complicated tasks, it is desired to acquire the state information of the neurons. Accordingly, the state estimation issue for ANNs has been an active research topic and received considerable research interest in the past few decades., see [118,128,152].

An implicit assumption made in most of the literature considering the state estimation problem for ANNs is that the measurement outputs of all the neurons are accessible to the state estimator. However, such a prerequisite requirement is unrealistic due either to the limited measurement technology or to the cost of acquiring all the measurements of a large scale of neuron nodes. As such, in engineering practice, it is neither possible nor necessary to collect all the neurons' measurement outputs, which gives rise to the partial-neurons-based state estimation problem and some relevant results have been reported in [14,68]. For example, in [98] the partial-neurons-based state estimation problem has been addressed for a class of neural networks in the presence of both state-dependent noises and time-varying delays. Moreover, for the complex networks with unbounded distributed

DOI: 10.1201/9781003534853-3

delays and energy-bounded measurement noises, the partial-nodes-based state estimation problem has been discussed in [98].

Serving as one of the essential steps of a networked control/state estimation process, data exchange has a major effect on the overall system performance. Generally speaking, the communication quality of a digital channel is heavily dependent on the bandwidth of the communication network, which is also known as the bit rate of the channel. In an ideal case, a communication channel with an infinite bit rate provides a reliable communication process without any distortion of the original data. However, this is only an extreme situation and can hardly be satisfied in real applications due mainly to the limitation of the network resource. In practice, it is often the case that the bit rate of the communication media is limited, which is particularly true for those complicated systems with a large scale of unit systems and the tight communication coupling among them, such as the ANNs, sensor networks and multi-agent system, etc. Therefore, how to develop an efficient data transmission mechanism for the ANNs in a bit-rate-constrained environment is of both theoretical significance and engineering importance.

It is worth pointing out that most of the existing literature concerning the state estimation problem for ANNs is based on the Lyapunov asymptotic stability and the system performance is evaluated over an infinite time horizon. However, such a performance metric is not suitable in certain practical scenarios since most of the real systems (e.g. the missile system, the robot control system, and the calculation of drug dose) require that the dynamics achieve the prescribed performance requirement in a finite time interval, which is referred to as the *finite time stability* [1]. Compared with the traditional asymptotic stability, the performance index of the finite-time stability is more appealing since the system trajectory is confined to a specified state space in a finite time, which would achieve a better transient performance [1]. So far, although some research efforts have been dedicated to the finite-time control/estimation problem [1,175], the corresponding results for the ANN with time-varying delays subject to partial neuron measurements have been scattered, not to mention the communication channel is of constrained bit rate, which constitutes the primary motivation of the investigation in this chapter.

3.1 Problem Formulation

The artificial neural network considered here is composed of n_u neurons and each of them is described by the following discrete-time nonlinear dynamics:

$$
\left\{
\begin{aligned}
u_i(t+1) &= a_i u_i(t) + \sum_{j=1}^{n_u} b_{ij} p_j(u_j(t)) + \sum_{j=1}^{n_u} d_{ij} q_j(u_j(t - \pi(t))) \\
&\quad + J_i(t) + l_i w_i(t), \quad i = 1, 2, \cdots, n, \\
u_i(r) &= \phi_i(r), r \in [-\pi_M, 0]
\end{aligned}
\right.
\tag{3.1}
$$

where $u_i(t) \in \mathbb{R}$ stands for the ith neuron's state information and a_i represents the state feedback weight. $p_j(\cdot)$ and $q_j(\cdot)$ are the activation functions of neuron j. $w_i(t)$ refers to a bounded exogenous disturbance satisfying $\|w_i(t)\| \le \bar{w}_i$ with $\bar{w}_i$ being a given scalar. For the ith neuron, the parameters b_{ij} and d_{ij} are the connection weights for the non-delayed term $p_j(u_j(t))$ and the delayed term $q_j(u_j(t - \pi_j(t)))$, respectively. $J_i(t)$ is the external input of the ith neuron. $\pi_j(t)$ is a positive integer denoting the time-varying delay of the jth neuron, which satisfies $\pi_m \le \pi_j(t) \le \pi_M$ with π_m and π_m being known scalars. Moreover, it is assumed that $\phi_i(r)$ is a given initial condition sequence.

In this chapter, among the n_u neuron nodes, only a small fraction of measurement outputs can be obtained. Without loss of generality, we employ a set $\mathcal{N}_a \triangleq \{1, 2, \ldots, n_s\} \subset \{1, 2, \ldots, n_u\}$ to denote the available measurements, in which the element $i \in \mathcal{N}_a$ corresponds to the ith neuron whose output is measurable with the following form:

$$
s_i(t) = c_i u_i(t) + r_i v_i(t), \quad i = 1, 2, \cdots, n_s
\tag{3.2}
$$

where $s_i(t)$ is the measurement output collected by the ith sensor. $v_i(t) \in \mathbb{R}$ is an amplitude bounded disturbance sequence, that is $\|v_i(t)\| \le \bar{v}_i$ with $\bar{v}_i$ being a given scalar. c_i and r_i are given constant scalars.

For the activation functions $p_i(\cdot)$ and $q_i(\cdot)$ of the ith neuron, the following assumption is given.

Assumption 3.1: *For the neuron i ($i \in \{i = 1, 2, \cdots, n_u\}$), the neuron activation functions $p_i(\cdot)$ and $q_i(\cdot)$ in (3.1) are constrained by the following conditions:*

$$
[p_i(s_1) - p_i(s_2) - h_{i1}(s_1 - s_2)][p_i(s_1) - p_i(s_2) - h_{i2}(s_1 - s_2)] \le 0,
$$

$$
[q_i(s_1) - q_i(s_2) - g_{i1}(s_1 - s_2)][q_i(s_1) - q_i(s_2) - g_{i2}(s_1 - s_2)] \le 0
\tag{3.3}
$$

for any $s_1, s_2 \in \mathbb{R}$, where h_{i1}, h_{i2}, g_{i1} and g_{i2} are known constants.

In order to facilitate the further development, we rewrite the component-based artificial neural network model (3.1) to the following compact form:

$$
\left\{
\begin{aligned}
u(t+1) &= Au(t) + Bp(u(t)) + Dq(u(t - \pi(t))) + J(t) + L\omega(t), \\
u(r) &= \phi(r), r \in [-\pi_M, 0]
\end{aligned}
\right.
\tag{3.4}
$$

where

$$u(t) \triangleq \mathrm{vec}_{n_u}\{u_i(t)\}, \; p(u(t)) \triangleq \mathrm{vec}_{n_u}\{p_i(u_i(t))\},$$

$$\phi(r) \triangleq \mathrm{vec}_{n_u}\{\phi_i(r)\}, \; q(u(t)) \triangleq \mathrm{vec}_{n_u}\{q_i(u_i(t))\},$$

$$A \triangleq \mathrm{diag}_{n_u}\{a_i\}, \; q_{\pi(t)}(u(t)) \triangleq \mathrm{vec}_{n_u}\{q(u(t - \pi_i(t)))\},$$

$$B \triangleq (b_{ij})_{n_u \times n_u}, \; D \triangleq (d_{ij})_{n_u \times n_u},$$

$$J(t) \triangleq \mathrm{vec}_{n_u}\{J_i(t)\}, \; L \triangleq \mathrm{vec}_{n_u}\{l_i\}.$$

Similarly, the augmented measurement output model of (3.2) is given as follows:

$$s(t) = Cu(t) + Rv(t) \tag{3.5}$$

where

$$s(t) \triangleq \mathrm{vec}_{n_s}\{s_i(t)\}, v(t) \triangleq \mathrm{vec}_{n_s}\{v_i(t)\},$$

$$C \triangleq \begin{bmatrix} C_1 & 0_{n_s \times (n_u - n_s)} \end{bmatrix}, R \triangleq \begin{bmatrix} R_1 & 0_{n_s \times (n_u - n_s)} \end{bmatrix},$$

$$C_1 \triangleq \mathrm{diag}\{c_1, c_2, \cdots, c_{n_s}\}, R_1 \triangleq \mathrm{diag}\{r_1, r_2, \cdots, r_{n_s}\}.$$

In this chapter, due to the scarcity of the communication resource, the data delivery from the sensor to the estimator is implemented through a channel with limited bandwidth. To be more specific, it is assumed that the total available bit rate of the communication media is $\mathcal{R}$. According to the average allocation protocol [68], the bit rate of each sensor-to-estimator transmission channel is assigned as $\mathcal{R}_i$, which means

$$\mathcal{R} = \sum_{i=1}^{n_s} \mathcal{R}_i. \tag{3.6}$$

In order to comply with the digitalized data transmission under the bandwidth constraint, a uniform-quantization-based encoding-decoding mechanism is introduced.

3.1.1 Encoding-Decoding Mechanism under Bit Rate Constraint

One of the purposes of the encoding-decoding mechanism is to convert the raw analogue signal to a digital one with less bits. To do this, in this subsection, we shall develop an easy-to-implement encoding-decoding mechanism for the data exchange between the sensor to the state estimator. First, a brief introduction of the uniform quantization scheme is provided. For the measurement output $s(t) \in \mathbb{R}^{n_s}$, we introduce a hyperrectangle

$$\mathcal{S}_b = \{|s_j(t)| \le b, j = 1, 2, \ldots, n_s\} \tag{3.7}$$

in which $s(t)$ is contained, where b is the scaling parameter, which is determined in terms of the practical requirement. Next, we divide the interval $[-b\ b]$ into N subsection uniformly and each of the subsection can be represented by

$$\mathcal{I}_{1j}(b) \triangleq \left\{ s_j \middle| -b \le s_j < -b + \frac{2b}{N} \right\}$$

$$\mathcal{I}_{2j}(b) \triangleq \left\{ s_j \middle| -b + \frac{2b}{N} \le s_j < -b + \frac{4b}{N} \right\}$$

$$\vdots$$

$$\mathcal{I}_{Nj}(b) \triangleq \left\{ s_j \middle| b - \frac{2b}{N} \le s_j \le b \right\}. \tag{3.8}$$

Accordingly, the hyperrectangle $\mathcal{S}_b$ is divided into N^{n_s} sub-hyperrectangles $\mathcal{I}_{\beta_1}(b) \times \mathcal{I}_{\beta_2}(b) \times \cdots \times \mathcal{I}_{\beta_{n_s}}(b)$ where $\beta_1,\ \beta_2,\ \ldots,\ \beta_{n_s} \in \{1, 2, \cdots, N\}$. Because an individual artificial neuron is allocated $\mathcal{R}_i$ bit rate for the data communication, according to the average allocation protocol, the quantization level for each $\mathcal{I}_{\beta_i}(b)$ $(i = 1, 2, \ldots, n_s)$ is determined as $\bar{N}_i = 2^{\mathcal{R}_i}$. Thus, the center of the hyperrectangle $\mathcal{S}_b$ is given by

$$\theta_b(\beta_1, \beta_2, \ldots, \beta_{n_s}) \triangleq \begin{bmatrix} g_1 & g_2 & \cdots & g_{n_s} \end{bmatrix}^T \tag{3.9}$$

where $g_i = -b + \frac{(2\beta_i - 1)b}{\bar{N}_i}$, $i = 1, 2, \ldots, n_s$. It is easy to find that the measurement output $s(t)$ uniquely corresponds to a sub-hyperrectangle $\mathcal{I}_{\beta_1}(b) \times \mathcal{I}_{\beta_2}(b) \times \cdots \times \mathcal{I}_{\beta_{n_s}}(b)$. Moreover, the distance between $s(t)$ and the center of the hyperrectangle $\mathcal{S}_b$ satisfies

$$\left\| s(t) - \theta_b(\beta_1, \beta_2, \ldots, \beta_{n_s}) \right\| \le \sqrt{\sum_{i=1}^{n_s} \frac{1}{2^{2\mathcal{R}_i}}} \, b. \tag{3.10}$$

After introducing the uniform quantization scheme and giving related discussions (3.7)–(3.10), we are now in a position to employ the following encoding-decoding mechanism which will be utilized for the digitalized data transmission.

Encoding Scheme:

Given the measurement output $s(t)$, if $s(t) \in \mathcal{I}_{\beta_1}(b) \times \mathcal{I}_{\beta_2}(b) \times \cdots \times \mathcal{I}_{\beta_{n_s}}(b)$, then $s(t)$ is encoded with the following codewords

$$\chi_{\mathcal{R}} = \begin{bmatrix} \beta_1 & \beta_2 & \cdots & \beta_{n_s} \end{bmatrix}. \tag{3.11}$$

Decoding Scheme:

Having received the generated codewords $\chi_{\mathcal{R}}$, the measurement output $s(t)$ is correspondingly decoded as

$$\bar{s}(t) = \eta_b(\chi_{\mathcal{R}})$$
$$= \left[-b + \frac{(2\beta_1-1)b}{2^{\mathcal{R}_1}} \quad -b + \frac{(2\beta_2-1)b}{2^{\mathcal{R}_2}} \quad \cdots \quad -b + \frac{(2\beta_{n_s}-1)b}{2^{\mathcal{R}_{n_s}}} \right]^T . \tag{3.12}$$

Denote the decoding error as $\tilde{s}(t) \triangleq s(t) - \bar{s}(t)$ and it follows from (3.9)–(3.12) that the decoding error $\tilde{s}(t)$ is bounded by

$$\|\tilde{s}(t)\| \leq \sqrt{\sum_{i=1}^{n_s} \frac{1}{2^{2\mathcal{R}_i}} b} \triangleq \bar{b}. \tag{3.13}$$

3.1.2 Estimator Structure

After obtaining the decoded measurement output $\bar{s}(t)$, the following state estimator is constructed for the dynamical system (3.4)

$$\begin{cases} \hat{u}(t+1) = A\hat{u}(t) + Bp(\hat{u}(t)) + Dq(\hat{u}(t - \pi(t))) + J(t) \\ \qquad\qquad + K(\bar{s}(t) - C\hat{u}(t)), \\ \hat{u}(r) = \hat{\phi}(r), r \in [-\pi_M, 0] \end{cases} \tag{3.14}$$

where $\hat{u}(t)$ is the estimations of $u(t)$ and $K \in \mathbb{R}^{n_u \times n_s}$ is the estimator gain matrix to be determined later.

Setting $\tilde{u}(t) \triangleq u(t) - \hat{u}(t)$, one derives the following estimation error dynamics:

$$\begin{cases} \tilde{u}(t+1) = \tilde{A}\tilde{u}(t) + B\tilde{p}(\tilde{u}(t)) + D\tilde{q}(\tilde{u}(t - \pi(t))) + E\xi(t), \\ \tilde{u}(r) = \tilde{\phi}(r), r \in [-\pi_M, 0] \end{cases} \tag{3.15}$$

where

$$\tilde{A} \triangleq A - KC, \quad \tilde{\phi}(r) \triangleq \phi(r) - \hat{\phi}(r),$$
$$E \triangleq [L \quad -KR \quad -K],$$
$$\xi(t) \triangleq [w^T(t) \quad v^T(t) \quad \tilde{s}^T(t)]^T,$$
$$\tilde{p}(\tilde{u}(t)) \triangleq p(u(t)) - \hat{p}(\hat{u}(t)),$$
$$\tilde{q}(\tilde{u}(t - \pi(t))) \triangleq q(u(t - \pi(t))) - \hat{q}(\hat{u}(t - \pi(t))).$$

Before proceeding further, we introduce the following definition and assumption.

Definition 3.1: *The estimation error dynamics* (3.15) *is said to be finite-time bounded with respect to* $(\varepsilon_1, \varepsilon_2, \mathcal{S}, \bar{\bar{\xi}}, k_s)$, *if, for* $\forall t = 1, 2, \ldots, k_s$, *the following relationship is true*

$$\begin{cases} \tilde{u}^T(k)\mathcal{S}\tilde{u}(k) \leq \varepsilon_1, k \in [-\pi_M, 0] \\[2ex] \displaystyle\sum_{t=0}^{k_s} \xi^T(t)\xi(t) \leq \bar{\bar{\xi}} \end{cases} \implies \tilde{u}^T(t)\mathcal{S}\tilde{u}(t) \leq \varepsilon_2 \qquad (3.16)$$

where $\bar{\bar{\xi}} \triangleq (k_s + 1)(n_u\bar{w}^2 + n_s\bar{v}^2 + \bar{b})$, $0 \leq \varepsilon_1 < \varepsilon_2$ *are known positive scalars and* $\mathcal{S} > 0$ *is a prescribed weighted matrix.*

Assumption 3.2: *The initial values of the error dynamics* (3.15) *meet the following requirements*

$$\begin{cases} \tilde{u}^T(i)\mathcal{S}\tilde{u}(i) \leq \varepsilon_1 \\[2ex] (\tilde{u}(i+1) - \tilde{u}(i))^T \mathcal{S}(\tilde{u}(i+1) - \tilde{u}(i)) \leq \varpi \end{cases} \qquad (3.17)$$

for $i = -\pi_M, -\pi_M + 1, \cdots, -1, 0$.

The objective of this chapter is to design a filter structure (3.14) based on partial measurable neuron outputs subject to constrained bit-rates such that the estimation error dynamics (3.15) is finite-time bounded with respect to $(\varepsilon_1, \varepsilon_2, \mathcal{S}, \bar{\bar{\xi}}, k_s)$.

3.2 Main Results

Theorem 3.1: *For the ANN* (3.4) *with the bit-rate constraint* (3.6), *the estimator gain matrix K is given. The corresponding state estimation error dynamics* (3.15) *is finite-time bounded with respect to* $(\varepsilon_1, \varepsilon_2, \mathcal{S}, \bar{\bar{\xi}}, k_s)$ *if there exist positive matrices* $\mathcal{Q}_i$ $(i = 1, 2, 3, 4)$, *some positive scalars* $\alpha \in (0, 1)$, $\tau_i > 0$ $(i = 1, 2)$ *and* $\sigma_i > 0$ $(i = 0, 1)$ *such that the following inequality constraints*

$$\Gamma \triangleq \begin{bmatrix} \Gamma_{11} & 0 & 0 & 0 & \Gamma_{15} & \Gamma_{16} & \Gamma_{17} & \Gamma_{18} \\ * & \Gamma_{22} & \Gamma_{23} & 0 & 0 & 0 & 0 & 0 \\ * & * & \Gamma_{33} & \Gamma_{34} & 0 & \Gamma_{36} & 0 & 0 \\ * & * & * & \Gamma_{44} & 0 & 0 & 0 & 0 \\ * & * & * & * & \Gamma_{55} & 0 & 0 & \Gamma_{58} \\ * & * & * & * & * & \Gamma_{66} & 0 & \Gamma_{68} \\ * & * & * & * & * & * & \Gamma_{77} & \Gamma_{78} \\ * & * & * & * & * & * & * & \Gamma_{88} \end{bmatrix} < 0, \qquad (3.18)$$

$$\sigma_0 \mathcal{S} \leq \mathcal{Q}_j \leq \sigma_1 \mathcal{S}, \quad j = 1, 2, 3, \tag{3.19}$$

$$\frac{(1-\alpha)^t \rho_2 + (1-\alpha)\lambda_{\max}\{\mathcal{Q}_4\}\bar{\xi}}{\sigma_0} \leq \varepsilon_2 \tag{3.20}$$

hold, where

$$\Gamma_{11} \triangleq \left(A - KC - \left(1 - \frac{\alpha}{2}\right)I\right)^T \mathcal{Q}_1 + \mathcal{Q}_1 \left(A - KC - \left(1 - \frac{\alpha}{2}\right)I\right)$$
$$+ (\pi + 1)\mathcal{Q}_2 - \tau_1 \tilde{\mathcal{H}}_1,$$

$$\Gamma_{15} \triangleq \mathcal{Q}_1 B - \tau_1 \tilde{\mathcal{H}}_2, \Gamma_{16} \triangleq \mathcal{Q}_1 D, \Gamma_{17} \triangleq \mathcal{Q}_1 E,$$

$$\Gamma_{18} \triangleq (A - KC - I)^T, \Gamma_{22} = \Gamma_{23} = \Gamma_{34} = \Gamma_{44} \triangleq -\frac{(1-\alpha)^{\tau_M}}{\pi}\mathcal{Q}_3,$$

$$\Gamma_{33} \triangleq -\frac{2(1-\alpha)^{\tau_M}}{\pi}\mathcal{Q}_3 - \tau_2 \tilde{\mathcal{G}}_1, \Gamma_{36} \triangleq -\tau_2 \tilde{\mathcal{G}}_2, \Gamma_{55} \triangleq -\tau_1 I,$$

$$\Gamma_{58} \triangleq B^T, \Gamma_{66} \triangleq -\tau_2 I, \Gamma_{68} \triangleq D^T, \Gamma_{77} \triangleq -\mathcal{Q}_4, \Gamma_{78} \triangleq E^T,$$

$$\Gamma_{88} \triangleq -\mathscr{Q}^{-1}, \mathscr{Q} \triangleq \mathcal{Q}_1 + \pi\mathcal{Q}_3, \pi \triangleq \pi_M - \pi_m, \mu \triangleq \frac{(1-\alpha)^{\pi_M}}{\pi},$$

$$\rho_2 \triangleq (1 + \pi_M + \pi_M(\pi_M - \pi_m))\sigma_1\varepsilon_1 + (\pi_M - \pi_m)\pi_M\varpi,$$

$$\tilde{\mathcal{H}}_1 \triangleq \frac{1}{2}(\mathcal{H}_1^T \mathcal{H}_2 + \mathcal{H}_2^T \mathcal{H}_1), \tilde{\mathcal{H}}_2 \triangleq -\frac{1}{2}(\mathcal{H}_1^T + \mathcal{H}_2^T),$$

$$\tilde{\mathcal{G}}_1 \triangleq \frac{1}{2}(\mathcal{G}_1^T \mathcal{G}_2 + \mathcal{G}_2^T \mathcal{G}_1), \tilde{\mathcal{G}}_2 \triangleq -\frac{1}{2}(\mathcal{G}_1^T + \mathcal{G}_2^T),$$

$$\mathcal{H}_1 \triangleq \mathrm{diag}\{h_{11}, h_{21}, h_{31}\}, \mathcal{H}_2 \triangleq \mathrm{diag}\{h_{12}, h_{22}, h_{32}\},$$

$$\mathcal{G}_1 \triangleq \mathrm{diag}\{g_{11}, g_{21}, g_{31}\}, \mathcal{G}_2 \triangleq \mathrm{diag}\{g_{12}, g_{22}, g_{32}\}.$$

Proof: *Construct a Lyapunov functional as*

$$V(t) \triangleq V_1(t) + V_2(t) + V_3(t) \tag{3.21}$$

where

$$V_1(t) \triangleq \tilde{u}^T(t)\mathcal{Q}_1\tilde{u}(t),$$

$$V_2(t) \triangleq \sum_{i=t-\pi(t)}^{t-1} (1-\alpha)^{t-i-1}\tilde{u}^T(i)\mathcal{Q}_2\tilde{u}(i),$$

$$V_3(t) \triangleq \sum_{j=t-\pi_M+1}^{t-\pi_m} \sum_{i=j}^{t-1} (1-\alpha)^{t-i-1}\tilde{u}^T(i)\mathcal{Q}_2\tilde{u}(i),$$

$$V_4(t) \triangleq \sum_{j=t-\pi_M+1}^{t-\pi_m} \sum_{i=j}^{t} (1-\alpha)^{t-i}(\tilde{u}(i) - \tilde{u}(i-1))^T \mathcal{Q}_3(\tilde{u}(i) - \tilde{u}(i-1)).$$

Then, calculating the difference along the trajectory of (3.15), one obtains

$$\Delta V(t) = \sum_{i=1}^{4} \Delta V_i(t) \tag{3.22}$$

where $\Delta V_i(t) \triangleq V_i(t+1) - V_i(t)$. *Next, terms* $\Delta V_i(t) + \alpha V_i(t)$ $(i = 1, 2, 3, 4)$ *are, respectively, calculated as follows:*

$$\begin{aligned}
\Delta V_1(t) + \alpha V_1(t) &= V_1(t+1) - (1-\alpha)V_1(t) \\
&= \tilde{u}^T(t+1)\mathcal{Q}_1\tilde{u}(t+1) - (1-\alpha)\tilde{u}^T(t)\mathcal{Q}_1\tilde{u}(t)\} \\
&= (\tilde{u}(t+1) - \tilde{u}(t))^T \mathcal{Q}_1(\tilde{u}(t+1) - \tilde{u}(t)) \\
&\quad + 2\tilde{u}^T(t)\mathcal{Q}_1\left(\tilde{u}(t+1) - \left(1 - \frac{\alpha}{2}\right)\tilde{u}(t)\right),
\end{aligned} \tag{3.23}$$

$$\begin{aligned}
\Delta V_2(t) + \alpha V_2(t) &= \sum_{i=t-\pi(t+1)+1}^{t} (1-\alpha)^{t-i}\tilde{u}^T(i)\mathcal{Q}_2\tilde{u}(i) \\
&\quad - \sum_{i=t-\pi(t)}^{t-1} (1-\alpha)^{t-i-1}\tilde{u}^T(i)\mathcal{Q}_2\tilde{u}(i) \\
&\quad + \alpha \sum_{i=t-\pi(t)}^{t-1} (1-\alpha)^{t-i-1}\tilde{u}^T(i)\mathcal{Q}_2\tilde{u}(i) \\
&= \sum_{i=t-\pi(t+1)+1}^{t} (1-\alpha)^{t-i}\tilde{u}^T(i)\mathcal{Q}_2\tilde{u}(i) \\
&\quad - \sum_{i=t-\pi(t)}^{t-1} (1-\alpha)^{t-i}\tilde{u}^T(i)\mathcal{Q}_2\tilde{u}(i) \\
&= \tilde{u}^T(t)\mathcal{Q}_2\tilde{u}(t) - (1-\alpha)^{\pi(t)}\tilde{u}^T(t-\pi(t))\mathcal{Q}_2 \\
&\quad \times \tilde{u}(t-\pi(t)) + \sum_{i=t-\pi(t+1)+1}^{t-1} (1-\alpha)^{t-i}\tilde{u}^T(i)\mathcal{Q}_2\tilde{u}(i) \\
&\quad - \sum_{i=t-\pi(t)+1}^{t-1} (1-\alpha)^{t-i}\tilde{u}^T(i)\mathcal{Q}_2\tilde{u}(i) \\
&\leq \tilde{u}^T(t)\mathcal{Q}_2\tilde{u}(t) - (1-\alpha)^{\pi_M}\tilde{u}^T(t-\pi(t))\mathcal{Q}_2\tilde{u}(t-\pi(t)) \\
&\quad + \sum_{i=t-\pi(t+1)+1}^{t-\pi_m} (1-\alpha)^{t-i}\tilde{u}^T(i)\mathcal{Q}_2\tilde{u}(i),
\end{aligned} \tag{3.24}$$

$$\Delta V_3(t) + \alpha V_3(t) = \sum_{j=t-\pi_M+2}^{t-\pi_m+1} \sum_{i=j}^{t} (1-\alpha)^{t-i} \tilde{u}^T(i) \mathcal{Q}_2 \tilde{u}(i)$$

$$- \sum_{j=t-\pi_M+1}^{t-\pi_m} \sum_{i=j}^{t-1} (1-\alpha)^{t-i-1} \tilde{u}^T(i) \mathcal{Q}_2 \tilde{u}(i)$$

$$+ \alpha \sum_{j=t-\pi_M+1}^{t-\pi_m} \sum_{i=j}^{t-1} (1-\alpha)^{t-i-1} \tilde{u}^T(i) \mathcal{Q}_2 \tilde{u}(i) \qquad (3.25)$$

$$= \sum_{j=t-\pi_M+1}^{t-\pi_m} \left(\sum_{i=j+1}^{t} - \sum_{i=j}^{t-1} \right) (1-\alpha)^{t-i} \tilde{u}^T(i) \mathcal{Q}_2 \tilde{u}(i)$$

$$= \pi \tilde{u}^T(t) \mathcal{Q}_2 \tilde{u}(t) - \sum_{i=t-\pi_M+1}^{t-\pi_m} (1-\alpha)^{t-i} \tilde{u}^T(i) \mathcal{Q}_2 \tilde{u}(i),$$

$$\Delta V_4(t) + \alpha V_4(t)$$

$$= \sum_{j=t-\pi_M+2}^{t-\pi_m+1} \sum_{i=j}^{t+1} (1-\alpha)^{t-i+1} (\tilde{u}(i) - \tilde{u}(i-1))^T \mathcal{Q}_3 (\tilde{u}(i) - \tilde{u}(i-1)) \qquad (3.26)$$

$$- \sum_{j=t-\pi_M+1}^{k-\pi_m} \sum_{i=j}^{t} (1-\alpha)^{t-i} (\tilde{u}(i) - \tilde{u}(i-1))^T \mathcal{Q}_3 (\tilde{u}(i) - \tilde{u}(i-1))$$

$$+ \alpha \sum_{j=t-\pi_M+1}^{t-\pi_m} \sum_{i=j}^{t} (1-\alpha)^{t-i} (\tilde{u}(i) - \tilde{u}(i-1))^T \mathcal{Q}_3 (\tilde{u}(i) - \tilde{u}(i-1))$$

$$= \sum_{j=t-\pi_M+1}^{t-\pi_m} \left(\sum_{i=j+1}^{t+1} - \sum_{i=j}^{t} \right) (1-\alpha)^{t-i+1}$$

$$\times (\tilde{u}(i) - \tilde{u}(i-1))^T \mathcal{Q}_3 (\tilde{u}(i) - \tilde{u}(i-1))$$

$$= \pi (\tilde{u}(t+1) - \tilde{u}(t))^T \mathcal{Q}_3 (\tilde{u}(t+1) - \tilde{u}(t))$$

$$- \sum_{i=t-\pi_M+1}^{t-\pi_m} (1-\alpha)^{t-i+1} (\tilde{u}(i) - \tilde{u}(i-1))^T \mathcal{Q}_3 (\tilde{u}(i) - \tilde{u}(i-1))$$

$$\leq \pi (\tilde{u}(t+1) - \tilde{u}(t))^T \mathcal{Q}_3 (\tilde{u}(t+1) - \tilde{u}(t))$$

$$- \mu (\tilde{u}(t-\pi(t)) - \tilde{u}(t-\pi_M))^T \mathcal{Q}_3 (\tilde{u}(t-\pi(t)) - \tilde{u}(t-\pi_M))$$

$$- \mu (\tilde{u}(t-\pi_m) - \tilde{u}(t-\pi(t)))^T \mathcal{Q}_3 (\tilde{u}(t-\pi_m) - \tilde{u}(t-\pi(t))). \qquad (3.27)$$

Noting the nonlinear constraint for $p(u(t))$ and $q(u(t))$ in Assumption 3.1, one further obtains

$$\begin{bmatrix} \tilde{u}(t) \\ \tilde{p}(\tilde{u}(t-\pi(t))) \end{bmatrix}^T \begin{bmatrix} \tilde{\mathcal{H}}_1 & \tilde{\mathcal{H}}_2 \\ * & I \end{bmatrix} \begin{bmatrix} \tilde{u}(t) \\ \tilde{p}(\tilde{u}(t)) \end{bmatrix} \le 0, \tag{3.28}$$

$$\begin{bmatrix} \tilde{u}(t-\pi(t)) \\ \tilde{q}(\tilde{u}(t-\pi(t))) \end{bmatrix}^T \begin{bmatrix} \tilde{\mathcal{G}}_1 & \tilde{\mathcal{G}}_2 \\ * & I \end{bmatrix} \begin{bmatrix} \tilde{u}(t-\pi(t)) \\ \tilde{q}(\tilde{u}(t-\pi(t))) \end{bmatrix} \le 0. \tag{3.29}$$

Based on the discussions from (3.22)–(3.29), one derives

$$\begin{aligned}
\Delta V(t) + \alpha V(t) \le{}& (\tilde{u}(t+1) - \tilde{u}(t))^T \mathscr{Q}(\tilde{u}(t+1) - \tilde{u}(t)) \\
& + 2\tilde{u}^T(t)\mathcal{Q}_1 \left(\tilde{u}(t+1) - \left(1 - \frac{\alpha}{2}\right)\tilde{u}(t) \right) + (\pi+1)\tilde{u}^T(t)\mathcal{Q}_2\tilde{u}(t) \\
& - (1-\alpha)^{\pi_M}\tilde{u}^T(t-\pi(t))\mathcal{Q}_2\tilde{u}(t-\pi(t)) \\
& - \mu\big(\tilde{u}(t-\pi_m) - \tilde{u}(t-\pi(t))\big)^T \mathcal{Q}_3\big(\tilde{u}(t-\pi_m) - \tilde{u}(t-\pi(t))\big) \\
& - \mu\big(\tilde{u}(t-\pi(t)) - \tilde{u}(t-\pi_M)\big)^T \mathcal{Q}_3\big(\tilde{u}(t-\pi(t)) - \tilde{u}(t-\pi_M)\big) \\
& - \tau_1 \begin{bmatrix} \tilde{u}(t) \\ \tilde{p}(\tilde{u}(t)) \end{bmatrix}^T \begin{bmatrix} \tilde{\mathcal{H}}_1 & \tilde{\mathcal{H}}_2 \\ * & I \end{bmatrix} \begin{bmatrix} \tilde{u}(t) \\ \tilde{p}(\tilde{u}(t)) \end{bmatrix} \\
& - \tau_2 \begin{bmatrix} \tilde{u}(t-\pi(t)) \\ \tilde{q}(\tilde{u}(t-\pi(t))) \end{bmatrix}^T \begin{bmatrix} \tilde{\mathcal{G}}_1 & \tilde{\mathcal{G}}_2 \\ * & I \end{bmatrix} \begin{bmatrix} \tilde{u}(t-\pi(t)) \\ \tilde{q}(\tilde{u}(t-\pi(t))) \end{bmatrix}.
\end{aligned} \tag{3.30}$$

In light of (3.30), one has

$$\Delta V(t) + \alpha V(t) - \xi^T(t)\mathcal{Q}_4\xi(t) \le \eta^T(t)\tilde{\Gamma}\eta(t) \tag{3.31}$$

where

$$\eta(k) \triangleq \Big[\tilde{u}^T(t) \quad \tilde{u}^T(t-\pi_m) \quad \tilde{u}^T(t-\pi(t)) \quad \tilde{u}^T(t-\pi_M)$$

$$\tilde{p}^T(\tilde{u}(t)) \quad \tilde{q}^T(\tilde{u}(t-\pi(t))) \quad \xi^T(t) \Big]^T,$$

$$\Omega \triangleq \begin{bmatrix} \hat{A} & 0 & 0 & 0 & B & D & E \end{bmatrix},$$

$$\hat{A} \triangleq \left(A - KC - \left(1 - \frac{\alpha}{2}\right)I \right), \quad \tilde{\Gamma} \triangleq \bar{\Gamma} + \Omega^T \mathscr{Q}\Omega,$$

$$\bar{\Gamma} \triangleq \begin{bmatrix}
\Gamma_{11} & 0 & 0 & 0 & \Gamma_{15} & \Gamma_{16} & \Gamma_{17} \\
* & \Gamma_{22} & \Gamma_{23} & 0 & 0 & 0 & 0 \\
* & * & \Gamma_{33} & \Gamma_{34} & 0 & \Gamma_{36} & 0 \\
* & * & * & \Gamma_{44} & 0 & 0 & 0 \\
* & * & * & * & \Gamma_{55} & 0 & 0 \\
* & * & * & * & * & \Gamma_{66} & 0 \\
* & * & * & * & * & * & \Gamma_{77}
\end{bmatrix}.$$

In terms of the Schur Complement Lemma, it follows from the condition (3.18) in Theorem 3.1 that $\tilde{\Gamma} < 0$ and thus one finds $\Delta V(t) + \alpha V(t) - \xi^T(t)\mathcal{Q}_4\xi(t) \leq 0$, which further implies

$$V(t+1) \leq (1-\alpha)V(t) + \lambda_{\max}\{\mathcal{Q}_4\}\xi^T(t)\xi(t)$$

$$\leq (1-\alpha)^2 V(t-1) + (1-\alpha)\lambda_{\max}\{\mathcal{Q}_4\}\xi^T(t-1)\xi(t-1)$$

$$+ \lambda_{\max}\{\mathcal{Q}_4\}\xi^T(t)\xi(t)$$

$$\vdots \tag{3.32}$$

$$\leq (1-\alpha)^{t+1}V(0) + \sum_{i=0}^{t}(1-\alpha)^{t-i}\lambda_{\max}\{\mathcal{Q}_4\}\xi^T(i)\xi(i)$$

$$\leq (1-\alpha)^{t+1}V(0) + (1-\alpha)\lambda_{\max}\{\mathcal{Q}_4\}\bar{\xi}.$$

Based on the Assumption 3.2, the initial condition $V(0)$ satisfies

$$V(0) \leq (1 + \pi_M + \pi_M(\pi_M - \pi_m))\sigma_1\varepsilon_1 + (\pi_M - \pi_m)\pi_M\varpi$$

$$\triangleq \rho_2.$$

It is easy to see from (3.22) that $V(t) \geq \sigma_0\tilde{u}^T(t)\mathcal{S}\tilde{u}(t)$. On the basis of (3.19) and the discussions made above, we have

$$\sigma_0\tilde{u}^T(t)\mathcal{S}\tilde{u}(t) \leq V(t) \leq (1-\alpha)^t\rho_2 + (1-\alpha)\lambda_{\max}\{\mathcal{Q}_4\}\bar{\xi}, \tag{3.33}$$

which further indicates

$$\tilde{u}^T(t)\mathcal{S}\tilde{u}(t) \leq \frac{(1-\alpha)^t\rho_2 + (1-\alpha)\lambda_{\max}\{\mathcal{Q}_4\}\bar{\xi}}{\sigma_0} = \varepsilon_2. \tag{3.34}$$

Consequently, it follows from the definition (3.1) that the estimation error dynamics (3.15) is finite-time bounded and the proof of this theorem is complete.

Theorem 3.2: *For the ANN (3.4) with the bit-rate constraint (3.6), the corresponding state estimation error dynamics (3.15) is finite-time bounded with respect to $(\varepsilon_1, \varepsilon_2, \mathcal{S}, \bar{\xi}, k_s)$ if there exist positive matrices $\mathcal{Q}_i$ ($i = 1, 2, 3, 4$), a matrix $\mathcal{X}$, some positive scalars $\alpha \in (0, 1)$, $\tau_i > 0$ ($i = 1, 2$) and $\sigma_i > 0$ ($i = 0, 1$) such that the following inequality constraints*

$$\Gamma \triangleq \begin{bmatrix} \hat{\Gamma}_{11} & 0 & 0 & 0 & \Gamma_{15} & \Gamma_{16} & \hat{\Gamma}_{17} & \hat{\Gamma}_{18} \\ * & \Gamma_{22} & \Gamma_{23} & 0 & 0 & 0 & 0 & 0 \\ * & * & \Gamma_{33} & \Gamma_{34} & 0 & \Gamma_{36} & 0 & 0 \\ * & * & * & \Gamma_{44} & 0 & 0 & 0 & 0 \\ * & * & * & * & \Gamma_{55} & 0 & 0 & \hat{\Gamma}_{58} \\ * & * & * & * & * & \Gamma_{66} & 0 & \Gamma_{68} \\ * & * & * & * & * & * & \Gamma_{77} & \hat{\Gamma}_{78} \\ * & * & * & * & * & * & * & \hat{\Gamma}_{88} \end{bmatrix} < 0, \tag{3.35}$$

$$\sigma_0 \mathcal{S} \leq \mathcal{Q}_j \leq \sigma_1 \mathcal{S}, \quad j = 1, 2, 3 \tag{3.36}$$

$$\frac{(1-\alpha)^t \rho_2 + (1-\alpha)\lambda_{\max}\{\mathcal{Q}_4\}\bar{\xi}}{\sigma_0} \leq \varepsilon_2 \tag{3.37}$$

hold, where

$$\hat{\Gamma}_{11} \triangleq \left(\mathcal{Q}_1 A - \mathcal{X}C - \left(1 - \frac{\alpha}{2}\right)\mathcal{Q}_1\right)^T + \left(\mathcal{Q}_1 A - \mathcal{X}C - \left(1 - \frac{\alpha}{2}\right)\mathcal{Q}_1\right)$$
$$+ (\pi + 1)\mathcal{Q}_2 - \tau_1 \tilde{\mathcal{H}}_1$$

$$\Gamma_{15} \triangleq \mathcal{Q}_1 B - \tau_1 \tilde{\mathcal{H}}_2, \Gamma_{16} \triangleq \mathcal{Q}_1 D, \hat{\Gamma}_{17} \triangleq [\mathcal{Q}_1 L \quad -\mathcal{X}R \quad -\mathcal{X}],$$

$$\hat{\Gamma}_{18} \triangleq (\mathcal{Q}_1 A - \mathcal{X}_1 C - \mathcal{Q}_1)^T,$$

$$\Gamma_{22} = \Gamma_{23} = \Gamma_{34} = \Gamma_{44} \triangleq -\frac{(1-\alpha)^{\tau_M}}{\pi}\mathcal{Q}_3,$$

$$\Gamma_{33} \triangleq -\frac{2(1-\alpha)^{\tau_M}}{\pi}\mathcal{Q}_3 - \tau_2 \tilde{\mathcal{G}}_1, \Gamma_{36} \triangleq -\tau_2 \tilde{\mathcal{G}}_2, \Gamma_{55} \triangleq -\tau_1 I,$$

$$\hat{\Gamma}_{58} \triangleq (\mathcal{Q}_1 B)^T, \Gamma_{66} \triangleq -\tau_2 I, \hat{\Gamma}_{68} \triangleq (\mathcal{Q}_1 D)^T, \Gamma_{77} \triangleq -\mathcal{Q}_4,$$

$$\hat{\Gamma}_{78} \triangleq [\mathcal{Q}_1 L \quad -\mathcal{X}R \quad -\mathcal{X}]^T, \Gamma_{88} \triangleq \mathcal{Q} - \mathcal{Q}_1^T - \mathcal{Q}_1.$$

Moreover, the estimator gain matrix K is determined by

$$K = \mathcal{Q}_1^{-1}\mathcal{X}. \tag{3.38}$$

Proof: *Implementing the congruence transformation* $\mathrm{diag}\{I, I, I, I, I, I, I, \mathcal{Q}_1^{-1}\}$ *to (3.35) results in*

$$\begin{bmatrix} \hat{\Gamma}_{11} & 0 & 0 & 0 & \Gamma_{15} & \Gamma_{16} & \hat{\Gamma}_{17} & \Gamma_{18} \\ * & \Gamma_{22} & \Gamma_{23} & 0 & 0 & 0 & 0 & 0 \\ * & * & \Gamma_{33} & \Gamma_{34} & 0 & \Gamma_{36} & 0 & 0 \\ * & * & * & \Gamma_{44} & 0 & 0 & 0 & 0 \\ * & * & * & * & \Gamma_{55} & 0 & 0 & \Gamma_{58} \\ * & * & * & * & * & \Gamma_{66} & 0 & \Gamma_{68} \\ * & * & * & * & * & * & \Gamma_{77} & \Gamma_{78} \\ * & * & * & * & * & * & * & \hat{\Gamma}_{88}^1 \end{bmatrix} < 0 \tag{3.39}$$

where $\hat{\Gamma}_{88}^1 \triangleq \mathcal{Q}_1^{-T}\mathcal{Q}\mathcal{Q}_1^{-1} - \mathcal{Q}_1^{-T} - \mathcal{Q}_1^{-1}$ *and the other parameters are as the same as given in Theorem 3.1. Applying the useful inequality* $-\mathcal{Q}^{-1} \leq \mathcal{Q}_1^{-T}\mathcal{Q}\mathcal{Q}_1^{-1} - \mathcal{Q}_1^{-T} - \mathcal{Q}_1^{-1}$ *and (3.38), it is not difficult to see that (3.18) is true provided that (3.35) is met. Therefore, according to the proof of Theorem 3.1, the desired finite-time bounded requirement is accordingly achieved and the state estimator gain matrix can be designed as* $K = \mathcal{Q}_1^{-1}\mathcal{X}$. *The proof of this theorem is now ended.*

3.3 An Illustrative Example

This section presents an illustrative numerical example to show the effectiveness of the proposed partial-neurons-based finite-time state estimation problem for a class of discrete-time delayed neural networks subject to constrained bit rates.

In this example, the considered neural network is assumed to be composed of three neurons and the relevant parameters (e.g. the system matrices, the connection weights, etc.) are in the simulation are given as follows:

$$A = \begin{bmatrix} 0.81 & 0 & 0 \\ 0 & 0.85 & 0 \\ 0 & 0 & 0.75 \end{bmatrix}, B = \begin{bmatrix} 0.2 & 0.1 & 0.4 \\ 0.3 & 0.2 & 0 \\ 0 & 0.3 & 0.4 \end{bmatrix},$$

$$C = \begin{bmatrix} 2.2 & 0 & 0 \\ 0 & 2.8 & 0 \end{bmatrix}, D = \begin{bmatrix} 0.5 & 0.4 & 0.13 \\ 0.5 & 0.1 & 0.1 \\ 0.5 & 0.1 & 0.5 \end{bmatrix},$$

$$R = \begin{bmatrix} 0.05 & 0 \\ 0 & 0.04 \end{bmatrix}, L = \begin{bmatrix} 0.1 & 0 & 0 \\ 0 & 0.3 & 0.3 \\ 0 & 0 & 0.2 \end{bmatrix}.$$

It is seen from the expression of matrix C that only the first and the second neurons' state information can be measured and the status of the third neuron is unavailable, which is consistent with the theoretical analysis for the partial-neuron-based state estimation problem. The nonlinear activation function $p(\cdot)$ and $q(\cdot)$ are chosen as

$$p(u(t)) = \begin{bmatrix} -0.5u_1(t) + \tanh(0.2u_1(t)) \\ -\tanh(0.45u_2(t)) + 0.65u_2(t) \\ \tanh(0.7u_3(t)) + 0.3u_3(t) \end{bmatrix},$$

$$q(u(t - \pi(t))) = \begin{bmatrix} -0.5u_1(t - \pi(t)) + \tanh(0.2u_1(t - \pi(t))) \\ -\tanh(0.45u_2(t - \pi(t))) + 0.65u_2(t - \pi(t)) \\ \tanh(0.7u_3(t - \pi(t))) + 0.3u_3(t - \pi(t)) \end{bmatrix}.$$

Set the varying time delay as $\pi(t) = 2 + \frac{1+(-1)^t}{2}$ and consequently the upper and lower bounds of $\pi(t)$ are, respectively, chosen as $\pi_m = 2$ and $\pi_M = 3$. The input signal is set as $J(t) = [0.1\exp(-t) \quad 0 \quad 0.08\sin(t/2)]^T$. Let the upper bound of the process disturbance $w_i(t)$ and the measurement noise $v_i(t)$ take values over the interval $[-0.5\ 0.5]$ uniformly and thus their upper bounds are, respectively, given as $\bar{w}_i = 0.5$ and $\bar{v}_i = 0.5$. The total available bit rate of the complex network is given as 24bps. According to the average allocation protocol in [68], the individual bit rate for each neuron node is $\mathcal{R}_1 = \mathcal{R}_2 = \mathcal{R}_3 = 8$bps and the quantization level is consequently calculated as $\bar{q}_i = 256$.

The initial condition of the system state $u(t)$ and its estimation are given as $u(t) = [-1 \quad 1.4 \quad -0.8]^T$ and $\hat{u}(t) = [0 \quad 0 \quad 0]^T$. Let the matrix S in Definition 3.1 be $S = 0.1I$. With the help of the Matlab toolbox, the matrix inequalities in (3.35) can be solved and the estimator gain matrix is determined as

$$
K = \begin{bmatrix} 0.2683 & 0.1564 \\ 0.0788 & 0.2607 \\ 0.3369 & 0.2755 \end{bmatrix}.
$$

The corresponding simulation results are exhibited in Figures 3.1–3.7. Figures 3.1–3.3 show the curves of the states of three neurons and their

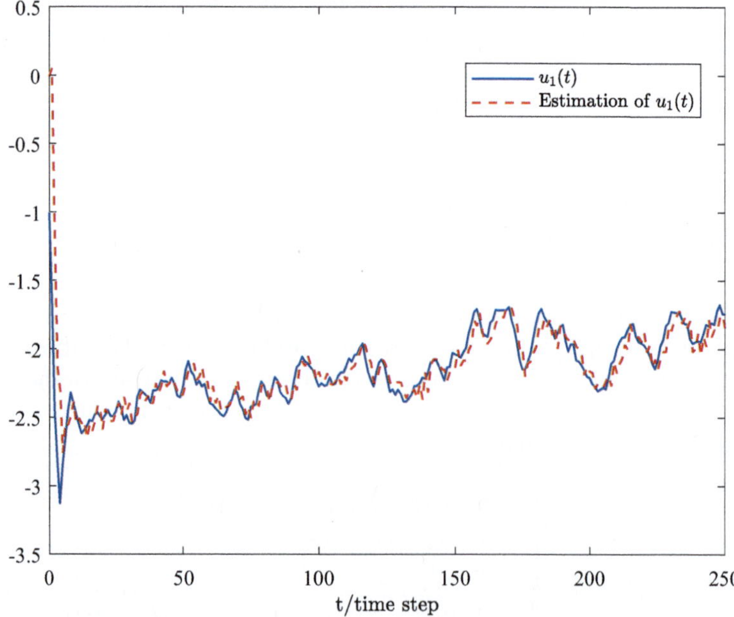

FIGURE 3.1

State trajectory $u_1(t)$ and its estimation.

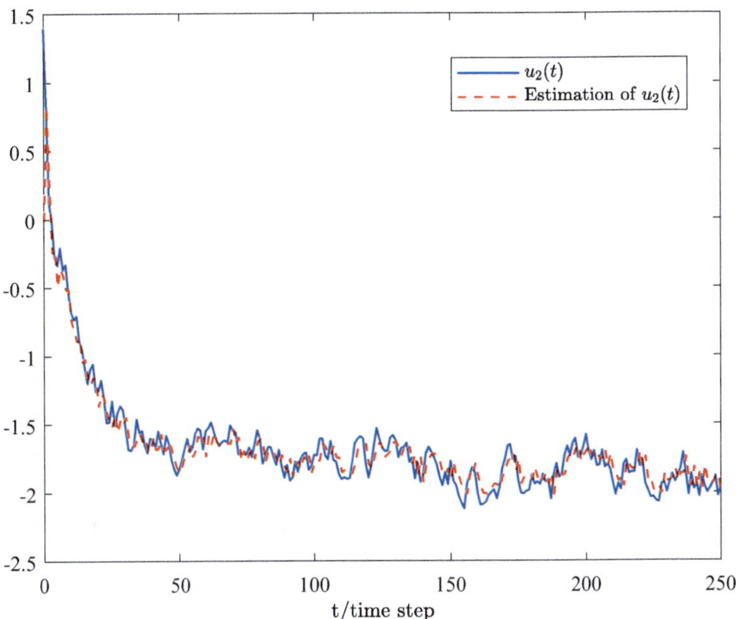

FIGURE 3.2

State trajectory $u_2(t)$ and its estimation.

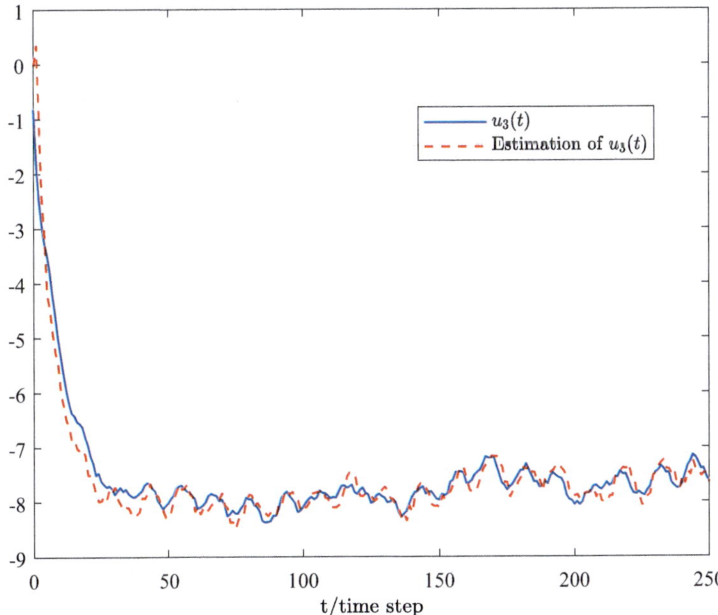

FIGURE 3.3

State trajectory $u_3(t)$ and its estimation.

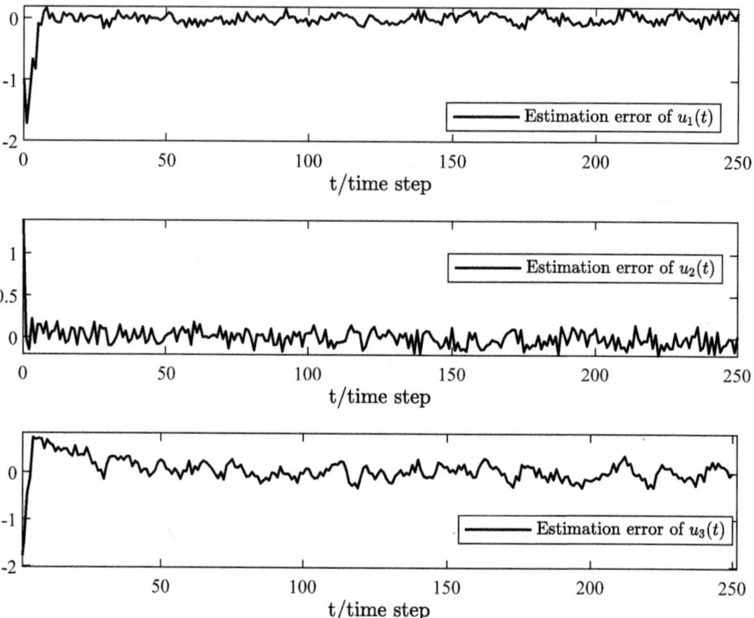

FIGURE 3.4

Estimation error of neuron i ($i = 1, 2, 3$).

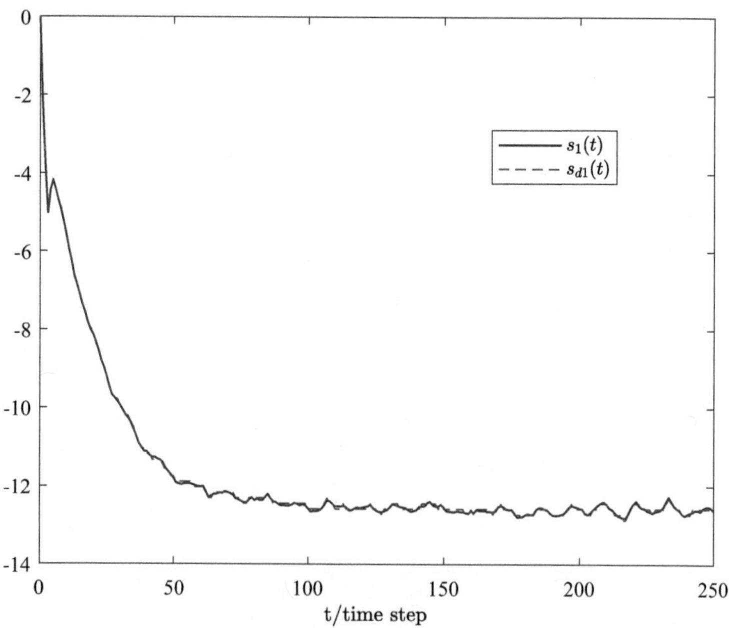

FIGURE 3.5

The measurement output $s_1(t)$ of the first neuron and its decoded value.

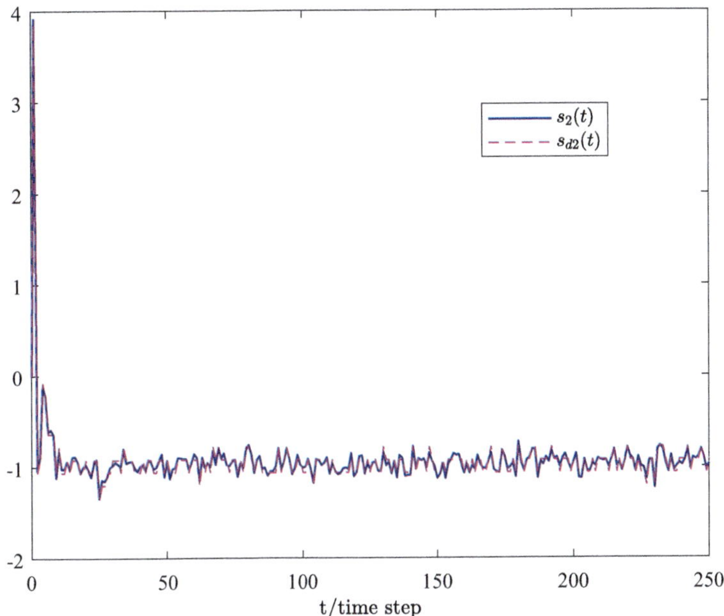

FIGURE 3.6

The measurement output $s_2(t)$ of the second neuron and its decoded value.

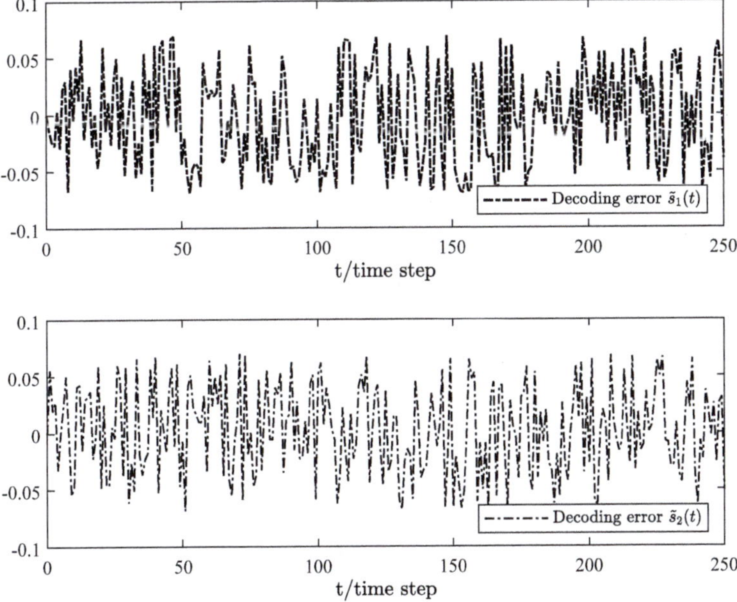

FIGURE 3.7

Decoding error $\tilde{s}_i(t)$ ($i = 1, 2$).

estimates. Figure 3.4 gives the estimation error of each neuron state. Figures 3.5–3.6 describe the measurement output $s_i(t)$ ($i = 1, 2$) of the first and second neurons and their decoded values. Figure 3.7 plots the decoding error for the measurement output $s(t)$ of each neuron. It can be observed from the above simulation results that the proposed partial-neurons-based state estimation strategy achieves an acceptable estimation performance in a bandwidth resource-constrained communication scenario with the aid of the proposed encoding-decoding mechanism. Therefore, all the simulation results well support the validness of the derived theoretical results.

3.4 Summary

In this chapter, the finite-time state estimation issue has been addressed for a class of time-delayed ANNs subject to certain bit rate constraint. For the purpose of meeting the network bandwidth requirement, a uniform-quantization-based encoding-decoding mechanism has been applied to encode the raw measurement outputs into specific digital codewords which occupy fewer communication resource. During the design of the state estimation scheme, only a small fraction of the neurons' outputs have been assumed to be measurable and utilized. By constructing a time-delay-dependent Lyapunov functional, sufficient conditions have been established to ensure the finite-time boundedness of the estimation error dynamics. The desired estimator gain matrix has been obtained by solving certain matrix inequalities. In addition, the effectiveness of the developed state estimator has been further demonstrated by a numerical simulation.

4

Synchronization Control for a Class of Discrete-Time Dynamical Networks with Packet Dropouts: A Coding-Decoding-Based Approach

The synchronization problem, which is concerned with a typical collective behavior in nature, has long been serving as an important research topic for dynamical networks because of its practical background and potential applications. However, in reality, there may be a case that the synchronization phenomenon cannot be attained for autonomous networks by their local connections, that is, the dynamic evolutions exhibit the asynchronous nature. For such kind of asynchronous networks, there is a vital need to develop effective synchronization control approaches in order to enforce the trajectories of all nodes to the desired synchronization manifold.

Concerning the synchronization control problems of dynamical networks, the available control strategies include, but are not limited to, the sampled-data control [146], the impulsive control [42], the pinning control [105] and the intermittent control [48]. Note that most reported synchronization control protocols have been assumed to work within the *analog* communication framework where the information is translated into electric pulses of varying amplitude. Compared with the traditional data transmission scheme in an analog manner, up to now, very little attention has been devoted to the synchronization control issue of dynamical networks with a *digital* communication mechanism, despite the fact that the digital information interaction has already become a dominant communication fashion for its obvious advantages of strong disturbance rejection capacity, low power consumption as well as high reliability. It is worth mentioning that, in the digitalization era, the digital communication strategy has been widely utilized in the areas of signal processing and control engineering.

In the real world, almost all the physical signals are virtually of analog nature and it would be improper to directly transmit them through digital channels. As such, before transmitting such analog signals in a digital pattern, one should first transform them into the digital form by an analog-to-digital (A/D) converter with the steps including sampling, reservation,

DOI: 10.1201/9781003534853-4

quantization and encoding. To some extent, the data coding stage is most crucial in realizing the digitization transmission since it is closely associated with the detectability issue, that is, whether and how the coded data could be recovered from the codewords with a prescribed accuracy requirement.

Compared with the traditional analog communication schemes, the distinct features of the digital communication protocols lie in that they a) have a good noise immunity; b) possess a high communication reliability; and c) are convenient for the data encryption. Here, the features a) and b) facilitate the long distant data transmission, and the feature c) prevents the transmitted data from being vulnerable to attacks launched by adversaries in response to the increasingly important security requirement of the information transmission. In particular, it is significantly important to design an appropriate coding-decoding procedure so as to ensure a satisfactory control performance via digital communication channels.

In the past few years, the coding-decoding communication scheme has gained some initial research attention. Up to now, almost all results concerning the coding-decoding issues have been on linear/nonlinear systems without network topology specifications, and the corresponding results on dynamical networks have not been addressed yet despite their wide applications in engineering practice.

The coding-decoding-based synchronization control problem for discrete-time dynamical networks appears to be an essentially difficult issue with three fundamental challenges identified as follows. 1) The first challenge stems from the mathematical difficulty in analyzing the addressed problem. As a typical dynamical network usually comes with both high dimension and tight coupling among the nodes, rather complicated/difficult mathematical analysis is expected and this is particularly true when the random packet dropout phenomenon is taken into account during the codeword transmissions [154]. On the other hand, due to the existence of the decoding error between the decoded system state and actual system state, it is fairly challenging to research into the issue of how to utilize imperfect decoded information to attain the desired control objective. In addition, since the performance of the coding-decoding protocol is influenced by a number of factors (e.g. the coding period, the size of the coding alphabet and the probabilistic packet dropouts), the third challenge is how to examine the effects of these combined factors on the prescribed system performance indices. It is, therefore, the main aim of this chapter to deal with the listed challenges by addressing the synchronization control problem for dynamical networks using a coding-decoding-based approach.

4.1 Problem Formulation

Consider a discrete-time dynamical network consisting of N coupled nodes described by:

$$
\left\{
\begin{aligned}
x_i(k+1) &= f(x_i(k)) + \sum_{j=1}^{N} w_{ij}\Gamma x_j(k) + u_i(k), \\
y_i(k) &= C_i x_i(k), \\
x_i(0) &= x_{i0} \in \mathcal{X}_0, \ i = 1, 2, \ldots, N
\end{aligned}
\right.
\tag{4.1}
$$

where, for the ith node, $x_i(k) \in \mathbb{R}^n$ is the state vector which is inaccessible directly, and $y_i(k) \in \mathbb{R}^m$ and $u_i(k) \in \mathbb{R}^n$ are the measurement output and the control input, respectively. $\Gamma = \mathrm{diag}\{\gamma_1, \gamma_2, \ldots, \gamma_n\}$ is the inner coupling matrix linking the jth state variable if $\gamma_j \neq 0$. $W = [w_{ij}]_{N \times N}$ is the coupled configuration matrix of the network with $w_{ij} \geq 0$ $(i \neq j)$ but not all zero. x_{i0} is the initial value of the ith node belonging to a known set $\mathcal{X}_0$. C_i is a constant matrix of appropriate dimensions. The nonlinear vector-valued function $f(\cdot) : \mathbb{R}^n \mapsto \mathbb{R}^n$ is continuous which satisfies $f(0) = 0$ and the following condition:

$$
[f(x) - f(y) - U_1(x - y)]^T [f(x) - f(y) - U_2(x - y)] \leq 0
\tag{4.2}
$$

for all $x, y \in \mathbb{R}^n$, where U_1 and U_2 are known real matrices of appropriate dimensions.

Note that full state information of the dynamical network (4.1) may not be obtained directly and only the network output is available. In this case, we construct the following state estimator for node i:

$$
\left\{
\begin{aligned}
\tilde{x}_i(k+1) &= f(\tilde{x}_i(k)) + \sum_{j=1}^{N} w_{ij}\Gamma \tilde{x}_j(k) \\
&\quad + L_i(y_i(k) - C_i\tilde{x}_i(k)) + u_i(k), \\
\tilde{x}_i(0) &= \tilde{x}_{i0} \in \mathcal{X}_0
\end{aligned}
\right.
\tag{4.3}
$$

where $\tilde{x}_i(k) \in \mathbb{R}^n$ is the estimator state for the ith node, $L_i \in \mathbb{R}^{n \times m}$ is the estimator gain matrix to be designed and $\tilde{x}_{i0}$ is the ith estimator's initial condition.

Let $s(k) \in \mathbb{R}^n$ be the solution to the unforced isolate node

$$
\left\{
\begin{aligned}
s(k+1) &= f(s(k)), \\
s(0) &= s_0 \in \mathcal{X}_0
\end{aligned}
\right.
\tag{4.4}
$$

where s_0 is the initial condition.

Denote by $e_{si}(k) \triangleq x_i(k) - s(k)$ the synchronization error vector. Then, for node i, we have the following synchronization error dynamics:

$$\begin{cases} e_{si}(k+1) = f(e_{si}(k)) + \sum_{j=1}^{N} w_{ij}\Gamma e_{sj}(k) + u_i(k), \\[2mm] e_{si}(0) = e_{si0} \in \mathcal{X}_0 \end{cases} \qquad (4.5a)$$

where $f(e_{si}(k)) \triangleq f(x_i(k)) - f(s(k))$ and $e_{si0} \triangleq x_{i0} - s_0$ is the initial value.

To comply with the digital communication fashion, the purpose of this chapter is to address the synchronization control problem for the dynamical network (4.1) by using a coding-decoding communication protocol. To be exact, by developing an efficient coding-decoding procedure based on the estimated network states, we shall design a decoder-based control protocol such that all the network nodes can be synchronized to the dynamics (4.4).

Before proceeding, a general form of the coding-decoding procedure for dynamical networks (4.1) is given as follows:

Coder for Node i:

$$g^i(lh) = \mathcal{F}_l^i(\tilde{x}_i(h), \, \tilde{x}_i(2h), \dots, \tilde{x}_i(lh)), \qquad (4.6)$$

Decoder for Node i:

$$\hat{X}^i(lh) = \mathcal{G}_l^i(\alpha_i(h)g^i(h), \, \alpha_i(2h)g^i(2h), \dots, \alpha_i(lh)g^i(lh)) \qquad (4.7)$$

for $l = 1, 2, \dots$, where $g^i(lh)$ is the codeword generated from the coder i at the coding instant lh, and $\hat{X}^i(lh)$ is defined as

$$\hat{X}^i(lh) \triangleq \{\hat{x}_i(lh), \, \hat{x}_i(lh+1), \dots, \hat{x}_i((l+1)h-1)\}$$

with $\hat{x}_i(k)$ being the decoded state of $\tilde{x}_i(k)$ for $k \in [lh, (l+1)h)$. $\mathcal{F}_l^i(\cdot)$ and $\mathcal{G}_l^i(\cdot)$ are coder and decoder functions to be designed, respectively. $\alpha_i(lh)$ is a Bernoulli distributed random variable which characterizes the packet dropouts and satisfies the following probability distribution:

$$\text{Prob}\{\alpha_i(lh) = 0\} = \bar{\alpha}_i, \quad \text{Prob}\{\alpha_i(lh) = 1\} = 1 - \bar{\alpha}_i$$

where $\bar{\alpha}_i \in [0, 1]$ is a given scalar. In addition, $\alpha_i(lh)$ and $\alpha_j(lh)$ are mutually independent random variables for $i \neq j$ with $i, j = 1, 2, \dots, N$.

For node i, in terms of the decoded state $\hat{x}_i(k)$, the *decoder-based synchronization controller* is given as

$$u_i(k) = K_c(\hat{x}_i(k) - s(k)) \qquad (4.8)$$

where K_c is the controller parameter to be designed.

Remark 4.1: *In the coding-decoding procedure (4.6)–(4.7), by employing the Bernoulli distributed white sequence $\alpha_i(lh)$, the phenomenon of probabilistic packet loss occurring in the transmission of codewords is well described, thereby better reflecting the reality. Also, because the coding period is h times of the system sampling period, it can be observed from (4.7) that, for a fixed coding period $[lh, (l+1)h)$, the decoding function $G_l^i(\,\cdot\,)$ of node i applies the codeword $g^i(lh)$ to generate a set of decoded state $\hat{x}_i(k)$ $(k \in [lh, (l+1)h))$ which indicates that the decoder can generate the decoded states at both the coding instant lh and those non-coding time instants $lh+1, \cdots, (l+1)h - 1$. Here, the decoded state $\hat{x}_i(k)$ $k \in (lh, (l+1)h)$ generated at non-coding instants can be viewed as a "prediction" of the actual system state.*

For notation simplicity, we denote

$$
\begin{aligned}
x(k) &= \mathrm{vec}_N\{x_i(k)\}, \; \tilde{x}(k) = \mathrm{vec}_N\{\tilde{x}_i(k)\}, \\
\hat{x}(k) &= \mathrm{vec}_N\{\hat{x}_i(k)\}, \; e_s(k) = \mathrm{vec}_N\{e_{si}(k)\}, \\
x_0 &= \mathrm{vec}_N\{x_{i0}\}, \; \tilde{x}_0 = \mathrm{vec}_N\{\tilde{x}_{i0}\}, \\
e_{s0} &= \mathrm{vec}_N\{e_{si0}\}, \; F(x(k)) = \mathrm{vec}_N\{f(x_i(k)\}, \\
F(\tilde{x}(k)) &= \mathrm{vec}_N\{f(\tilde{x}_i(k)\}, \; C = \mathrm{diag}_N\{C_i\}, \\
F(e_s(k)) &= \mathrm{vec}_N\{f(e_{si}(k))\}, \; L = \mathrm{diag}_N\{L_i\}.
\end{aligned}
\tag{4.9}
$$

By recurring to the Kronecker product, the original dynamical network (4.1) can be rewritten in the following compact form:

$$
\begin{cases}
x(k+1) = F(x(k)) + (W \otimes \Gamma)x(k) + u(k), \\
\quad y(k) = Cx(k), \\
\quad x(0) = x_0.
\end{cases}
\tag{4.10}
$$

Accordingly, we develop the following state estimator for the dynamical network (4.10):

$$
\begin{cases}
\tilde{x}(k+1) = F(\tilde{x}(k)) + (W \otimes \Gamma)\tilde{x}(k) \\
\qquad\qquad\quad + LC(x(k) - \tilde{x}(k)) + u(k), \\
\tilde{x}(0) = \tilde{x}_0
\end{cases}
\tag{4.11}
$$

and obtain the compact form of synchronization error dynamics (4.5a) as follows:

$$
\begin{cases}
e_s(k+1) = F(e_s(k)) + (W \otimes \Gamma)e_s(k) + u(k), \\
\quad e_s(0) = e_{s0}.
\end{cases}
\tag{4.12}
$$

To facilitate the subsequent developments, we introduce the following definitions which are needed for stating the problem to be investigated.

Definition 4.1: *Consider the discrete-time dynamical network of the compact form (4.10) subject to packet dropouts during the codeword transmissions. The network (4.10) is said to be detectable if there exist families of coder-decoder pairs (4.6) and (4.7) with a coding alphabet $\mathcal{H}$ of size χ such that*

$$\lim_{k \to \infty} \mathbb{E}\{\|x(k) - \hat{x}(k)\|_2\} = 0 \tag{4.13}$$

holds for any solution of (4.10).

Definition 4.2: *The discrete-time dynamical network (4.10) is said to be synchronized to the isolated node (4.4) if there exist families of decoder-controller pairs (4.8) such that*

$$\lim_{k \to \infty} \mathbb{E}\{\|e_s(k)\|_2\} = 0 \tag{4.14}$$

holds for any solution of closed-loop system (4.12).

In this chapter, we investigate the synchronization control problem for the dynamical network (4.10) by using a decoder-based control protocol. For the sake of realizing the synchronization criterion, one has to examine the detectability of the whole dynamical networks first. In other words, we are interested in developing an efficient coding-decoding procedure (4.6) and (4.7) for each node such that

a) The dynamical network (4.10) subject to the packet dropouts is detectable.

b) Based on the established detectability criterion, the dynamical network (4.10) is synchronized to the prescribed dynamics $s(k)$ by the decoder-based control protocol (4.8).

4.2 Main Results

In this section, our attention is focused on the analysis and design problems of the coding-decoding procedure for dynamical network (4.10). Then, based on the established coding-decoding procedure, we plan to deal with the decoder-based synchronization control problem for the addressed dynamical network to achieve the desired performance requirements.

4.2.1 Preliminaries

To start with, we give the following useful lemma which will be needed for the subsequent derivation of our main results.

Lemma 4.1: *Let the scalar $\mu_1 > 0$ be given. If there exist a positive definite matrix $P > 0$ and a scalar $\varepsilon_1 > 0$ satisfying the following linear matrix inequality (LMI)*

$$\Pi_1 = \begin{bmatrix} -\Pi_{11} & -\varepsilon_1 \Phi_{2\Lambda}^T & (W \otimes \Gamma)^T P \\ * & -\varepsilon_1 I_{Nn} & P \\ * & * & -P \end{bmatrix} < 0, \tag{4.15}$$

then we have

$$\|v_1(k+1) - v_2(k+1)\|_2 \le c_0 \|v_1(k) - v_2(k)\|_2 \tag{4.16}$$

where $\Pi_{11} = (1 + \mu_1)P - \varepsilon_1 \Phi_{1\Lambda}$, $c_0 = \sqrt{\beta}$, $\beta = \frac{(1+\mu_1)\lambda_{max}\{P\}}{\lambda_{min}\{P\}}$, and $v_1(k)$ and $v_2(k)$ are any two solutions of (4.10).

Proof: *First, denoting $z(k) \triangleq v_1(k) - v_2(k)$ and $F(z(k)) \triangleq F(v_1(k)) - F(v_2(k))$, it can be inferred immediately from (4.10) that $z(k+1) = F(z(k)) + (W \otimes \Gamma)z(k)$. Then, consider the Lyapunov function $V(k) = z^T(k)Pz(k)$. The term $\Delta V(k) - \mu_1 V(k)$, where $\Delta V(k)$ is the difference of $V(k)$ along the dynamics of $z(k+1) = F(z(k)) + (W \otimes \Gamma)z(k)$, can be subsequently calculated as*

$$\Delta V(k) - \mu_1 V(k) = z^T(k)[(W \otimes \Gamma)^T P(W \otimes \Gamma) - (1 + \mu_1)P]z(k)$$
$$+ 2F^T(z(k))P(W \otimes \Gamma)z(k) + F^T(z(k))PF(z(k)). \tag{4.17}$$

It is easy to see from (4.2) that

$$\begin{bmatrix} z(k) \\ F(z(k)) \end{bmatrix}^T \begin{bmatrix} \Phi_{1\Lambda} & \Phi_{2\Lambda}^T \\ * & I_{Nn} \end{bmatrix} \begin{bmatrix} z(k) \\ F(z(k)) \end{bmatrix} \le 0 \tag{4.18}$$

where $\Phi_{1\Lambda} = I_N \otimes \Phi_1$, $\Phi_{2\Lambda} = I_N \otimes \Phi_2$, $\Phi_1 = \frac{U_1^T U_2 + U_2^T U_1}{2}$ and $\Phi_2 = -\frac{U_1 + U_2}{2}$. Then, substituting (4.18) into (4.17) yields

$$\Delta V(k) - \mu_1 V(k) \le z^T(k)[(W \otimes \Gamma)^T P(W \otimes \Gamma) - (1 + \mu_1)P]z(k)$$
$$+ 2F^T(z(k))P(W \otimes \Gamma)z(k) + F^T(z(k))PF(z(k))$$
$$- \varepsilon_1 \begin{bmatrix} z(k) \\ F(z(k)) \end{bmatrix}^T \begin{bmatrix} \Phi_{1\Lambda} & \Phi_{2\Lambda}^T \\ * & I_{Nn} \end{bmatrix} \begin{bmatrix} z(k) \\ F(z(k)) \end{bmatrix} \tag{4.19}$$
$$= \xi_1^T(k)\bar{\Pi}_1\xi_1(k)$$

where

$$\xi_1(k) = \begin{bmatrix} z^T(k) & F^T(z(k)) \end{bmatrix}^T,$$

$$\bar{\Pi}_1 = \begin{bmatrix} \bar{\Pi}_{11} & (W \otimes \Gamma)^T P - \varepsilon_1 \Phi_{2\Lambda}^T \\ * & P - \varepsilon_1 I_{Nn} \end{bmatrix}$$

with $\bar{\Pi}_{11} = (W \otimes \Gamma)^T P (W \otimes \Gamma) - (1 + \mu_1)P - \varepsilon_1 \Phi_{1\Lambda}$.

By using the Schur Complement Lemma, it can be obtained from (4.15) that $\bar{\Pi}_1 < 0$, and therefore we have that $V(k + 1) \leq (1 + \mu_1)V(k)$. In addition, it can be readily seen from the definition of $V(k)$ that $\lambda_{\min}\{P\}\|z(k + 1)\|_2^2 \leq V(k + 1) \leq (1 + \mu_1)\lambda_{\max}\{P\}\|z(k)\|_2^2$, which implies that $\|z(k + 1)\|_2 \leq \sqrt{\beta}\|z(k)\|_2$, namely, $\|v_1(k + 1) - v_2(k + 1)\|_2 \leq c_0\|v_1(k) - v_2(k)\|_2$. The proof is complete.

4.2.2 State Estimator Design

The following lemma provides a sufficient condition for the design problem of the state estimator (4.11) to ensure that the estimation error dynamics is convergent. Furthermore, the estimator gain matrix for each node is explicitly characterized.

Lemma 4.2: Let the positive scalar $0 < \mu_2 < 1$ be given and consider the dynamical network (4.10) with estimator (4.11). It is assumed that there exist a positive definite matrix $Q = diag_N\{Q_i\} > 0$, a matrix $X = diag_N\{X_i\}$ and a positive scalar $\varepsilon_2 > 0$ satisfying

$$\Pi_2 = \begin{bmatrix} -(1 - \mu_2)Q - \varepsilon_2 \Phi_{1\Lambda} & -\varepsilon_2 \Phi_{2\Lambda}^T & \bar{Q} \\ * & -\varepsilon_2 I_{Nn} & Q \\ * & * & -Q \end{bmatrix} < 0 \qquad (4.20)$$

where $\bar{Q} = (Q(W \otimes \Gamma) - XC)^T$. Then, for the augmented dynamical network (4.10) and estimator (4.11), there always exist an integer $h \in \mathbb{Z}^+$ and a scalar c_1 $(0 < c_1 < 1)$ such that

$$\|x(k + h) - \tilde{x}(k + h)\|_2 < c_1\|x(k) - \tilde{x}(k)\|_2. \qquad (4.21)$$

Moreover, the desired estimator gain matrix for each node can be obtained in the form of

$$L_i = Q_i^{-1}X_i, \quad i = 1, 2, \ldots, N. \qquad (4.22)$$

Proof: Following the similar line in the proof of Lemma 4.1, we denote the estimation error as $e(k) \triangleq x(k) - \tilde{x}(k)$. Then, subtracting (4.11) from (4.10) results in the estimation error dynamics

$$e(k + 1) = F(e(k)) + (W \otimes \Gamma - LC)e(k),$$

where $F(e(k)) \triangleq F(x(k)) - F(\tilde{x}(k))$. Choosing the Lyapunov function as $V(k) = e^T(k)Qe(k)$, we have

$$\Delta V(k) + \mu_2 V(k) = [F(e(k)) + (W \otimes \Gamma - LC)e(k)]^T Q[F(e(k))$$
$$+ (W \otimes \Gamma - LC)e(k)] - (1 - \mu_2)e^T(k)Qe(k)$$
$$= e^T(k)\left\{[(W \otimes \Gamma) - LC]^T Q[(W \otimes \Gamma) - LC]\right.$$
$$\left. - (1 - \mu_2)Q\right\}e(k) + 2F^T(e(k))Q[(W \otimes \Gamma) - LC]e(k)$$
$$+ F^T(e(k))QF(e(k)).$$
$$(4.23)$$

Similarly, by noting (4.2), one has

$$\begin{bmatrix} e(k) \\ F(e(k)) \end{bmatrix}^T \begin{bmatrix} \Phi_{1\Lambda} & \Phi_{2\Lambda}^T \\ * & I_{Nn} \end{bmatrix} \begin{bmatrix} e(k) \\ F(e(k)) \end{bmatrix} \leq 0. \qquad (4.24)$$

Then, it can be easily obtained from (4.23) and (4.24) that

$$\Delta V(k) + \mu_2 V(k) \leq e(k)^T \left[(W \otimes \Gamma - LC)^T Q(W \otimes \Gamma - LC) \right.$$
$$\left. - (1 - \mu_2)Q\right] e(k) + 2F^T(e(k))Q(W \otimes \Gamma - LC)e(k)$$
$$+ F^T(e(k))QF(e(k))$$
$$- \varepsilon_2 \begin{bmatrix} e(k) \\ F(e(k)) \end{bmatrix}^T \begin{bmatrix} \Phi_{1\Lambda} & \Phi_{2\Lambda}^T \\ * & I_{Nn} \end{bmatrix} \begin{bmatrix} e(k) \\ F(e(k)) \end{bmatrix} \qquad (4.25)$$
$$= \xi_2^T(k)\bar{\Pi}_2 \xi_2(k)$$

where

$$\xi_2(k) - \begin{bmatrix} e^T(k) & F^T(e(k)) \end{bmatrix}^T,$$
$$\bar{\Pi}_2 = \begin{bmatrix} \bar{\Pi}_{21} & (W \otimes \Gamma - LC)^T Q - \varepsilon_2 \Phi_{2\Lambda}^T \\ * & Q - \varepsilon_2 I_{Nn} \end{bmatrix}$$

with $\bar{\Pi}_{21} = (W \otimes \Gamma - LC)^T Q(W \otimes \Gamma - LC) - (1 - \mu_2)Q - \varepsilon_2 \Phi_{1\Lambda}$.

By noticing (4.22), it follows from (4.20) and the Schur Complement Lemma that $\Delta V(k) + \mu_2 V(k) \leq 0$, which implies that $V(k + h) \leq (1 - \mu_2)V(k + h - 1) \leq (1 - \mu_2)^2 V(k + h - 2) \leq \cdots \leq (1 - \mu_2)^h V(k)$. Based on such a fact, one can further obtain that

$$\lambda_{\min}\{Q\}\|e(k + h)\|_2^2 \leq V(k + h) \leq (1 - \mu_2)^h V(k)$$
$$\leq (1 - \mu_2)^h \lambda_{\max}\{Q\}\|e(k)\|_2^2, \qquad (4.26)$$

which means that

$$\|x(k + h) - \tilde{x}(k + h)\|_2 < c_1 \|x(k) - \tilde{x}(k)\|_2 \qquad (4.27)$$

where $c_1 = \sqrt{\frac{(1-\mu_2)^h \lambda_{\max}\{Q\}}{\lambda_{\min}\{Q\}}}$. Therefore, it is not difficult to verify that there always exists a proper integer h to ensure $0 < c_1 < 1$. The proof of Lemma 4.2 is complete.

4.2.3 Detectability Analysis

Motivated by the uniform quantization approach proposed in [140,190], in this chapter, a modified uniform quantization approach is put forward, which would play an important role in the development of the coding-decoding procedure. For such a procedure, at each coding instant lh, since the current decoded state $\hat{x}_i(lh)$ of the ith node is inaccessible at this stage, we introduce an auxiliary state $\bar{x}_i(lh)$ which is determined by the previously decoded state $\hat{x}_i(lh-1)$ and will be defined later, see (4.32). Then, a modified uniform quantization approach is applied to the error between the estimated state $\tilde{x}_i(lh)$ and the auxiliary state $\bar{x}_i(lh)$. In this regard, denote by $\delta_i(lh) \triangleq \tilde{x}_i(lh) - \bar{x}_i(lh)$ the error vector for the ith node at the coding instant lh. In the following, a brief introduction of the proposed quantization method is given to facilitate the readers. For presentation convenience, sometimes, the arguments of a function or a matrix will be omitted in the analysis when no confusion can arise.

For the given scaling parameters $a_i > 0$ $(i = 1, 2, \ldots, N)$ and integer q, we can partition the hyperrectangles $\mathcal{B}_{a_i} = \{\delta_i \in \mathbb{R}^n : |\delta_i^{(j)}| \le a_i, j = 1, \ldots, n\}$ into q^n hyperrectangles $I_{s_1^i}^{i1}(a_i) \times I_{s_2^i}^{i2}(a_i) \times \cdots \times I_{s_n^i}^{in}(a_i)$, where $s_1^i, s_2^i, \ldots, s_n^i \in \{1, 2, \ldots, q\}$ and

$$I_1^{ij}(a_i) \triangleq \left\{ \delta_i^{(j)} \middle| -a_i \le \delta_i^{(j)} < -a_i + \frac{2a_i}{q} \right\},$$

$$I_2^{ij}(a_i) \triangleq \left\{ \delta_i^{(j)} \middle| -a_i + \frac{2a_i}{q} \le \delta_i^{(j)} < -a_i + \frac{4a_i}{q} \right\},$$

$$\vdots$$

$$I_q^{ij}(a_i) \triangleq \left\{ \delta_i^{(j)} \middle| a_i - \frac{2a_i}{q} \le \delta_i^{(j)} \le a_i \right\}$$

$$(4.28)$$

where $\delta_i^{(j)}$ is the jth element of the vector δ_i. As a result, for each $\mathcal{B}_{a_i}$, the center of the hyperrectangle $I_{s_1^i}^{i1}(a_i) \times I_{s_2^i}^{i2}(a_i) \times \cdots \times I_{s_n^i}^{in}(a_i)$ can be defined as

$$\eta_{a_i}^i(s_1^i, s_2^i, \ldots, s_n^i) \triangleq \begin{bmatrix} b_{i1} & b_{i2} & \cdots & b_{in} \end{bmatrix}^T \qquad (4.29)$$

where $b_{ij} = -a_i + \frac{(2s_j^i - 1)a_i}{q}$, $j = 1, 2, \ldots, n$.

Hence, for any $\delta_i \in \mathcal{B}_{a_i}$, there exist unique integers $s_1^i, s_2^i, \ldots, s_n^i \in \{1, 2, \ldots, q\}$ such that $\delta_i \in I_{s_1^i}^{i1}(a_i) \times I_{s_2^i}^{i2}(a_i) \times \cdots \times I_{s_n^i}^{in}(a_i)$, which amounts to the following

$$\|\delta_i - \eta_{a_i}^i(s_1^i, s_2^i, \ldots, s_n^i)\|_2 \leq \frac{\sqrt{n}a_i}{q}. \tag{4.30}$$

In the coding-decoding procedure to be proposed, each segment $I_{s_j^i}^{ij}(a_i)$

$(j = 1, 2, \ldots, n)$ of the hyperrectangle $\mathcal{B}_{a_i}$ determines an integer s_j^i, which is one component of the codeword. So, the transmitted codeword corresponds to a certain integer sequence $s_1^1, \ldots, s_n^1, s_1^2, \ldots, s_n^2, \ldots, s_1^N, \ldots, s_n^N$, which is vitally important to the decoding procedure.

Assume that Lemma 4.1 and Lemma 4.2 hold for some proper constants c_0 and c_1, respectively. The following specific form of coding-decoding procedure is proposed to handle the detectability analysis problem for the ith node.

Coder for Node i: For $\delta_i(lh) \triangleq \tilde{x}_i(lh) - \bar{x}_i(lh) \in I_{s_1^i}^{i1}(a_i(lh)) \times I_{s_2^i}^{i2}(a_i(lh)) \times \cdots \times I_{s_n^i}^{in}(a_i(lh)) \subset \mathcal{B}_{a_i(lh)}$, we have

$$g^i(lh) = [s_1^i, \ s_2^i, \ldots, s_n^i] \tag{4.31}$$

where $\bar{x}_i(lh)$ is defined by

$$\bar{x}_i(0) = 0,$$
$$\bar{x}_i(k) = \hat{x}_i(k), \ k \neq lh,$$
$$\bar{x}_i(lh) = f(\hat{x}_i(lh - 1)) + \sum_{j=1}^{N} w_{ij}\Gamma\hat{x}_j(lh - 1) + u_i(lh - 1). \tag{4.32}$$

Decoder for Node i:

$$\hat{x}_i(0) = 0,$$
$$\hat{x}_i(k + 1) = f(\hat{x}_i(k)) + \sum_{j=1}^{N} w_{ij}\Gamma\hat{x}_j(k) + u_i(k), \ k \neq lh - 1, \tag{4.33}$$
$$\hat{x}_i(lh) = \bar{x}_i(lh) + \alpha_i(lh)\eta_{a_i(lh)}^i(s_1^i, s_2^i, \ldots, s_n^i).$$

By utilizing the Kronecker product, we reformulate the compact coding-decoding procedure as follows.

Coder: For $\delta(lh) \triangleq \tilde{x}(lh) - \bar{x}(lh) \in I_{s_1^1}^{11}(a_1(lh)) \times \cdots \times I_{s_n^1}^{1n}(a_1(lh)) \times I_{s_1^2}^{21}(a_2(lh)) \times \cdots \times I_{s_n^2}^{2n}(a_2(lh)) \times \cdots \times I_{s_1^N}^{N1}(a_N(lh)) \times \cdots \times I_{s_n^N}^{Nn}(a_N(lh)) \subset \mathcal{B}_{a(lh)}$, we have

$$g(lh) = [s_1^1, \ldots, s_n^1, s_1^2, \ldots, s_n^2, \ldots, s_1^N, \ldots, s_n^N] \tag{4.34}$$

where $\mathcal{B}_{a(lh)} = \{x(lh) \in \mathbb{R}^{Nn} : |x^{(j)}(lh)| < a(lh), j = 1, \ldots, Nn\}$ and $\bar{x}(lh) = \text{vec}_N\{\bar{x}_i(lh)\}$ is defined by

$$\begin{aligned}
\bar{x}(0) &= 0, \\
\bar{x}(k) &= \hat{x}(k), \; k \neq lh, \\
\bar{x}(lh) &= F(\hat{x}(lh-1)) + (W \otimes \Gamma)\hat{x}(lh-1) + u(lh-1).
\end{aligned} \tag{4.35}$$

Decoder:

$$\begin{aligned}
\hat{x}(0) &= 0, \\
\hat{x}(k+1) &= F(\hat{x}(k)) + (W \otimes \Gamma)\hat{x}(k) + u(k), \quad k \neq lh-1, \\
\hat{x}(lh) &= \bar{x}(lh) + \alpha(lh)\eta_{a(lh)}(s_1^1, \ldots, s_n^N)
\end{aligned} \tag{4.36}$$

where $a(lh) = \max_{1 \leq i \leq N}\{a_i(lh)\}$, $\alpha(lh) = \text{diag}_N\{\alpha_i(lh)I_n\}$ and $\eta_{a(lh)}(s_1^1, \ldots, s_n^N) = \text{vec}_N\{\eta_{a_i(lh)}^i(s_1^i, s_2^i, \ldots, s_n^i)\}$.

In the following, some definitions are given to facilitate the subsequent developments.

$$\begin{aligned}
s_0 &\triangleq \sup_{x_{i0} \in \mathcal{X}_0} \|x_{i0}\|_2^2, \\
a(h) &\triangleq (2c_1 + c_0^h)\sqrt{Ns_0}, \\
\mathbb{E}\{a((l+1)h)\} &\triangleq 2c_1^l\sqrt{Ns_0}(c_1 + c_0^h) + \tilde{\alpha}\sqrt{Nn}c_0^h a(lh)
\end{aligned} \tag{4.37}$$

where $\tilde{\alpha} = \frac{\bar{\alpha}}{q} + (1 - \bar{\alpha})$ and $\bar{\alpha} = \prod_{i=1}^N (1 - \bar{\alpha}_i)$.

The following lemma shows that the coding-decoding protocol developed in this chapter is well posed. That is, the decoding condition $\mathbb{E}\{\tilde{x}(lh) - \bar{x}(lh)\} \in \mathcal{B}_{a(lh)}$ holds for all $l = 1, 2, \ldots$.

Lemma 4.3: *The coding-decoding procedure (4.34)–(4.36) satisfies the following constraint*

$$\mathbb{E}\{\|\tilde{x}(lh) - \bar{x}(lh)\|_\infty\} \leq a(lh), \; l = 1, 2, \ldots. \tag{4.38}$$

Proof: *This lemma can be proved by using the mathematical induction. First, for $l = 1$, by considering Lemma 4.1, Lemma 4.2 and the property of vector norm, we have*

$$\mathbb{E}\{\|\tilde{x}(h) - \bar{x}(h)\|_2\} \leq \mathbb{E}\{\|\tilde{x}(h) - x(h)\|_2\} + \mathbb{E}\{\|x(h) - \bar{x}(h)\|_2\}$$
$$\leq c_1\|e(0)\|_2 + c_0\|x(h-1) - \hat{x}(h-1)\|_2 \qquad (4.39)$$
$$\leq 2c_1\sqrt{Ns_0} + c_0^h\|x(0) - \hat{x}(0)\|_2,$$

which guarantees

$$\mathbb{E}\{\|\tilde{x}(lh) - \bar{x}(lh)\|_\infty\} \leq a(lh). \qquad (4.40)$$

Subsequently, assuming that $\mathbb{E}\{\|\tilde{x}(jh) - \bar{x}(jh)\|_\infty\} \leq \sqrt{Nn}\mathbb{E}\{a(jh)\}$ *for all* $j = 2,$
$\ldots, l$, we arrive at

$$\mathbb{E}\{\|\tilde{x}((l+1)h) - \bar{x}((l+1)h)\|_2\}$$
$$\leq \mathbb{E}\{\|\tilde{x}((l+1)h) - x((l+1)h)\|_2\}$$
$$+ \mathbb{E}\{\|x((l+1)h) - \bar{x}((l+1)h)\|_2\} \qquad (4.41)$$
$$\leq c_1^{l+1}\|e(0)\|_2 + c_0\mathbb{E}\{\|x(lh + h - 1) - \hat{x}(lh + h - 1)\|_2\}$$
$$\leq 2c_1^{l+1}\sqrt{Ns_0} + c_0^h\mathbb{E}\{\|x(lh) - \hat{x}(lh)\|_2\}.$$

Furthermore, taking the packet dropout phenomenon into account, we obtain

$$\|x(lh) - \hat{x}(lh)\|_2 = \|x(lh) - \bar{x}(lh) - \alpha(lh)\eta_{a(lh)}(s_1^1, \ldots, s_n^N)\|_2$$
$$\leq \|x(lh) - \tilde{x}(lh)\|_2 + \|\tilde{x}(lh) - \bar{x}(lh) \qquad (4.42)$$
$$- \alpha(lh)\eta_{a(lh)}(s_1^1, \ldots, s_n^N)\|_2$$
$$\leq 2c_1^l\sqrt{Ns_0} + \|\tilde{x}(lh) - \bar{x}(lh) - \alpha(lh)\eta_{a(lh)}(s_1^1, \ldots, s_n^N)\|_2.$$

Since the codeword loss occurs in a random way for each node and is governed by a set of mutually independent Bernoulli distributed sequence $\alpha_i(lh)$, *the term* $\|\tilde{x}(lh) - \bar{x}(lh) - \alpha(lh)\eta_{a(lh)}(s_1^1, \ldots, s_n^N)\|_2$ *in (4.42) can be further rewritten us*

$$\|\tilde{x}(lh) - \bar{x}(lh) - \alpha(lh)\eta_{a(lh)}(s_1^1, \ldots, s_n^N)\|_2$$
$$= \prod_{i=1}^{N}\alpha_i(lh)\left\|\tilde{x}(lh) - \bar{x}(lh) - \eta_{a(lh)}(s_1^1, \ldots, s_n^N)\right\|_2$$
$$+ \sum_{j=1}^{N}\prod_{i=1, i\neq j}^{N}(1 - \alpha_i(lh))\alpha_j(lh)\left\|\tilde{x}(lh) - \bar{x}(lh) \right.$$
$$\left. - diag\{\underbrace{I_n, \ldots, I_n}_{j-1}, 0, \underbrace{I_n, \ldots, I_n}_{N-j}\}\eta_{a(lh)}(s_1^1, \ldots, s_n^N)\right\|_2$$
$$+ \cdots + \prod_{i=1}^{N}(1 - \alpha_i(lh))\|\tilde{x}(lh) - \bar{x}(lh)\|_2 \qquad (4.43)$$

where the first term on the right-hand side of (4.43) implies the case that there is no packet dropout occurring during the codeword transmissions, the second term describes that only one node is subject to the codeword loss and the last term indicates that all the nodes lose the codewords at the transmission stage.

For presentation convenience, only some cases of packet dropouts for the dynamical network are described in (4.43). Specifically, without loss of generality, we assume that the first m nodes suffer from the packet dropouts where $m \in \{1, 2, \ldots, N\}$. Then, we have

$$\left\| \tilde{x}(lh) - \bar{x}(lh) - diag\{\underbrace{0, \ldots, 0}_{m}, \underbrace{I_n, \ldots, I_n}_{N-m}\} \eta_{a(lh)}(s_1^1, \ldots, s_n^N) \right\|_2$$

$$= \left\| \tilde{x}(lh) - \bar{x}(lh) - \eta_{a(lh)}(s_1^1, \ldots, s_n^N) \right.$$

$$\left. + diag\{\underbrace{I_n, \ldots, I_n}_{m}, \underbrace{0, \ldots, 0}_{N-m}\} \eta_{a(lh)}(s_1^1, \ldots, s_n^N) \right\|_2$$

$$\leq \left\| \tilde{x}(lh) - \bar{x}(lh) - \eta_{a(lh)}(s_1^1, \ldots, s_n^N) \right\|_2 \tag{4.44}$$

$$+ \left\| [\underbrace{\eta_{a^1(lh)}^{1T}, \ldots, \eta_{a^m(lh)}^{mT}}_{m}, \underbrace{0, \ldots, 0}_{N-m}]^T \right\|_2$$

$$\leq \sqrt{Nn} \frac{a(lh)}{q} + \sqrt{Nn} \left(a(lh) - \frac{a(lh)}{q} \right)$$

$$= \sqrt{Nn} a(lh).$$

It is not difficult to verify that, if we consider the case of packet dropouts for arbitrary m nodes, the above result still holds. Accordingly, it can be derived from (4.42)–(4.44) that

$$\mathbb{E}\{\|x(lh) - \hat{x}(lh)\|_2\} \leq 2c_1^l \sqrt{Ns_0} + \sqrt{Nn} \frac{\bar{\alpha}}{q} \mathbb{E}\{a(lh)\}$$

$$+ \sqrt{Nn}(1 - \bar{\alpha}) \mathbb{E}\{a(lh)\}. \tag{4.45}$$

Thus, it can be concluded from (4.37), (4.41) and (4.45) that

$$\mathbb{E}\left\{\|\tilde{x}((l+1)h) - \bar{x}((l+1)h)\|_2\right\} \leq a((l+1)h), \tag{4.46}$$

which means

$$\mathbb{E}\left\{\|\tilde{x}((l+1)h) - \bar{x}((l+1)h)\|_\infty\right\} \leq a((l+1)h). \tag{4.47}$$

Consequently, it follows that the Lemma 4.3 holds for all $l \geq 1$, and the proof of this lemma is complete.

The following theorem offers a sufficient condition to guarantee the detectability of the dynamical network (4.10) by utilizing the proposed coding-decoding protocol.

Theorem 4.1: *The dynamical network (4.10) is detectable with the coding-decoding procedure (4.34)-(4.36) if the inequality*

$$\left(\frac{\bar{\alpha}}{q} + 1 - \bar{\alpha}\right)\sqrt{Nn}c_0^h < 1 \tag{4.48}$$

holds subject to (4.15) and (4.20) for some positive integer q, where c_0 and h satisfy Lemma 4.1 and Lemma 4.2, respectively.

Proof: *It follows from (4.37), (4.48) and $0 < c_1 < 1$ that $\lim_{l\to+\infty} \mathbb{E}\{a(lh)\} = 0$, which indicates*

$$\lim_{l\to\infty} \mathbb{E}\{\|\tilde{x}(lh) - \bar{x}(lh)\|_2\} = 0. \tag{4.49}$$

Then, it can be easily seen from (4.45) that, at each coding instant lh, we have

$$\lim_{l\to\infty} \mathbb{E}\{\|x(lh) - \hat{x}(lh)\|_2\} = 0. \tag{4.50}$$

On the other hand, when $lh < k < (l+1)h$, noticing that $x(k)$ and $\hat{x}(k)$ can be viewed as two solutions of (4.10), one finds $\mathbb{E}\{\|x(k) - \hat{x}(k)\|_2\} \leq c_0^{k-lh}\mathbb{E}\{\|x(lh) - \hat{x}(lh)\|_2\}$ by Lemma 4.1, which implies that $\mathbb{E}\{\|x(k) - \hat{x}(k)\|_2\}$ is also bounded at non-coding instants. Therefore, it follows immediately from the above analysis that

$$\lim_{k\to\infty} \mathbb{E}\{\|x(k) - \hat{x}(k)\|_2\} = 0, \tag{4.51}$$

which ends the proof.

Remark 4.2: *In Theorem 4.1, a criterion has been established to ensure the detectability of the dynamical network (4.10). It is not difficult to find that all the information (including the system matrices, the coding period, the size of coding alphabet and the statistics characteristics of packet dropouts) are reflected in the condition (4.48). Hence, the detectability of the dynamical network (4.10) depends heavily on the combined influences of those factors mentioned above. Let us now evaluate the impact on the performance of the proposed coding-decoding protocol from the following three aspects. 1) From (4.48), one has $q > \frac{\bar{\alpha}\sqrt{Nn}c_0^h}{1-(1-\bar{\alpha})\sqrt{Nn}c_0^h}$, which requires that the size of the coding alphabet $\mathcal{H}$ should satisfy $\chi > \frac{(\bar{\alpha}\sqrt{Nn}c_0^h)^{Nn}}{(1-(1-\bar{\alpha})\sqrt{Nn}c_0^h)^{Nn}}$. Especially, when $\bar{\alpha} = 1$, namely, there is no packet dropout occurring, (4.48) reduces to $q > \sqrt{Nn}c_0^h$, then the size χ is required to satisfy $\chi > (\sqrt{Nn}c_0)^{Nnh}$. 2) The* *inequality (4.48) also indicates $1 - \bar{\alpha} < \frac{q\sqrt{N}nc_0^{-h}-1}{q-1}$. Recalling the definition of $\bar{\alpha}$, it is noticed that the term $1 - \bar{\alpha}$ can be regarded as the total packet dropout probability of the whole network. Therefore, it is easily seen that the upper bound of admissible data packet dropout probability is $\frac{q\sqrt{N}nc_0^{-h}-1}{q-1}$. 3) For a given size χ of the coding alphabet $\mathcal{H}$, with increased packet dropout probabilities $\bar{\alpha}_i$, the $\bar{\alpha}$ becomes smaller and thus inevitably weakens the feasibility of the inequality (4.48) and even leads to the possible divergence of the coding-decoding algorithm. Accordingly, this gives a trade-off between the packet dropout probabilities and the convergence of the proposed algorithm.*

4.2.4 Synchronization Control

Based on the established results of the detectability analysis problem in the last subsection, now let us tackle the decoder-based synchronization control problem for the dynamical network (4.10). Due to the existence of control input error induced by the decoding error, the ISS analysis approach is employed.

For the convenience of the readers, let us recall some basic notions and necessary foundations on the ISS theory before driving our main results.

Consider the following discrete-time nonlinear system

$$x(k+1) = f(x(k), u(k)) \tag{4.52}$$

where $x(k) \in \mathbb{R}^n$ is the system state vector, $u(k) \in \mathbb{R}^p$ is the control input and $f(\cdot, \cdot) : \mathbb{R}^n \times \mathbb{R}^p \mapsto \mathbb{R}^n$ is a continuous function satisfying $f(0,0) = 0$.

First, let us introduce some important function classes which will be used in the subsequent developments. $\mathcal{K}$, $\mathcal{K}_\infty$ and $\mathcal{KL}$ denote the different classes of functions. A function $\gamma(\cdot) : \mathbb{R}^+ \mapsto \mathbb{R}^+$ belongs to class $\mathcal{K}$ if it is a continuous strictly increasing function with $\gamma(0) = 0$ and belongs to class $\mathcal{K}_\infty$ if $\gamma(\cdot) \in \mathcal{K}$ with $\gamma(r) \mapsto \infty$ as $r \mapsto \infty$. Also, a function $\sigma(s, k) : \mathbb{R}^+ \times \mathbb{R}^+ \mapsto \mathbb{R}^+$ belongs to class $\mathcal{KL}$ if, for each fixed k, the function $\sigma(\cdot, k) \in \mathcal{K}$, and for each fixed s, the function $\sigma(s, \cdot)$ is decreasing and $\sigma(\cdot, k) \mapsto 0$ as $k \mapsto \infty$. A function is said to be a $\mathcal{K}$ ($\mathcal{K}_\infty$ or $\mathcal{KL}$) class function if it belongs to the class $\mathcal{K}$ ($\mathcal{K}_\infty$ or $\mathcal{KL}$).

Definition 4.3: *The system (4.52) is said to be input-to-state stable if there exist a $\mathcal{KL}$ class function $\beta(\cdot, \cdot)$ and a $\mathcal{K}$ class function $\gamma(\cdot)$ such that the evolution of the system state $x(k)$ satisfies*

$$\|x(k)\|_2 \leq \beta(\|x(0)\|_2, k) + \gamma(\|u(k)\|_\infty) \tag{4.53}$$

for $\forall k \geq 0$ and $\forall x(0) \in \mathbb{R}^n$, where $\|u(k)\|_\infty \triangleq \sup_k\{\|u(k)\|_2\}$.

Remark 4.3: *It is clear from (4.53) that the bound of the state $x(k)$ is closely associated with the initial condition $x(0)$ and the system input $u(k)$. To be specific, if the system input $u(k)$ is bounded, then the state evolution trajectory $x(k)$ is naturally bounded. Also, it follows from the properties of functions $\beta(\cdot,\cdot)$ and $\gamma(\cdot)$ that the state $x(k)$ is ultimately bounded as k increases. In addition, as pointed out in [58], the ISS property further implies the case of "converging-input" to "converging-state", that is, the system state $x(k)$ converges to 0 if the input $u(k)$ achieves 0 as $k \mapsto \infty$.*

To proceed, we present the following lemma.

Lemma 4.4: *[58] The nonlinear discrete-time system (4.52) is said to be input-to-state stable if there exist a positive definite function $V(k, x(k)) : [0, +\infty) \times \mathbb{R}^n \mapsto \mathbb{R}$ (called an ISS-Lyapunov function), three $\mathcal{K}_\infty$ class functions $\alpha_1(\cdot)$, $\alpha_2(\cdot)$ and $\alpha_3(\cdot)$, and a $\mathcal{K}$ class function $\varpi(\cdot)$ such that the following two inequalities*

$$\alpha_1(\|x(k)\|_2) \le V(k, x(k)) \le \alpha_2(\|x(k)\|_2), \tag{4.54}$$
$$V(k+1, x(k+1)) - V(k, x(k)) \le -\alpha_3(\|x(k)\|_2) + \varpi(\|u(k)\|_2) \tag{4.55}$$

hold for all $x(k) \in \mathbb{R}^n$ and $u(k) \in \mathbb{R}^p$. Furthermore, if the above two inequalities are met simultaneously, then the functions $\beta(\cdot,\cdot)$ and $\gamma(\cdot)$ in Definition 4.3 can be selected as $\beta(\cdot,k) = \alpha_1^{-1}(\phi^k \alpha_2(\cdot))$ $(0 < \phi < 1)$ and $\gamma(\cdot) = \alpha_1^{-1}(\alpha_2(\alpha_3^{-1}(\varpi(\cdot))))$, respectively, where $\alpha_1^{-1}(\cdot)$ expresses the inverse function of the monotone function $\alpha_1(\cdot)$ and so does $\alpha_3^{-1}(\cdot)$.

Now, we rewrite the decoder-based control protocol (4.8) with the following compact form:

$$u(k) = \bar{K}_c(\hat{x}(k) - \bar{s}(k)) \tag{4.56}$$

where $\bar{K}_c \triangleq I_N \otimes K_c$ and $\bar{s}(k) \triangleq \mathbb{1}_N \otimes s(k)$.

Next, let $w_i(k) \triangleq \hat{x}_i(k) - x_i(k)$ be the decoding error vector for the ith node and set $w(k) \triangleq \text{vec}_N\{w_i(k)\}$. In light of (4.56) and noting $\hat{x}(k) = w(k) + x(k)$, the closed-loop system (4.12) can be rewritten as

$$e_s(k+1) = (\bar{K}_c + (W \otimes \Gamma))e_s(k) + F(e_s(k)) + \bar{K}_c w(k). \tag{4.57}$$

Definition 4.3 and Lemma 4.4 reveal the relationship between the system state evolution and the bound of inputs. On the other hand, based on the obtained detectability analysis results, we have that the decoding error $w(k)$ is bounded. In this sense, $w(k)$ can be viewed as a bounded input of (4.57) and therefore the introduced ISS theory can be utilized to investigate the coding-decoding-based synchronization control problem.

Theorem 4.2: *Under the condition in Theorem 4.1, the dynamical network (4.10) can be synchronized by the decoder-based controller (4.56) if there exist a positive definite matrix $R = I_N \otimes R_0 > 0$, a matrix $Y = I_N \otimes Y_0$ and a positive scalar $\varepsilon_3 > 0$ satisfying*

$$\Pi_3 = \begin{bmatrix} -R - \varepsilon_3 \Phi_{1\Lambda} & -\varepsilon_3 \Phi_{2\Lambda}^T & 0 & \bar{R}^T \\ * & -\varepsilon_3 I_{Nn} & 0 & R \\ * & * & -R & Y^T \\ * & * & * & -R \end{bmatrix} < 0 \tag{4.58}$$

where $\bar{R} = (R(W \otimes \Gamma) + Y)$. Moreover, the desired controller gain matrix is given as

$$K_c = R_0^{-1} Y_0. \tag{4.59}$$

Proof: *Choosing the ISS-Lyapunov function as $V(k) = e_s^T(k) R e_s(k)$ and calculating its difference along the trajectory of (4.57), one obtains*

$$\begin{aligned} V(k+1) - V(k) &\le \left[(\bar{K}_c + (W \otimes \Gamma))e_s(k) + F(e_s(k)) + \bar{K}_c w(k) \right]^T R \\ &\quad \times \left[(\bar{K}_c + (W \otimes \Gamma))e_s(k) + F(e_s(k)) + \bar{K}_c w(k) \right] \\ &\quad - e_s^T(k) R e_s(k) - \varepsilon_3 \begin{bmatrix} e_s(k) \\ F(e_s(k)) \end{bmatrix}^T \begin{bmatrix} \Phi_{1\Lambda} & \Phi_{2\Lambda}^T \\ * & I_{Nn} \end{bmatrix} \\ &\quad \times \begin{bmatrix} e_s(k) \\ F(e_s(k)) \end{bmatrix} \\ &= \xi_3^T(k) \bar{\Pi}_3 \xi_3(k) + w^T(k) R w(k) \end{aligned} \tag{4.60}$$

where

$$\xi_3(k) = \begin{bmatrix} e_s^T(k) & F^T(e_s(k)) & w^T(k) \end{bmatrix}^T,$$

$$\bar{\Pi}_3 = \begin{bmatrix} \bar{\Pi}_{31} & \mathbb{K}^T R - \varepsilon_3 \Phi_{3\Lambda}^T & \mathbb{K}^T \bar{K}_c \\ * & R - \varepsilon_3 I_{Nn} & \bar{K}_c \\ * & * & \bar{K}_c^T R \bar{K}_c - R \end{bmatrix}$$

with $\bar{\Pi}_{31} = \mathbb{K}^T R \mathbb{K} - \varepsilon_3 \Phi_{3\Lambda}$ and $\mathbb{K} = \bar{K}_c + W \otimes \Gamma$.

By the Schur Complement and (4.59), inequality (4.58) means $\bar{\Pi}_3 < 0$, which yields that $V(k+1) - V(k) \le -\lambda_{\min}\{-\bar{\Pi}_3\} \|e_s(k)\|_2^2 + \lambda_{\max}\{R\} \|w(k)\|_2^2$. Choosing $\alpha_1(\|e_s(k)\|_2) = \lambda_{\min}\{R\} \|e_s(k)\|_2^2$, $\alpha_2(\|e_s(k)\|_2) = \lambda_{\max}\{R\} \|e_s(k)\|_2^2$, $\alpha_3(\|e_s(k)\|_2) = \lambda_{\min}\{-\bar{\Pi}_3\} \|e_s(k)\|_2^2$ and $\varpi(\|w(k)\|_2) = \lambda_{\max}\{R\} \|w(k)\|_2^2$, it is inferred from Lemma 4.4 that the synchronization error system (4.57) is input-to-state stable. So, let us choose $\beta(\|e_{s0}\|_2, k) = \sqrt{\frac{\lambda_{\max}\{R\}}{\lambda_{\min}\{R\}}} \phi^k \|e_{s0}\|_2$ and $\gamma(\|w(k)\|_2) = \sqrt{\frac{\lambda_{\max}^2\{R\}}{c \lambda_{\min}\{R\} \lambda_{\min}\{-\bar{\Pi}_3\}}} \|w(k)\|_2$, where $0 < c < 1$. Then we have from Definition 4.3 that

$$\|e_s(k)\|_2 \leq \sqrt{\frac{\lambda_{\max}\{R\}}{\lambda_{\min}\{R\}}}\phi^k\|e_{s0}\|_2$$

$$+ \sqrt{\frac{\lambda_{\max}^2\{R\}}{c\lambda_{\min}\{R\}\lambda_{\min}\{-\bar{\Pi}_3\}}}\|w(k)\|_2. \tag{4.61}$$

Taking the mathematical expectation of both sides of (4.61), it can be immediately found from the detectability analysis result (4.51) and the property of "converging-input" to "converging state" discussed in Remark 4.3 that $\lim_{k\to\infty}\mathbb{E}\{\|e_s(k)\|_2\} = 0$, which completes the proof.

4.3 An Illustrative Example

In this section, two simulation examples are presented to demonstrate the detectability and the effectiveness of the proposed decoder-based synchronization control protocol for the dynamical network (4.1).

Consider a dynamical network (4.1) with three nodes which are connected according to the following coupling configuration matrix

$$W = \begin{bmatrix} -0.2 & 0.1 & 0.1 \\ 0.1 & -0.2 & 0.1 \\ 0.1 & 0.1 & -0.2 \end{bmatrix}$$

and the inner-coupling matrix $\Gamma = \mathrm{diag}\{0.1, 0.1\}$.

The nonlinear functions are assumed to be

$$f(x_i(k)) = \begin{bmatrix} -0.5x_{i1}(k) + \tanh(0.65x_{i1}(k)) - 0.15x_{i2}(k) \\ 1.1x_{i2}(k) - \tanh(0.95x_{i2}(k)) \end{bmatrix}.$$

Then, it is easy to see that the nonlinear function $f(\,\cdot\,)$ satisfies the sector bounded condition (4.2) with

$$U_1 = \begin{bmatrix} -0.5 & -0.15 \\ 0 & 1.1 \end{bmatrix}, \quad U_2 = \begin{bmatrix} 0.15 & -0.15 \\ 0 & 0.25 \end{bmatrix}.$$

The available measurements of the dynamical network described above are modeled in (4.1) with the following parameters

$$C_1 = \begin{bmatrix} 0.41 & 0.52 \end{bmatrix}, C_2 = \begin{bmatrix} 0.33 & 0.44 \end{bmatrix}, C_3 = \begin{bmatrix} 0.32 & 0.60 \end{bmatrix}.$$

In the following, we shall deal with the detectability as well as synchronization control problems for the dynamical network (4.1) with given parameters.

Example 4.1: *In this example, according to our established criterion, we shall test the detectability problem for the dynamical network (4.1). In this case, without loss of generality, it is assumed that $u(k) \equiv 0$. Choosing $\mu_1 = 0.3$ and solving the matrix inequality (4.15) by the Matlab software, we have that (4.16) holds with $c_0 = 1.2701$. For the given scalar $\mu_2 = 0.1$ and the coding period $h = 3$, $c_1 = 0.9435$ is obtained such that (4.21) holds. Moreover, the estimator gain matrices L_i are obtained as $L_1 = \begin{bmatrix} -0.3476 & 0.8462 \end{bmatrix}^T$, $L_2 = \begin{bmatrix} -0.4161 & 1.0285 \end{bmatrix}^T$ and $L_3 = \begin{bmatrix} -0.3223 & 0.8768 \end{bmatrix}^T$. According to Theorem 4.1, the packet probability $\bar{\alpha}_i$ $(i = 1, 2, 3)$ is taken as 0.1 for $q = 20$.*

To validate our analysis results and make our simulation non-trivial, we consider the above dynamical network which is set to be unstable. The simulation results are shown in Figures 4.1–4.3, where Figures 4.1–4.2 plot the actual states and their decoded values for all the nodes of the network (4.1) and Figure 4.3 shows the corresponding decoding errors $w_{ij}(k)$ $(i = 1, 2, 3; j = 1, 2)$. It is noticed from Figure 4.3 that the errors between the actual states of the whole network and their decoded states asymptotically approach zero. Therefore,

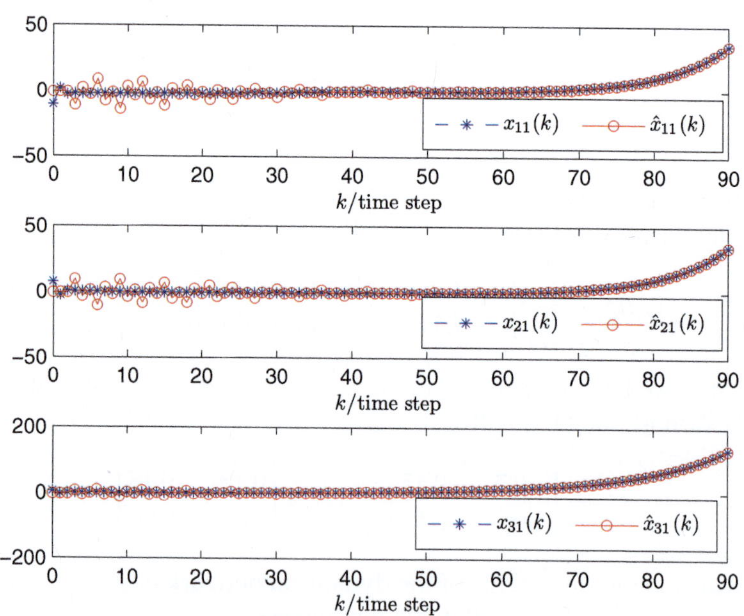

FIGURE 4.1

State trajectories $x_{i1}(k)$ and their decoded values $\hat{x}_{i1}(k)$ $(i = 1, 2, 3)$.

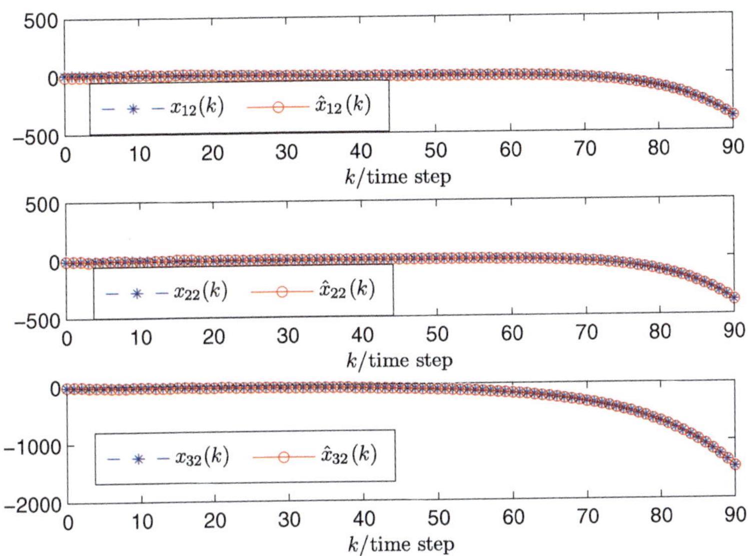

FIGURE 4.2
State trajectories $x_{i2}(k)$ and their decoded values $\hat{x}_{i2}(k)$ ($i = 1, 2, 3$).

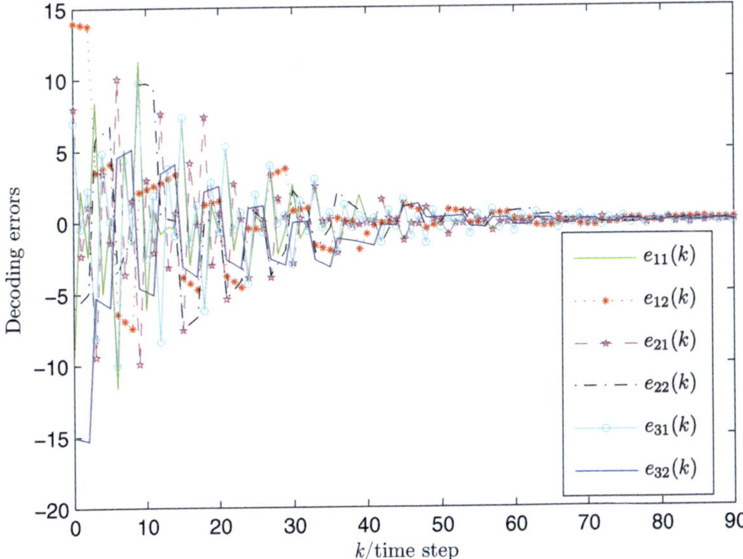

FIGURE 4.3
Decoding errors $w_{ij}(k)$ ($i = 1, 2, 3; j = 1, 2$).

the desired detectability performance of the addressed dynamical networks is well attained.

Remark 4.4: *For comparison, if we use the infinity-vector norm-based uniform quantization approach that has been widely used in the existing literature, we can calculate that $c_0 = 3.1110$ and $c_1 = 2.3111$ for this example, which violates the condition $0 < c_1 < 1$ and the feasibility of the coding-decoding procedure cannot be guaranteed. Such a comparison result embodies the superiority of our proposed modified uniform quantization approach.*

Example 4.2: *In the second example, we consider the decoder-based synchronization control problem for the dynamical network (4.1). The control input signal is taken as the form in (4.8) for each node. By solving (4.58) in Theorem 4.2, we obtain the desired controller parameter as*

$$K_c = R_0^{-1} Y_0 = \begin{bmatrix} 0.1935 & 0.1521 \\ -0.0003 & -0.6601 \end{bmatrix}.$$

As stated in Theorem 4.2, the considered dynamical network can be synchronized with the designed controller parameters given above. Simulation results are presented in Figures 4.4–4.5. From Figure 4.4, it can be observed that the state response curves of all nodes diverge from the isolate node drastically when there is no

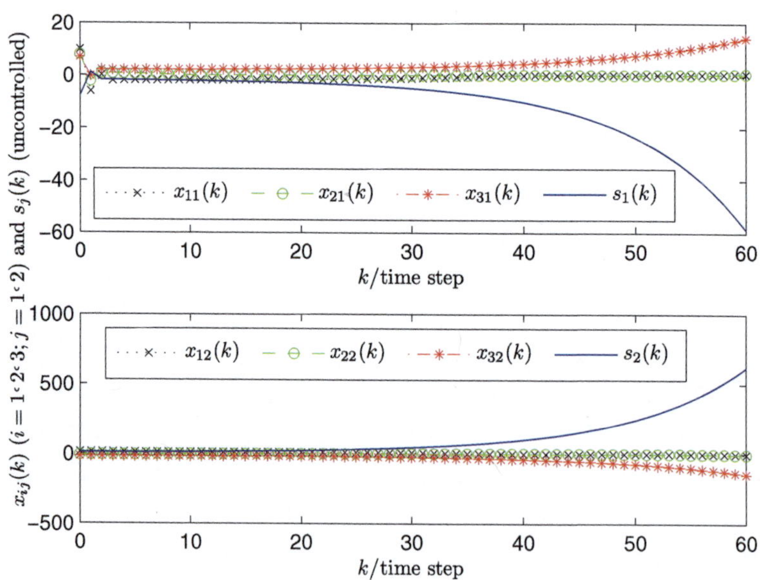

FIGURE 4.4

State trajectories $x_{ij}(k)$ of the uncontrolled nodes i ($i = 1, 2, 3; j = 1, 2$).

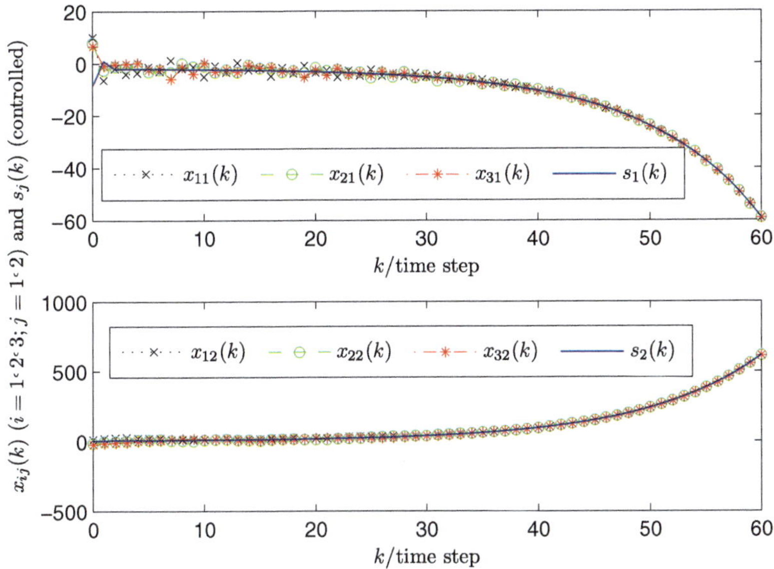

FIGURE 4.5
State trajectories $x_{ij}(k)$ of the controlled nodes i ($i = 1, 2, 3; j = 1, 2$).

control signal to the dynamical network. Figure 4.5 shows that all the state response curves of the controlled network can synchronize to the isolate node according to the proposed control protocol. Therefore, the simulation examples have confirmed our theoretical analysis very well.

4.4 Summary

This chapter has dealt with the synchronization control problem for a class of dynamical networks in a coding-decoding framework. By employing a modified uniform quantization approach, a set of coder-decoder-controller pairs has been designed to comply with the digital communication fashion. With the help of ISS theory, sufficient conditions have been derived for the closed-loop network to achieve the desired synchronization performance and the explicit expression of the controller gain matrix has been characterized in terms of the solution to certain LMIs. Finally, two illustrative examples have highlighted the detectability of the addressed dynamical network and the effectiveness of the synchronization control protocol presented in this chapter.

5

Observer-Based Consensus Control for Discrete-Time Multi-Agent Systems with Coding-Decoding Communication Protocol

The past few decades have witnessed particular research interests in the problems of collective behaviors (e.g. consensus, flocking and swarming) of networked multiple autonomous agents for their broad applications on various aspects of engineering systems including cooperative control for multi-robot systems, coordination of distributed sensor networks, formation control of unmanned air vehicles and satellite flying, to name just a few. In order to reach the goal of the coordination control for the individual agent in a networked environment, a pivotal issue is how to develop certain simple yet efficient distributed control protocols (or algorithms) only by sharing the local information between the adjacent agents. Because of its theoretical importance in distributed computing and application potentials in practical engineering, the consensus problem has long been an active research topic with a great many results reported in the literature.

It should be mentioned that almost all existing literature on the consensus control problems for MASs has been based on the assumption that the signals can be directly transmitted in an analog manner with infinite precision. Such an assumption is, however, sometimes too stringent because the signals are usually digitalized before being sent in a *networked environment*. As opposed to the traditional analog communication scheme, the digital communication strategy enjoys the distinct merits of well interference rejection capability, high security as well as reduced power consumption, and has therefore become increasingly popular in nowadays prevalent networked control systems (NCSs). Roughly speaking, there are four steps in an analog-to-digital (A/D) conversion, namely, sampling, reservation, quantization and encoding. It is worth pointing out that, compared with sampling and quantization issues, the signal encoding problem has received much less research attention in the context of NCSs.

In NCSs, the limited communication resource (e.g. channel bandwidth) is well known to be the main cause for quality deterioration of signal transmissions and performance degradation of control systems . Especially,

DOI: 10.1201/9781003534853-5

in networked distributed systems (e.g. MASs, sensor networks and complex networks), the subsystems usually share the communication network cables or wireless media, and it is therefore a critical issue to make adequate use of the limited communication capacity through exploiting energy-saving communication protocols.

To this end, another important communication protocol, namely, coding-decoding communication protocol (CDCP), which sends the *symbolic* data through communication channels and therefore occupies less resource as compared to the *original* data, has largely been overlooked despite its popularity in the environment of digital communication.

Motivated by the above discussion, in this chapter, our objective is to design an observer-based consensus controller for a class of discrete-time networked MASs with CDCPs. In doing so, three challenges are identified as follows: 1) how to establish a unified theoretical framework that takes both the observer-based control scheme and the CDCP into consideration? 2) how to obtain sufficient conditions for the existence of the desired easy-to-implement consensus controllers? and 3) how to examine the impact from the directed communication topology and the proposed CDCP on the consensus performance? To address these three identified challenges, we shall first focus our attention on the analysis/synthesis issues of the CDCP for the discrete-time MASs, which is of vital importance to the further investigation on the consensus performance. Then, based on the designed communication protocol, the observer-based consensus controller is obtained to guarantee the expected consensus performance of the closed-loop MAS.

5.1 Problem Formulation

In this chapter, it is assumed that the MAS has M agents which communicate with each other according to a fixed network topology represented by a directed graph $\mathcal{G} = (\mathcal{V}, \mathcal{E}, \mathcal{A})$ with the set of agents $\mathcal{V} = \{1, 2, \ldots, M\}$, the set of edges $\mathcal{E} \subseteq \mathcal{V} \times \mathcal{V}$, and the weighted adjacency matrix $\mathcal{A} = [a_{ij}]_{M \times M}$ with nonnegative adjacency element a_{ij}. An edge of $\mathcal{G}$ is denoted by the ordered pair (i, j). The adjacency elements associated with the edges of the graph are positive, that is, $a_{ij} > 0 \iff (i, j) \in \mathcal{E}$, which means that agent i can obtain information from agent j. The neighborhood of agent i is denoted by $\mathcal{N}_i = \{j \in \mathcal{V} : (j, i) \in \mathcal{E}\}$. An element of $\mathcal{N}_i$ is called a neighbor of agent i. In this chapter, self-edges (i, i) are not allowed, that is, $(i, i) \notin \mathcal{E}$ for any $i \in \mathcal{V}$. The in-degree of agent i is defined as $\deg_{\text{in}}^i = \sum_{j \in \mathcal{N}_i} a_{ij}$. Denote the diagonal matrix $\mathcal{D} = \text{diag}_M\{\deg_{\text{in}}^i\}$. The Laplacian matrix of $\mathcal{G}$ is given by $\mathcal{L}_{\mathcal{G}} = [l_{ij}]_{M \times M} = \mathcal{D} - \mathcal{A}$.

Consider a class of discrete-time MASs consisting of M agents where the dynamics of the ith agent is described by

$$\begin{cases} x_i(k+1) = Ax_i(k) + Bu_i(k) \\ y_i(k) = Cx_i(k) \\ x_i(0) = x_{i0} \end{cases} \tag{5.1}$$

where $x_i(k) \in \mathbb{R}^n$, $y_i(k) \in \mathbb{R}^m$ and $u_i(k) \in \mathbb{R}^p$ are, respectively, the state vector, the measurement output and the control input of the ith agent. $x_{i0} \in \mathbb{R}^n$ is a given initial condition. A, B and C are known real matrices with appropriate dimensions.

In most existing results regarding the MASs, a common assumption is that the relative full state information for each agent is available from itself and its adjacent agents, which is defined by

$$\xi_i(k) \triangleq \sum_{j \in \mathcal{N}_i} a_{ij}(x_j(k) - x_i(k)) \tag{5.2}$$

with the initial condition $\xi_i(0) = s_{i0}$, where $s_{i0} = \sum_{j \in \mathcal{N}_i} a_{ij}(x_{j0} - x_{i0})$. However, from an engineering point of view, the relative full state information is often unavailable in reality. Hence, for agent i, it is assumed that only the relative measurement output of neighboring agents is obtained as follows:

$$\tilde{y}_i(k) \triangleq \sum_{j \in \mathcal{N}_i} a_{ij}(y_j(k) - y_i(k)). \tag{5.3}$$

5.1.1 Traditional Observer Structure

With the available measurement output $\tilde{y}_i(k)$, we would be able to estimate the relative full state information $\xi_i(k)$ by using the following Luenberger-like type observer for agent i:

$$\hat{\xi}_i(k+1) = A\hat{\xi}_i(k) + B \sum_{j \in \mathcal{N}_i} a_{ij}(u_j(k) - u_i(k)) + L(\tilde{y}_i(k) - C\hat{\xi}_i(k)) \tag{5.4}$$

with the initial condition $\hat{\xi}_i(0) = \hat{s}_{i0}$, where $\hat{s}_{i0}$ is a known vector. Moreover, the control input $u_i(k)$ in (5.4) is given as $u_i(k) = K\hat{\xi}_i(k)$, where K is the controller gain to be determined.

5.1.2 CDCP-Based Observer Structure

In order to realize the digital transmission, we introduce a CDCP between the observer and the controller for each agent. To be more specific, a general form of coder-decoder scheme for the MAS (5.1) is given as follows:

Coder for Agent i:

$$g^i(dh) = \mathcal{F}_d^i(\hat{\xi}_i(h),\ \hat{\xi}_i(2h),\ldots,\hat{\xi}_i(dh)),$$ (5.5)

Decoder for Agent i:

$$\Xi^i(dh) = \mathcal{G}_d^i(g^i(h),\ g^i(2h),\ldots,g^i(dh))$$ (5.6)

for $d = 1,\ 2\ldots,$ where $g^i(dh)$ is the encoded data (codeword) generated at the coding instant dh by coder i and $\Xi^i(dh)$ is defined by $\Xi^i(dh) \triangleq [\check{\xi}_i^T(dh)\ \ \check{\xi}_i^T(dh+1)\ \ \cdots\ \ \check{\xi}_i^T((d+1)h-1)]^T$ with $\check{\xi}_i(k)$ being the decoded state of $\hat{\xi}_i(k)$ for $k \in [dh, (d+1)h)$. $\mathcal{F}_d^i(\cdot)$ and $\mathcal{G}_d^i(\cdot)$ are, respectively, the coder and decoder functions to be designed for agent i.

Instead of using the estimated state $\hat{\xi}_i(k)$ directly, only the decoded state $\check{\xi}_i(k)$ is available at the controller side due to the introduction of the CDCP, and thus the following decoder-based consensus control scheme is employed for agent i:

$$u_i(k) = K\check{\xi}_i(k)$$ (5.7)

where $\check{\xi}_i(k) \in \mathbb{R}^n$ is the decoded signal generated by decoder i. Based on such a fact, the control input $u_i(k) = K\hat{\xi}_i(k)$ in the observer (5.4) should be replaced by (5.7). On the other hand, it should be pointed out that, owing to the constraint of the network bandwidth, it is unrealistic and impossible to feed the decoded signal $\check{\xi}_i(k)$ back to the state estimator side, which means that the decoded state $\check{\xi}_i(k)$ cannot be directly utilized to construct state observer (5.4). Therefore, an auxiliary control input $\breve{u}_i(k)$ defined in the next section (see (5.31)) is introduced to establish the following observer:

$$\hat{\xi}_i(k+1) = A\hat{\xi}_i(k) + B\sum_{j\in\mathcal{N}_i} a_{ij}(\breve{u}_j(k) - \breve{u}_i(k)) + L(\tilde{y}_i(k) - C\hat{\xi}_i(k))$$ (5.8)

subject to the initial condition $\hat{\xi}_i(0) = \hat{s}_{i0}$, where $\hat{s}_{i0}$ is a known vector. As will be shown in the later developments, the auxiliary control input $\breve{u}_i(k)$ equals $u_i(k)\ \forall k \geq 0$. Therefore, by letting $\tilde{\xi}_i(k) \triangleq \xi_i(k) - \hat{\xi}_i(k)$, it follows from (5.2) and (5.8) that the following estimation error dynamics for agent i is derived:

$$\tilde{\xi}_i(k+1) = (A - LC)\tilde{\xi}_i(k).$$ (5.9)

The following lemma is needed in deriving our main results.

Lemma 5.1: *([135]). Given a matrix $A = [a_{ij}]_{n \times n}$, where $a_{ii} \leq 0$, $a_{ij} > 0$ ($\forall i \neq j$) and $\sum_{j=1}^{n} a_{ij} = 0$ for each i. Then, A has at least one zero eigenvalue and all of the nonzero eigenvalues are in the open left half plane. Furthermore, A has exactly one zero eigenvalue if and only if the directed graph associated with A has a spanning tree.*

For brevity, we set

$$x(k) = \mathrm{col}_M\{x_i(k)\}, \; w(k) = \mathrm{col}_M\{w_i(k)\},$$

$$v(k) = \mathrm{col}_M\{v_i(k)\}, \; \xi(k) = \mathrm{col}_M\{\xi_i(k)\},$$

$$\hat{\xi}(k) = \mathrm{col}_M\{\hat{\xi}_i(k)\}, \; \tilde{\xi}(k) = \mathrm{col}_M\{\tilde{\xi}_i(k)\}, \tag{5.10}$$

$$\check{\xi}(k) = \mathrm{col}_M\{\check{\xi}_i(k)\}, u(k) = \mathrm{col}_M\{u_i(k)\},$$

$$\bar{\mathscr{L}}_{\mathscr{G}} = \mathscr{L}_{\mathscr{G}}^T \mathscr{L}_{\mathscr{G}} = \left[\bar{l}_{ij}\right]_{M \times M}.$$

Subsequently, from (5.2) and (5.7), the augmentation of the dynamics $\xi_i(k)$ can be obtained with the following form:

$$\xi(k+1) = (I_M \otimes A)\xi(k) - (\mathscr{L}_{\mathscr{G}} \otimes BK)\check{\xi}(k). \tag{5.11}$$

Moreover, denote $\vec{\xi}(k) \triangleq \xi(k) - \check{\xi}(k)$ and rewrite $u(k) \triangleq (I_M \otimes K)(\xi(k) - \vec{\xi}(k))$. By augmenting the system state $x_i(k)$, we obtain the closed-loop system as follows:

$$x(k+1) = (I_M \otimes A - \mathscr{L}_{\mathscr{G}} \otimes BK)x(k) - (I_M \otimes BK)\vec{\xi}(k). \tag{5.12}$$

Similarly, we reformulate the observation error dynamics (5.9) as follows:

$$\tilde{\xi}(k+1) = (I_M \otimes (A - LC))\tilde{\xi}(k). \tag{5.13}$$

Next, introduce the weighted average state for all agents as $\bar{x}(k) = \sum_{i=1}^{M} \eta_{1i}x_i(k) = (\eta_1^T \otimes I_n)x(k)$, where η_{1i} is the ith element of $\eta_1 = \mathrm{col}_M\{\eta_{1i}\} \in \mathbb{R}^M$ with η_1 being the unique nonnegative left eigenvector of $\mathscr{L}_{\mathscr{G}}$ associated with eigenvalue 0 satisfying $\sum_{i=1}^{M} \eta_{1i} = 1$. Therefore, it derives from (5.12) and $\eta_1^T \mathscr{L}_{\mathscr{G}} = 0$ that

$$\begin{aligned}
\bar{x}(k+1) &= (\eta_1^T \otimes I_n)x(k+1) \\
&= A(\eta_1^T \otimes I_n)x(k) - (\eta_1^T \otimes BK)\vec{\xi}(k) \\
&= A\bar{x}(k) - (\eta_1^T \otimes BK)\vec{\xi}(k).
\end{aligned} \tag{5.14}$$

By setting $\tilde{x}_i(k) \triangleq x_i(k) - \bar{x}(k)$ and $\tilde{x}(k) \triangleq \mathrm{col}_M\{\tilde{x}_i(k)\}$, it is inferred from (5.12) and (5.14) that the deviation from each agent's state $x_i(k)$ to the weighted average state $\bar{x}(k)$ is calculated with a compact form as follows:

$$
\begin{aligned}
\tilde{x}(k+1) &= x(k+1) - (\mathbb{1}_M \otimes I_n)\bar{x}(k+1) \\
&= (I_M \otimes A - \mathscr{L}_{\mathscr{G}} \otimes BK)x(k) - (I_M \otimes BK)\vec{\xi}(k) \\
&\quad - (\mathbb{1}_M \otimes I_n)[A\bar{x}(k) - (\eta_1^T \otimes BK)\vec{\xi}(k)] \\
&= (I_M \otimes A - \mathscr{L}_{\mathscr{G}} \otimes BK)\tilde{x}(k) - (\mathcal{H} \otimes BK)\vec{\xi}(k) \\
&= (I_M \otimes A - \mathscr{L}_{\mathscr{G}} \otimes BK)\tilde{x}(k) + H\delta(k),
\end{aligned}
\tag{5.15}
$$

where $\mathcal{H} = [h_{ij}]_{M \times M}$ with $h_{ij} = 1 - \eta_{1j}$ for $i = j$ and $h_{ij} = -\eta_{1j}$ for $i \neq j$.

Subsequently, along the same line as in [84], denote Λ by the Jordan form associated with the Laplace matrix $\mathscr{L}_{\mathscr{G}}$, for example $P^{-1}\mathscr{L}_{\mathscr{G}}P = \Lambda$, where $P = [p_1 \quad p_2 \quad \cdots \quad p_M] \in \mathbb{C}^{M \times M}$ and $P^{-1} = [q_1 \quad q_2 \quad \cdots \quad q_M]^T \in \mathbb{C}^{M \times M}$ are nonsingular matrices with $p_i \in \mathbb{C}^M$ and $q_i \in \mathbb{C}^M$ being the M-dimensional complex vectors for $i = 1, 2, \ldots, M$. Different from the case of undirected graph $\mathscr{G}$ where $\mathscr{L}_{\mathscr{G}}$ is symmetric and Λ is a diagonal matrix, for a directed graph $\mathscr{G}$, $\mathscr{L}_{\mathscr{G}}$ could be a non-symmetric matrix whose eigenvalues are complex and Λ is no more diagonal but with the form $\Lambda = \mathrm{diag}\{\Lambda_1, \Lambda_2, \ldots, \Lambda_r\}$, where $\Lambda_i \in \mathbb{C}^{M_i \times M_i}$ is a Jordan block

$$
\Lambda_i = \begin{pmatrix} \lambda_i & 1 & 0 & 0 \\ 0 & \ddots & \ddots & 0 \\ 0 & 0 & \ddots & 1 \\ 0 & 0 & 0 & \lambda_i \end{pmatrix}_{M_i \times M_i}.
\tag{5.16}
$$

In (5.16), the diagonal elements λ_i are the eigenvalues of $\mathscr{L}_{\mathscr{G}}$ with multiplicity $M_i \leq M$ for $i = 1, 2, \ldots, r$ and $\sum_{i=1}^r M_i = M$. Specially, when $r = M$, we have $M_i = 1$ and Λ reduces to a diagonal matrix. It is assumed that the directed graph $\mathscr{G}$ contains a directed spanning tree and, therefore, according to Lemma 5.1, we denote Λ_1 by the simple eigenvalue 0 with the corresponding eigenvectors $p_1 = \mathbb{1}_M$ and $q_1 = \eta_1$, which also implies $M_1 = 1$.

Based on the above discussion, we define $\delta(k) \triangleq (P^{-1} \otimes I_n)\tilde{x}(k) \in \mathbb{C}^{Mn}$ and $W \triangleq [q_2 \quad \cdots \quad q_M]^T \in \mathbb{C}^{(M-1) \times M}$, and then make a partition to $\delta(k)$ as $\delta(k) \triangleq [\zeta_1^T(k) \tilde{\delta}^T(k)]^T$, where $\zeta_1(k) \in \mathbb{R}^n$ contains the first nth entries of $\delta(k)$ and $\tilde{\delta}(k)$ is of the form $\tilde{\delta}(k) \triangleq [\tilde{\delta}_2^T(k), \ldots, \tilde{\delta}_r^T(k)]^T$ with $\tilde{\delta}_i(k) \in \mathbb{C}^{M_i n}$ for $i = 2, \ldots, r$. Bearing in mind the facts that $\eta_1^T \mathcal{H} = 0$ and $W\mathcal{H} = W$, we immediately have that

$$
\begin{cases}
\zeta_1(k+1) = (\eta_1^T \otimes I_n)\tilde{x}(k+1) = 0 & \text{(5.17a)} \\
\tilde{\delta}(k+1) = (I_{M-1} \otimes A - \tilde{\Lambda} \otimes BK)\tilde{\delta}(k) - (W \otimes BK)\vec{\xi}(k) & \text{(5.17b)}
\end{cases}
$$

where $\tilde{\Lambda} = \mathrm{diag}\{\Lambda_2, \ldots, \Lambda_r\}$.

To proceed, we introduce the following definitions and lemmas which are useful for the later developments.

Definition 5.1: *The discrete-time dynamical system (5.11) is said to be detectable if there exist families of coder-decoder pairs (5.5) and (5.6) with a coding alphabet $\mathbb{H}$ of size χ such that*

$$\lim_{k \to \infty} |\xi(k) - \check{\xi}(k)| = 0. \tag{5.18}$$

Definition 5.2: *Suppose that the directed graph $\mathscr{G}$ of the communication network contains a directed spanning tree. The discrete-time MASs (5.1) is said to reach the asymptotic consensus if*

$$\lim_{k \to \infty} |x_i(k) - x_j(k)| = 0 \tag{5.19}$$

holds for $\forall i, j \in \mathscr{V}$.

Lemma 5.2: *Suppose that the directed graph $\mathscr{G}$ of the communication network contains a directed spanning tree. The MAS (5.1) reaches the asymptotic consensus if the component $\tilde{\delta}_i(k)$ in (5.17b) satisfies*

$$\lim_{k \to \infty} |\tilde{\delta}_i(k)| = 0 \tag{5.20}$$

for $i = 2, \ldots, r$.

Proof: *First, define $\tilde{p}_1 = p_1$ and $\tilde{p}_i \triangleq \underbrace{[p_{\alpha_i+1}\, p_{\alpha_i+2} \cdots p_{\alpha_i+M_i}]}_{M_i}$ with $\alpha_i = \sum_{s=1}^{i-1} M_s$ for $i = 2, \ldots, r$. Thus, the matrix P can be rewritten as $P = [\tilde{p}_1 \quad \tilde{p}_2 \quad \cdots \quad \tilde{p}_r]$, and then it follows from $\delta(k) = (P^{-1} \otimes I_n)\tilde{x}(k)$ and $\zeta_1(k) = 0$ that*

$$\lim_{k \to \infty} |\tilde{x}(k)| = \lim_{k \to \infty} |(P \otimes I_n)\delta(k)|$$

$$= \lim_{k \to \infty} \left|(p_1 \otimes I_n)\zeta_1(k) + \sum_{i=2}^{r} (\tilde{p}_i \otimes I_n)\tilde{\delta}_i(k)\right|$$

$$\leq \lim_{k \to \infty} \sum_{i=2}^{r} |\tilde{p}_i \otimes I_n|_F |\tilde{\delta}_i(k)|.$$

Consequently, it follows immediately from (5.20) that the conclusion $\lim_{k \to \infty} |\tilde{x}(k)| = 0$ is true, and the proof is complete.

Lemma 5.3: *Let the positive scalar ϵ be given. If there exists a positive definite matrix $R > 0$ satisfying*

$$\Pi_0 \triangleq (I_M \otimes A)^T R(I_M \otimes A) - \epsilon^2 R < 0, \tag{5.21}$$

then the following inequality

$$|\varphi(k+1) - \phi(k+1)| \leq c_0|\varphi(k) - \phi(k)| \tag{5.22}$$

is satisfied, where $c_0 > 0$ is a positive scalar, and $\varphi(k)$ and $\phi(k)$ are two trajectories of system (5.11).

Proof: *First, denoting $z(k) \triangleq \epsilon^{-k}(\varphi(k) - \phi(k))$, it is readily seen from (5.11) that $z(k+1) = \epsilon^{-1}(I_M \otimes A)z(k)$. Then, by choosing the quadratic function $V_1(k) = \epsilon^2 z^T(k)Rz(k)$, we calculate the difference along the trajectory of $z(k)$ as follows:*

$$\begin{aligned}
\Delta V_1(k) &\triangleq V_1(k+1) - V_1(k) \\
&= \epsilon^2 z^T(k+1)Rz(k+1) - \epsilon^2 z^T(k)Rz(k) \tag{5.23} \\
&\leq -\lambda_{\min}\{-\Pi_0\}|z(k)|^2.
\end{aligned}$$

Defining $\lambda_0 \triangleq \lambda_{\min}\{-\Pi_0\}$, $\bar{\lambda}_1 \triangleq \lambda_{\max}\{R\}$ and $\underline{\lambda}_1 \triangleq \lambda_{\min}\{R\}$, one has $\Delta V_1(k) \leq -\lambda_0\bar{\lambda}_1^{-1}\epsilon^{-2}V_1(k)$, which further indicates that $|\varphi(k+1)-\phi(k+1)| \leq c_0|\varphi(k)-\phi(k)|$ with $c_0 = \epsilon((1 - \lambda_0\bar{\lambda}_1^{-1}\epsilon^{-2})\bar{\lambda}_1\underline{\lambda}_1^{-1})^{1/2}$. The proof is now complete.

Lemma 5.4: *Let the positive scalar $0 < \mu < 1$ be given. Consider the augmented estimation error system (5.13). If there exist a positive definite matrix $\bar{S} > 0$ and a matrix $\bar{X}$ satisfying*

$$\Pi_1 \triangleq \begin{bmatrix} -(1-\mu)S & (I_M \otimes (\bar{S}A - \bar{X}C))^T \\ * & -S \end{bmatrix} < 0 \tag{5.24}$$

where $S = I_M \otimes \bar{S}$ and $X = I_M \otimes \bar{X}$, then there always exist an integer $h \subset \mathbb{Z}^+$ and a scalar c_1 $(0 < c_1 < 1)$ such that

$$|\tilde{\xi}(k+h)| < c_1|\tilde{\xi}(k)|. \tag{5.25}$$

Moreover, the desired observer gain matrix for (5.9) can be obtained with the form of

$$L = \bar{S}^{-1}\bar{X}. \tag{5.26}$$

Proof: *First, by selecting the quadratic function $V_2(k) = \tilde{\xi}^T(k)S\tilde{\xi}(k)$ for (5.13) and defining its difference as $\Delta V_2(k) \triangleq V_2(k+1) - V_2(k)$, we calculate the term $\Delta V_2(k) + \mu V_2(k)$ along the trajectory of (5.13) as*

$$\begin{aligned}
\Delta V_2(k) + \mu V_2(k) &= \tilde{\xi}^T(k+1)S\tilde{\xi}(k+1) - (1-\mu)\tilde{\xi}^T(k)S\tilde{\xi}(k) \\
&= \tilde{\xi}^T(k)(I_M \otimes (A - LC))^T S(I_M \otimes (A - LC))\tilde{\xi}(k) \\
&\quad - (1-\mu)\tilde{\xi}^T(k)S\tilde{\xi}(k).
\end{aligned}$$

It is inferred from (5.24), (5.26) and the Schur Complement Lemma that

$$\Delta V_2(k) + \mu V_2(k) \leq 0. \tag{5.27}$$

After some straightforward algebraic manipulations, it follows from inequality (5.27) that

$$\underline{\mu}_1 |\tilde{\xi}(k+h)|^2 \leq V_2(k+h) \leq (1-\mu)^h V_2(k) \leq (1-\mu)^h \bar{\mu}_1 |\tilde{\xi}(k)|^2$$

where $\bar{\mu}_1 = \lambda_{\max}\{S\}$ and $\underline{\mu}_1 = \lambda_{\min}\{S\}$. Therefore, one further has

$$|\tilde{\xi}(k+h)| \leq c_1 |\tilde{\xi}(k)|$$

with $c_1 = (\underline{\mu}_1^{-1} \bar{\mu}_1 (1-\mu)^h)^{1/2}$. Noting the fact that $0 < \mu < 1$, there must exist a positive integer h ensuring $0 < c_1 < 1$, which completes the proof of this lemma.

The objective of this chapter is to design an observer-based controller (5.7) for the discrete-time MAS (5.1). More specifically, for a given directed communication network topology, we are interested in looking for the controller parameter K such that the consensus performance (5.19) can be achieved under the CDCP.

5.2 Design of CDCP and Performance Analysis of MASs

In this section, we shall investigate the design issue of the CDCP (5.5)–(5.6) for discrete-time MAS (5.1). Then, based on the established communication protocol, we would like to handle the observer-based controller design problems for the addressed system.

5.2.1 Coding-Decoding Communication Protocol

In this subsection, we are ready to tackle the synthesis issue for the CDCP (5.5)–(5.6) by using the uniform quantization approach. For the convenience of the readers, to begin with, we provide a brief introduction regarding the quantization method which will be adopted later.

Let the scaling parameters $a_i > 0$ $(i = 1, 2, \ldots, M)$ and integer q be given. Then, the hyperrectangles $\mathcal{B}_{a_i} = \{\vartheta_i \in \mathbb{R}^n : |\vartheta_i^{(j)}| \leq a_i, j = 1, \ldots, n\}$ can be respectively partitioned into q^n hyperrectangles $\mathcal{I}_{s_1^i}^{i1}(a_i) \times \mathcal{I}_{s_2^i}^{i2}(a_i) \times \cdots \times \mathcal{I}_{s_n^i}^{in}(a_i)$, where $s_1^i, s_2^i, \ldots, s_n^i \in \{1, 2, \ldots, q\}$ and

$$\mathcal{I}_1^{ij}(a_i) \triangleq \left\{ \vartheta_i^{(j)} : -a_i \le \vartheta_i^{(j)} < -a_i + \frac{2a_i}{q} \right\};$$

$$\mathcal{I}_2^{ij}(a_i) \triangleq \left\{ \vartheta_i^{(j)} : -a_i + \frac{2a_i}{q} \le \vartheta_i^{(j)} < -a_i + \frac{4a_i}{q} \right\};$$

$$\vdots$$

$$\mathcal{I}_q^{ij}(a_i) \triangleq \left\{ \vartheta_i^{(j)} : a_i - \frac{2a_i}{q} \le \vartheta_i^{(j)} \le a_i \right\}.$$

Here, $\vartheta_i^{(j)}$ is the jth element of the vector ϑ_i.

For each $\mathcal{B}_{a_i}$, the center of the hyperrectangle $\mathcal{I}_{s_1^i}^{i1}(a_i) \times \mathcal{I}_{s_2^i}^{i2}(a_i) \times \cdots \times \mathcal{I}_{s_n^i}^{in}(a_i)$ can be determined by

$$\eta_{a_i}^i(s_1^i, s_2^i, \ldots, s_n^i) \triangleq \begin{bmatrix} \upsilon_{1i} & \upsilon_{2i} & \cdots & \upsilon_{ni} \end{bmatrix}^T$$

where $\upsilon_{1i} = -a_i + \frac{(2s_1^i-1)a_i}{q}$, $\upsilon_{2i} = -a_i + \frac{(2s_2^i-1)a_i}{q}$ and $\upsilon_{ni} = -a_i + \frac{(2s_n^i-1)a_i}{q}$. Consequently, for any $\vartheta_i \in \mathcal{B}_{a_i}$, there exist unique integers $s_1^i, s_2^i, \ldots, s_n^i \in \{1, 2, \ldots, q\}$ such that $\vartheta_i \in \mathcal{I}_{s_1^i}^{i1}(a_i) \times \mathcal{I}_{s_2^i}^{i2}(a_i) \times \cdots \times \mathcal{I}_{s_n^i}^{in}(a_i)$, which indicates that

$$|\vartheta_i - \eta_{a_i}^i(s_1^i, s_2^i, \ldots, s_n^i)| \le \frac{\sqrt{n}a_i}{q}. \tag{5.28}$$

In order to accomplish the control task for the MAS (5.1) under the digital transmission mechanism, we present the following design scheme of the CDCP: **Coder for Agent** i : For $\hat{\xi}_i(dh) - \bar{\xi}_i(dh) \in \mathcal{I}_{s_1^i}^{i1}(a^i(dh)) \times \mathcal{I}_{s_2^i}^{i2}(a^i(dh)) \times \cdots \times \mathcal{I}_{s_n^i}^{in}(a^i(dh)) \subset \mathcal{B}_{a^i(dh)}$, one obtains

$$g^i(dh) = \{s_1^i, s_2^i, \ldots, s_n^i\} \tag{5.29}$$

where $\bar{\xi}_i(dh)$ is governed by

$$\begin{cases} \bar{\xi}_i(0) = 0 \\ \bar{\xi}_i(k) = \check{\xi}_i(k), \ k \ne dh \\ \bar{\xi}_i(dh) = A\check{\xi}_i(dh-1) + B\check{u}_i(dh-1) \\ \check{u}_i(dh-1) = K\check{\xi}(dh-1) \end{cases} \tag{5.30}$$

and the dynamics of $\breve{\xi}_i(k)$ is determined by the following system

$$\begin{cases} \breve{\xi}_i(0) = 0 \\ \breve{\xi}_i(k+1) = A\breve{\xi}_i(k) + B\breve{u}_i(k), \; k \neq dh - 1 \\ \breve{\xi}_i(dh) = \bar{\xi}_i(dh) + \eta^i_{a^i(dh)}(s^i_1, s^i_2, \ldots, s^i_n) \\ \breve{u}_i(k) = K\breve{\xi}_i(k). \end{cases} \tag{5.31}$$

Decoder for Agent i :

$$\begin{cases} \check{\xi}_i(0) = 0 \\ \check{\xi}_i(k+1) = A\check{\xi}_i(k) + Bu_i(k), \; k \neq dh - 1 \\ \check{\xi}_i(dh) = \bar{\xi}_i(dh) + \eta^i_{a^i(dh)}(s^i_1, s^i_2, \ldots, s^i_n) \\ u_i(k) = K\check{\xi}_i(k). \end{cases} \tag{5.32}$$

Remark 5.1: *Two auxiliary systems (5.30) and (5.31) are constructed in the design of the coding-decoding protocol. The auxiliary system (5.30) generates a "prediction" $\bar{\xi}_i(dh)$ of the decoded state $\check{\xi}_i(dh)$ as the decoded state $\check{\xi}_i(dh)$ cannot be obtained at the coding instant dh. On the other hand, it follows from (5.31) and (5.32) that $\breve{\xi}_i(k) \equiv \check{\xi}_i(k)$ for $\forall k \geq 0$, and hence the auxiliary system (5.31) can be seen as a "copy" of the system (5.32). Different from the procedure in [190] where the decoded state is required to feed back to the coder side, in this chapter, such a requirement is removed due to the introduction of an additional auxiliary system (5.31), which is closer to the engineering practice.*

Denoting $\bar{\xi}(k) \triangleq \mathrm{col}_M\{\bar{\xi}_i(k)\}$ and $\breve{\xi}(k) \triangleq \mathrm{col}_M\{\breve{\xi}_i(k)\}$, we reformulate the coder-decoder procedure in a compact form as follows:

Coder : For $\hat{\xi}(dh) - \bar{\xi}(dh) \in \mathcal{I}^{11}_{s^1_1}(a(dh)) \times \cdots \times \mathcal{I}^{1n}_{s^1_n}(a(dh)) \times \mathcal{I}^{21}_{s^2_1}(a(dh)) \times \cdots \times \mathcal{I}^{2n}_{s^2_n}(a(dh)) \times \cdots \times \mathcal{I}^{M1}_{s^M_1}(a(dh)) \times \cdots \times \mathcal{I}^{Mn}_{s^M_n}(a(dh)) \subset \mathcal{B}_{a(dh)}$, we have

$$g(dh) = \{s^1_1, \ldots, s^1_n, s^2_1, \ldots s^2_n, \ldots, s^M_1, \ldots, s^M_n\} \tag{5.33}$$

where $\mathcal{B}_{a(dh)} = \{\vartheta(dh) \in \mathbb{R}^{Mn} : |\vartheta^{(j)}(dh)| \leq a(dh), j = 1, \ldots, Mn\}$ with $a(dh) = \max_{1 \leq i \leq M}\{a_i(dh)\}$ and $\bar{\xi}(dh) = \mathrm{col}_M\{\bar{\xi}_i(dh\}$ is constructed as

$$\begin{cases} \bar{\xi}(0) = 0 \\ \bar{\xi}(k) = \breve{\xi}(k), \; k \neq dh \\ \bar{\xi}(dh) = (I_M \otimes A)\breve{\xi}(dh - 1) - (\mathcal{L}_{\mathcal{G}} \otimes B)\breve{u}(dh - 1) \\ \breve{u}(dh - 1) = (I_M \otimes K)\breve{\xi}(dh - 1) \end{cases} \tag{5.34}$$

with the evolution of the dynamics $\check{\xi}(k)$ as follows

$$\begin{cases} \check{\xi}(0) = 0 \\ \check{\xi}(k+1) = A\check{\xi}(k) + B\check{u}(k), \ k \neq dh - 1 \\ \check{\xi}(dh) = \bar{\xi}(dh) + \eta_{a(dh)}(s_1^1, \ldots, s_n^M) \\ \check{u}(k) = (I_M \otimes K)\check{\xi}(k). \end{cases} \tag{5.35}$$

Decoder:

$$\begin{cases} \check{\xi}(0) = 0 \\ \check{\xi}(k+1) = (I_M \otimes A)\check{\xi}(k) - (\mathscr{L}_{\mathscr{G}} \otimes B)u(k), \ k \neq dh - 1 \\ \check{\xi}(dh) = \bar{\xi}(dh) + \eta_{a(dh)}(s_1^1, \ldots, s_n^M) \\ u(k) = (I_M \otimes K)\check{\xi}(k). \end{cases} \tag{5.36}$$

Before proceeding, let us give some notations as follows:

$$\begin{aligned} \bar{s}_0 &\triangleq |s_0|, \ s_0 \triangleq \mathrm{col}_M\{s_{i0}\}, \ \hat{s}_0 \triangleq \mathrm{col}_M\{\hat{s}_{i0}\}, \\ \tilde{s}_0 &\triangleq |s_0 - \hat{s}_0|, \ a(h) \triangleq c_1 \tilde{s}_0 + c_0^h \bar{s}_0, \\ a((d+1)h) &\triangleq c_1^d \tilde{s}_0(c_1 + c_0^h) + c_0^h \frac{\sqrt{Mn}}{q} a(dh). \end{aligned} \tag{5.37}$$

The following lemma is to verify that the CDCP proposed in this chapter is well defined. In other words, we would like to deal with the decoding condition $\hat{\xi}(dh) - \bar{\xi}(dh) \in \mathcal{B}_{a(dh)}$ for all $d = 1, 2, \ldots$.

Lemma 5.5: *The coding-decoding communication protocol (5.33) and (5.36) satisfies the following requirement:*

$$|\hat{\xi}(dh) - \bar{\xi}(dh)|_\infty \leq a(dh), \ d = 1, 2, \ldots. \tag{5.38}$$

Proof: *The mathematical induction method is utilized in this proof. First, for $d = 1$, it follows from Lemma 5.3, Lemma 5.4 and the property of vector norm that*

$$\begin{aligned} |\hat{\xi}(h) - \bar{\xi}(h)| &\leq |\hat{\xi}(h) - \xi(h)| + |\xi(h) - \bar{\xi}(h)| \\ &\leq c_1|\hat{\xi}(0) - \xi(0)| + |\xi(h) - \bar{\xi}(h)| \\ &\leq c_1 \tilde{s}_0 + c_0|\xi(h-1) - \check{\xi}(h-1)| \\ &\leq c_1 \tilde{s}_0 + c_0^2|\xi(h-2) - \check{\xi}(h-2)| \\ &\leq \cdots \leq c_1 \tilde{s}_0 + c_0^h \bar{s}_0 = a(h) \end{aligned} \tag{5.39}$$

or, equivalently,

$$|\hat{\xi}(dh) - \bar{\xi}(dh)|_\infty \le a(h). \tag{5.40}$$

Second, with the assumption that $|\hat{\xi}(jh) - \bar{\xi}(jh)|_\infty \le a(jh)$ for $j = 2, \ldots, d$, one has

$$
\begin{aligned}
|\hat{\xi}((d+1)h) &- \bar{\xi}((d+1)h)| \\
&\le |\hat{\xi}((d+1)h) - \xi((d+1)h)| + |\xi((d+1)h) - \bar{\xi}((d+1)h)| \\
&\le c_1 |\hat{\xi}(dh) - \xi(dh)| + c_0 |\xi((d+1)h - 1) - \check{\xi}((d+1)h - 1)| \\
&\le \cdots \le c_1^{d+1} \tilde{s}_0 + c_0^h |\xi(dh) - \check{\xi}(dh)|.
\end{aligned}
\tag{5.41}
$$

By considering the dynamics of $\check{\xi}(dh)$ in (5.36), it is further derived that

$$
\begin{aligned}
|\xi(dh) &- \check{\xi}(dh)| \\
&= |\xi(dh) - \bar{\xi}(dh) - \eta_{a(dh)}(s_1^1, \ldots, s_n^M)| \\
&\le |\xi(dh) - \hat{\xi}(dh)| + |\hat{\xi}(dh) - \bar{\xi}(dh) - \eta_{a(dh)}(s_1^1, \ldots, s_n^M)| \\
&\le c_1^d \tilde{s}_0 + \frac{\sqrt{Mn}}{q} a(dh).
\end{aligned}
\tag{5.42}
$$

Therefore, in combination with (5.41) and (5.42), we can draw the conclusion that

$$|\hat{\xi}((d+1)h) - \bar{\xi}((d+1)h)| \le c_1^d \tilde{s}_0 (c_1 + c_0^h) + c_0^h \frac{\sqrt{Mn}}{q} a(dh). \tag{5.43}$$

Moreover, by noting (5.37), (5.43) also implies

$$|\hat{\xi}((d+1)h) - \bar{\xi}((d+1)h)|_\infty \le a((d+1)h). \tag{5.44}$$

Finally, it follows readily from (5.39)–(5.44) that the requirement in Lemma 5.5 is satisfied for $d \ge 1$, which completes the proof of this lemma.

Theorem 5.1: *The discrete-time system (5.11) is detectable with a CDCP (5.33)–(5.36) if the following inequality*

$$c_0^h \frac{\sqrt{Mn}}{q} < 1 \tag{5.45}$$

holds subject to (5.21) and (5.24) for a certain proper positive integer q, where the parameters c_0 and h have been defined in Lemma 5.3 and Lemma 5.4, respectively.

Proof: *According to the definition of $a(dh)$ in (5.37) and $0 < c_1 < 1$, it is easy to see that*

$$\lim_{d \to +\infty} a(dh) = 0, \tag{5.46}$$

and then one can easily infer from (5.42) and (5.46) that

$$\lim_{d \to \infty} |\xi(dh) - \check{\xi}(dh)| = 0.$$

Moreover, for those non-coding instants $k \in (dh, (d+1)h)$, it is noticed that $\xi(k)$ and $\check{\xi}(k)$ are actually the trajectories of (5.11) and, therefore, it follows immediately from Lemma 5.3 that $|\xi(k) - \check{\xi}(k)|^2 \leq c_0^{k-dh}|\xi(dh) - \check{\xi}(dh)|^2$. So, we have that $|\xi(k) - \check{\xi}(k)|$ is bounded at the non-coding instants, which implies that the dynamical system (5.11) is detectable, namely, $\lim_{k \to \infty} |\xi(k) - \check{\xi}(k)| = 0$. The proof is now complete.

5.2.2 Consensus Analysis

Let us start with introducing some basic notions and background on input-to-state stability theory, which will be useful for the derivation of our main results later.

Consider the following discrete-time nonlinear system with the following form

$$\varsigma(k + 1) = f(\varsigma(k), \tau(k)) \tag{5.47}$$

where $\varsigma(k) \subset \mathbb{R}^n$, $\tau(k) \in \mathbb{R}^p$ and $f(\cdot, \cdot) : \mathbb{R}^n \times \mathbb{R}^p \mapsto \mathbb{R}^n$ are, respectively, the system state vector, the exogenous input and the continuous nonlinear function with $f(0, 0) = 0$.

Definition 5.3: *The system (5.47) is said to be input-to-state stable, if there exist a $\mathcal{KL}$ function $\beta(\cdot, \cdot)$ and a $\mathcal{K}$ class function $\gamma(\cdot)$ such that the dynamics of the system state $\varsigma(k)$ satisfies*

$$|\varsigma(k)| \leq \beta(|\varsigma(0)|, k) + \gamma(|\tau(k)|_\infty) \tag{5.48}$$

for $\forall k \geq 0$ and $\forall x(0) \in \mathbb{R}^n$, where $|u(k)|_\infty \triangleq \sup_k\{|u(k)|\}$..

Lemma 5.6: *([58]). The nonlinear discrete-time system (5.47) is said to be input-to-state stable if there exist a positive definite function $V(k, \varsigma(k)) : [0, +\infty) \times \mathbb{R}^n \mapsto \mathbb{R}$ (called an ISS-Lyapunov function), three $\mathcal{K}_\infty$ class functions $\alpha_1(\cdot)$, $\alpha_2(\cdot)$ and $\alpha_3(\cdot)$, and a $\mathcal{K}$ class function $\varpi(\cdot)$ such that the following two inequalities*

$$\alpha_1(|\varsigma(k)|) \leq V(k, \varsigma(k)) \leq \alpha_2(|\varsigma(k)|) \tag{5.49}$$

and

$$V(k+1, \varsigma(k+1)) - V(k, \varsigma(k)) \leq -\alpha_3(|\varsigma(k)|) + \varpi(|u(k)|) \tag{5.50}$$

hold for all $\varsigma(k) \in \mathbb{R}^n$ and $u(k) \in \mathbb{R}^p$. Furthermore, if the above two inequalities are met simultaneously, the functions $\beta(\cdot, \cdot)$ and $\gamma(\cdot)$ in Definition 5.3 can be selected as $\beta(\cdot, k) = \alpha_1^{-1}(\phi^k \alpha_2(\cdot))$ $(0 < \phi < 1)$ and $\gamma(\cdot) = \alpha_1^{-1}(\alpha_2(\alpha_3^{-1}(\varpi(\cdot))))$, respectively, where $\alpha_1^{-1}(\cdot)$ expresses the inverse function of the monotone function $\alpha_1(\cdot)$ and so does $\alpha_3^{-1}(\cdot)$.

Subsequently, recalling the special form of (5.17b), we decouple it into the following $r - 1$ parts

$$\tilde{\delta}_i(k+1) = (I_{M_i} \otimes A - \Lambda_i \otimes BK)\tilde{\delta}_i(k) - ([W]_{i-1} \otimes BK)\vec{\xi}(k), \quad i = 2, \ldots, r \tag{5.51}$$

where $[W]_j \in \mathbb{R}^{M_j \times M}$ stands for a matrix whose elements come from the 1st row to the M_2th row of matrix W for $j = 1$, and from the $(\sum_{s=2}^{j} M_s + 1)$th row to the $(\sum_{s=2}^{j+1} M_s)$th row of matrix W for $j \geq 2$.

As discussed in [182], from the properties of the Jordan form (5.16), the asymptotic property of system (5.51) is dominated by the diagonal terms, and therefore, the consensus issue of system (5.1) can be now converted into the problem of asymptotic stability of the following closed-loop system

$$\theta_i(k+1) = (A - \lambda_i BK)\theta_i(k) - (W_{i-1} \otimes BK)\vec{\xi}(k), \quad i = 2, \ldots, r \tag{5.52}$$

where θ_i stands for an n-dimensional vector and $W_j \in \mathbb{R}^{1 \times M}$ is a row vector whose elements come from the 1st row of matrix W for $j = 1$ and from the $(\sum_{s=2}^{j} M_s + 1)$th row of matrix W for $j \geq 2$.

It is worth mentioning that, since the topology structure of the communication network is directed, some eigenvalues of the Laplacian matrix $\mathscr{L}_{\mathscr{G}}$ may be complex values. To this end, for $i = 1, \ldots, r$, we denote $\lambda_i = a_i + b_i j$, $\theta_i(k) = \theta_i^{\mathcal{R}}(k) + \theta_i^{\mathcal{I}}(k)j$ and $W_i = W_{i-1}^{\mathcal{R}} + W_{i-1}^{\mathcal{I}} j$, where a_i, $\theta_i^{\mathcal{R}}(k)$, $W_{i-1}^{\mathcal{R}}$ and b_i, $\theta_i^{\mathcal{I}}(k)$, $W_i^{\mathcal{I}}$ are, respectively, the real and imaginary parts of λ_i, $\theta_i(k)$ and W_{i-1}. In addition, we assume that the real parts of the r different eigenvalues of $\mathscr{L}_{\mathscr{G}}$ in an ascending order are written as $0 = a_1 < a_2 \leq \cdots \leq a_r$. As such, separating the real and imaginary parts of (5.52) for $i = 2, \cdots, r$, one finds

$$\begin{cases} \theta_i^{\mathcal{R}}(k+1) &= (A - a_i BK)\theta_i^{\mathcal{R}}(k) + b_i BK\theta_i^{\mathcal{I}}(k) \\ &\quad -(W_{i-1}^{\mathcal{R}} \otimes BK)\vec{\xi}(k) \\ \theta_i^{\mathcal{I}}(k+1) &= (A - a_i BK)\theta_i^{\mathcal{I}}(k) - b_i BK\theta_i^{\mathcal{R}}(k) \\ &\quad -(W_{i-1}^{\mathcal{I}} \otimes BK)\vec{\xi}(k). \end{cases} \tag{5.53}$$

It is not difficult to see that the systems (5.51) are asymptotically stable as long as the systems (5.53) are asymptotically stable, which indicates from the Lemma 5.2 that the asymptotic consensus can be reached in MASs (5.1). Before providing the main results, we define

$$\theta^{\mathcal{R}}(k) \triangleq \mathrm{col}_{r-1}\{\theta_i^{\mathcal{R}}(k)\}, \quad \theta^{\mathcal{I}}(k) \triangleq \mathrm{col}_{r-1}\{\theta_i^{\mathcal{I}}(k)\},$$

$$\theta(k) \triangleq [(\theta^{\mathcal{R}}(k))^T (\theta^{\mathcal{I}}(k))^T]^T, \quad \mathcal{Q} \triangleq I_{2r} \otimes Q,$$

$$\tilde{W} \triangleq \max_{1 \le i \le r}\{(W_i^{\mathcal{R}})^T W_i^{\mathcal{R}} + (W_i^{\mathcal{I}})^T W_i^{\mathcal{I}}\}.$$

Theorem 5.2: *Suppose that the network topology $\mathscr{G}$ contains a directed spanning tree. Let the positive scalars ϱ_1, ϱ_2 and ϱ_3 be given. The asymptotic consensus of system (5.1) can be reached with the CDCP (5.29)-(5.32) if there exist a matrix K and a positive matrix Q satisfying*

$$\begin{cases} \bar{\Pi}_1 \triangleq A^T QBK + (BK)^T QA \\ \qquad - (a_2 + a_r)(BK)^T Q(BK) > 0 & (5.54a) \\ \bar{\Pi}_2 \triangleq (1 + \sigma_1)(A - a_2 BK)^T Q(A - a_2 BK) \\ \qquad + (1 + \sigma_2)b_m^2 (BK)^T Q(BK) - Q < 0 & (5.54b) \end{cases}$$

subject to (5.21), (5.24) and (5.45), where $\sigma_1 = \varrho_1 + \varrho_2$, $\sigma_2 = \varrho_1^{-1} + \varrho_3$ and $b_m = \max_{1 \le i \le r}\{b_i\}$.

Proof: *Choose $V(k) = \sum_{i=2}^{r}\left((\theta_i^{\mathcal{R}}(k))^T Q\theta_i^{\mathcal{R}}(k) + (\theta_i^{\mathcal{I}}(k))^T Q\theta_i^{\mathcal{I}}(k)\right)$ and calculate the difference along the trajectory of (5.53) as*

$$\begin{aligned}
V(k+1) - V(k) = \sum_{i=2}^{r} \Big\{ &((A - a_i BK)\theta_i^{\mathcal{R}}(k))^T Q(A - a_i BK)\theta_i^{\mathcal{R}}(k) \\
&+ b_i^2 (\theta_i^{\mathcal{I}}(k))^T (BK)^T QBK\theta_i^{\mathcal{I}}(k) \\
&+ \vec{\xi}^T(k)(W_{i-1}^{\mathcal{R}} \otimes BK)^T Q(W_{i-1}^{\mathcal{R}} \otimes BK)\vec{\xi}(k) \\
&+ 2b_i(\theta_i^{\mathcal{R}}(k))^T (A - a_i BK)^T QBK\theta_i^{\mathcal{I}}(k) \\
&- 2(\theta_i^{\mathcal{R}}(k))^T (A - a_i BK)^T Q(W_{i-1}^{\mathcal{R}} \otimes BK)\vec{\xi}(k) \\
&- 2b_i(\theta_i^{\mathcal{I}}(k))^T (BK)^T Q(W_{i-1}^{\mathcal{R}} \otimes BK)\vec{\xi}(k) \\
&+ (\theta_i^{\mathcal{I}}(k))^T (A - a_i BK)^T Q(A - a_i BK)\theta_i^{\mathcal{I}}(k) \\
&+ b_i^2 (\theta_i^{\mathcal{R}}(k))^T (BK)^T QBK\theta_i^{\mathcal{R}}(k)
\end{aligned}$$

$$
\begin{aligned}
&+ \vec{\xi}^T(k)(W^{\mathcal{I}}_{i-1} \otimes BK)^T Q(W^{\mathcal{I}}_{i-1} \otimes BK)\vec{\xi}(k) \\
&- 2b_i(\theta^{\mathcal{I}}_i(k))^T(A - a_iBK)^T QBK\theta^{\mathcal{R}}_i(k) \\
&- 2(\theta^{\mathcal{I}}_i(k))^T(A - a_iBK)^T Q(W^{\mathcal{I}}_{i-1} \otimes BK)\vec{\xi}(k) \\
&+ 2b_i(\theta^{\mathcal{R}}_i(k))^T(BK)^T Q(W^{\mathcal{I}}_{i-1} \otimes BK)\vec{\xi}(k) \\
&- (\theta^{\mathcal{R}}_i(k))^T Q\theta^{\mathcal{R}}_i(k) - (\theta^{\mathcal{I}}_i(k))^T Q\theta^{\mathcal{I}}_i(k) \Big\}.
\end{aligned}
$$

Then, considering the elementary inequality $2a^T b \le \delta a^T a + \delta^{-1} b^T b$, we obtain

$$
\begin{aligned}
V(k+1) - V(k) \le \sum_{i=2}^{r} \Big\{ &(\theta^{\mathcal{R}}_i(k))^T \Big((1 + \sigma_1)(A - a_iBK)^T Q(A - a_iBK) \\
&+ (1 + \sigma_2)b_i^2(BK)^T Q(BK) - Q \Big)\theta^{\mathcal{R}}_i(k) \\
&+ (\theta^{\mathcal{I}}_i(k))^T \Big((1 + \sigma_1)(A - a_iBK)^T Q(A - a_iBK) \\
&+ (1 + \sigma_2)b_i^2(BK)^T Q(BK) - Q \Big)\theta^{\mathcal{I}}_i(k) \Big\} + \vec{\xi}^T(k)\Gamma\vec{\xi}(k)
\end{aligned}
$$

where $\Gamma \triangleq (r - 1)(1 + \sigma_3)(\tilde{W} \otimes ((BK)^T Q(BK)))$ with $\sigma_3 = \varrho_2^{-1} + \varrho_3^{-1}$.

For $2 < i \le r$, denoting $\bar{\Pi}_i \triangleq (1 + \sigma_1)(A - a_iBK)^T Q(A - a_iBK) + (1 + \sigma_2) b_i^2(BK)^T Q(BK) - Q$ and noticing $0 < a_2 \le a_i$, it is easy to obtain from (5.54a) that

$$
\begin{aligned}
\bar{\Pi}_2 - \bar{\Pi}_i = (1 + \sigma_1)(&A^T QA - a_2 A^T QBK - a_2(BK)^T QA \\
&+ a_2^2(BK)^T QBK - A^T QA + a_i A^T QBK \\
&+ a_i(BK)^T QA - a_i^2(BK)^T QBK) \\
&+ (1 + \sigma_2)(b_m^2 - b_i^2)(BK)^T QBK \\
= (1 + \sigma_1)(a_i - a_2)(&A^T QBK + (BK)^T QA \\
&- (a_2 + a_i)(BK)^T QBK) \\
&+ (1 + \sigma_2)(b_m^2 - b_i^2)(BK)^T QBK \\
\ge (1 + \sigma_1)(a_i - a_2)(&A^T QBK + (BK)^T QA \\
&- (a_2 + a_r)(BK)^T QBK) \\
&+ (1 + \sigma_2)(b_m^2 - b_i^2)(BK)^T QBK \ge 0.
\end{aligned}
$$

(5.55)

Then, combining (5.54b) and (5.55), one has $\bar{\Pi}_i \leq \bar{\Pi}_2 < 0$. Consequently, in accordance with the definition of $\theta(k)$, we further derive that

$$V(k+1) - V(k) \leq \sum_{i=2}^{r} \left\{ (\theta_i^{\mathcal{R}}(k))^T \bar{\Pi}_2 \theta_i^{\mathcal{R}}(k) + (\theta_i^{\mathcal{I}}(k))^T \bar{\Pi}_2 \theta_i^{\mathcal{I}}(k) \right\}$$

$$+ \vec{\xi}^T(k) \Gamma \vec{\xi}(k)$$

$$\leq -\lambda_{\min}\{-\bar{\Pi}_2\} \sum_{i=2}^{r} \left\{ (\theta_i^{\mathcal{R}}(k))^T \theta_i^{\mathcal{R}}(k) + (\theta_i^{\mathcal{I}}(k))^T \theta_i^{\mathcal{I}}(k) \right\}$$

$$+ \vec{\xi}^T(k) \Gamma \vec{\xi}(k)$$

$$= -\lambda_{\min}\{-\bar{\Pi}_2\} |\theta(k)|^2 + \lambda_{\max}\{\Gamma\} |\vec{\xi}(k)|^2. \tag{5.56}$$

Noting that $V(k)$ can also be rewritten as $V(k) = \theta^T(k) \mathcal{Q} \theta(k)$, we choose

$$\alpha_1\left(|\theta(k)|\right) = \lambda_{\min}\{\mathcal{Q}\} |\theta(k)|^2,$$

$$\alpha_2\left(|\theta(k)|\right) = \lambda_{\max}\{\mathcal{Q}\} |\theta(k)|^2,$$

$$\alpha_3\left(|\theta(k)|\right) = \lambda_{\min}\{-\bar{\Pi}_2\} |\theta(k)|^2,$$

$$\varpi(|\vec{\xi}(k)|) = \lambda_{\max}\{\Gamma\} |\vec{\xi}(k)|^2.$$

Therefore, it is inferred from Lemma (5.53) is input-to-state stable and so does the system (5.52). Here, we can choose $\beta(|\theta(0)|, k) = \sqrt{\frac{\lambda_{\max}\{\mathcal{Q}\}}{\lambda_{\min}\{\mathcal{Q}\}}} \phi^k |\theta(0)|$ and $\gamma(|\vec{\xi}(k)|) = \sqrt{\frac{\lambda_{\max}\{\mathcal{Q}\} \lambda_{\max}\{\Gamma\}}{c \lambda_{\min}\{\mathcal{Q}\} \lambda_{\min}\{-\bar{\Pi}_2\}}} |\vec{\xi}(k)|$, where $0 < c < 1$. Furthermore, by using the property of "converging-input" to "converging state" in [58], it can be immediately found that $\lim_{k \to \infty} |\tilde{\delta}_i(k)| = 0$ for $i = 2, \ldots, r$, which completes the proof.

Remark 5.2: *Both the ISS theory and the algebraic graph method are utilized to derive the consensus conditions in this chapter. First, according to analysis results in Theorem 1, the decoding error $\vec{\xi}(k)$ can be regarded as the exogenous disturbance inputs which are bounded and asymptotically converge to zero, and hence the ISS theory can be directly used to analyze the coder-decoder-based consensus problem. In addition, the eigenvalue information (including both the real part and imaginary part) of the Laplacian matrix is reflected in the consensus conditions, which shows that the communication topology serving as an important factor has a non-negligible impact on the consensus of the MASs.*

Theorem 5.3: *Suppose that the network $\mathcal{G}$ contains a directed spanning tree. Let the positive scalars ϱ_1, ϱ_2 and ϱ_3 be given. The asymptotic consensus in system*

(5.1) can be reached with the CDCP (5.29)–(5.32) if there exist a positive matrix Q, a nonsingular matrix S and a matrix K satisfying

$$\begin{cases} \begin{bmatrix} -\tilde{a}^{-2}A^T QA & (\tilde{a}^{-1}SUA - \tilde{a}\tilde{K})^T \\ * & Q - (SU)^T - SU \end{bmatrix} < 0 & \text{(5.57a)} \\[4mm] \begin{bmatrix} -Q & \tilde{\sigma}_1(SUA - a_2\tilde{K})^T & \tilde{\sigma}_2 b_m \tilde{K}^T \\ * & Q - (SU)^T - SU & 0 \\ * & * & Q - (SU)^T - SU \end{bmatrix} \\[6mm] \qquad < 0 & \text{(5.57b)} \end{cases}$$

subject to (5.21), (5.24) and (5.45), where

$$U = [B((B^T B)^{-1})^T \quad B^{\perp}]^T, \ S = \begin{bmatrix} S_{11} & S_{12} \\ 0 & S_{22} \end{bmatrix},$$

$$\tilde{K} = [K^T \quad 0]^T, \ \tilde{a} = \sqrt{a_2 + a_r}, \ \tilde{\sigma}_1 = \sqrt{1 + \sigma_1},$$

$$\tilde{\sigma}_2 = \sqrt{1 + \sigma_2}.$$

Here, $B^{\perp}$ is an orthogonal basis of the null space for matrix B^T, that is $B^T B^{\perp} = 0$. $S_{11} \in \mathbb{R}^{p \times p}$, $S_{12} \in \mathbb{R}^{p \times (n-p)}$, and $S_{22} \in \mathbb{R}^{(n-p) \times (n-p)}$ are arbitrary matrices ensuring the non-singularity of S. Other parameters are defined as in Theorem 5.1 and Theorem 5.2. Furthermore, if inequalities (5.57a) and (5.57b) are feasible, the controller gain matrix can be determined by

$$K = S_{11}^{-1}K. \tag{5.58}$$

Proof: *First, we rewrite (5.54a) as*

$$-\tilde{a}^{-2}A^T QA + (\tilde{a}^{-1}A - \tilde{a}BK)^T Q(\tilde{a}^{-1}A - \tilde{a}BK) < 0. \tag{5.59}$$

By applying the Schur Complement Lemma, one can conclude that (5.54a) is true if and only if the following inequality

$$\begin{bmatrix} -\tilde{a}^{-2}A^T QA & (\tilde{a}^{-1}A - \tilde{a}BK)^T \\ * & -Q^{-1} \end{bmatrix} < 0 \tag{5.60}$$

holds.

On the other hand, it is worth mentioning that $-Q^{-1} \leq (SU)^{-1}Q(SU)^{-T} - (SU)^{-T} - (SU)^{-1}$, which means that (5.60) can be guaranteed by

$$\begin{bmatrix} -\tilde{a}^{-2}A^T QA & (\tilde{a}^{-1}A - \tilde{a}BK)^T \\ * & -W^T QW - W - W^T \end{bmatrix} < 0 \tag{5.61}$$

where $W = (\mathcal{SU})^{-T}$. Then, pre- and post-multiplying (5.61) by $\mathrm{diag}\{I, W^{-T}\}$ and its transpose leads to

$$\begin{bmatrix} -\tilde{a}^{-2}A^T Q A & (\tilde{a}^{-1}\mathcal{SU}A - \tilde{a}\mathcal{SU}BK)^T \\ * & Q - (\mathcal{SU})^T - \mathcal{SU} \end{bmatrix} < 0. \tag{5.62}$$

By noticing (5.58), we have $\mathcal{SU}BK = \tilde{K}$, which implies that (5.54a) can be guaranteed by (5.57a). Similarly, (5.54b) can also be guaranteed by (5.57b), and therefore the proof of this theorem is complete.

5.3 An Illustrative Example

In this section, a simulation example is presented to verify the effectiveness of the proposed observer-based consensus control scheme for the discrete-time MASs.

Consider the system (5.1) with the following parameters

$$A = \begin{bmatrix} 1 & 0.1 \\ 0.15 & 0.5 \end{bmatrix}, \ B = \begin{bmatrix} 0.2 \\ 0.25 \end{bmatrix}, \ C = [0.5 \quad 0.5].$$

In this example, we assume that there are five agents communicating with each other via a directed topology $\mathcal{G}$ with the adjacency matrix $\mathcal{A} = [a_{ij}]_{5\times5}$, where $a_{13} = a_{15} = a_{21} = a_{24} = a_{34} = a_{35} = a_{41} = a_{45} = a_{51} = a_{53} = 1$ and other elements are 0.

The eigenvalues of the associated Laplacian matrix $\mathcal{L}_{\mathcal{G}}$ are, respectively, calculated as 0, 2, $2.5 + 0.866j$, $2.5 - 0.866j$ and 3, where "j" stands for the imaginary unit.

For each agent, the initial state is set as $x_{i0} = [0.6 - 0.1i \quad 0.8 + 0.3i]^T$, $i = 1, 2, \ldots, 5$. Choose $\epsilon = 1.1$, $\mu = 0.3$, $h = 3$ and $q = 15$. Then, solving matrix inequality (5.21), (5.24), (5.57a) and (5.57b) by the Matlab software, we have $c_0 = 1.3247$ and $c_1 = 0.5556$. Moreover, the desired observer and controller gain matrices are obtained as $L = [1.3030 \quad 0.5608]^T$ and $K = [1.5918 \quad 0.1592]$, respectively.

The simulation results are shown in Figures 5.1–5.3. Figure 5.1 depicts the sate trajectories of each agent for the MAS (5.1) with the designed control scheme and the sate trajectories of each agent without the control input are plotted in Figure 5.2. In addition, in order to show the applicability of the coding-decoding-based communication protocol, Figure 5.3 displays the corresponding decoding error for each agent. It is inferred from the three figures that the closed-loop MAS (5.1) can attain the desired consensus performance while the open-loop one cannot. The simulation results have demonstrated that the developed consensus control scheme performs very well.

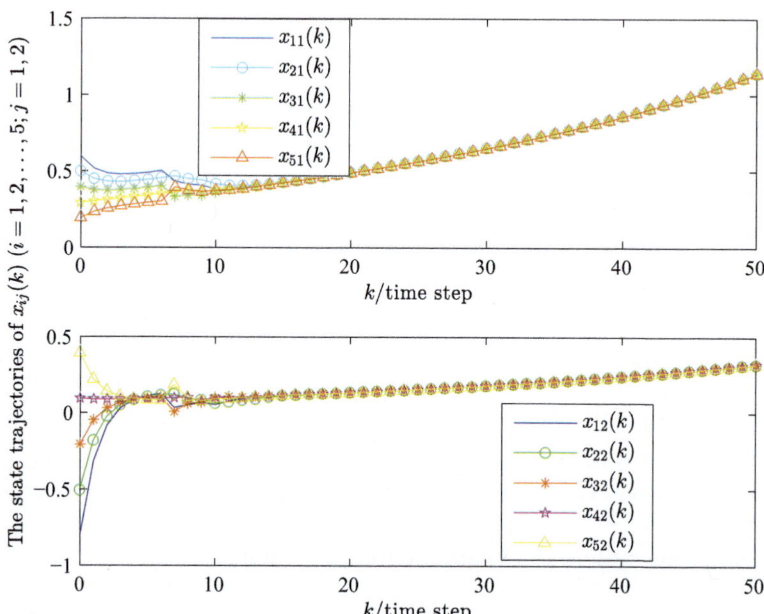

FIGURE 5.1

State trajectories $x_{ij}(k)$ $(i = 1, 2\ldots, 5; j = 1, 2)$ with control input.

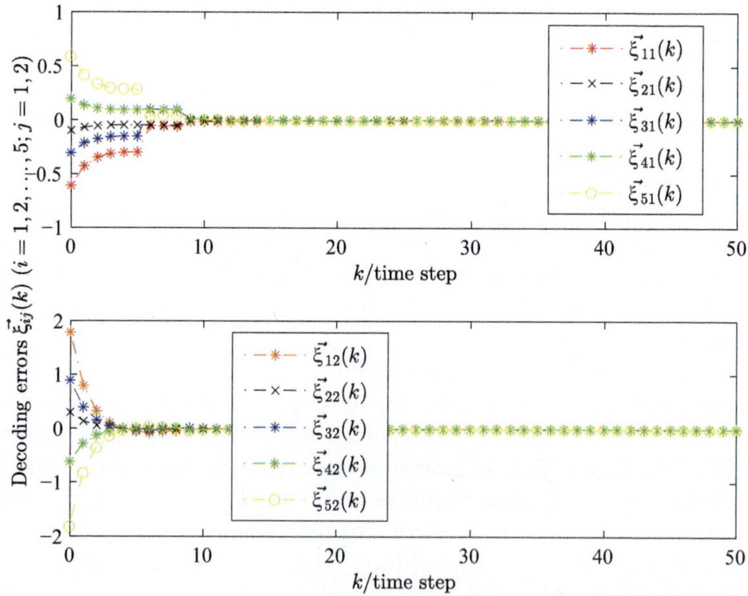

FIGURE 5.2

State trajectories $x_{ij}(k)$ $(i = 1, 2\ldots, 5; j = 1, 2)$ without control input.

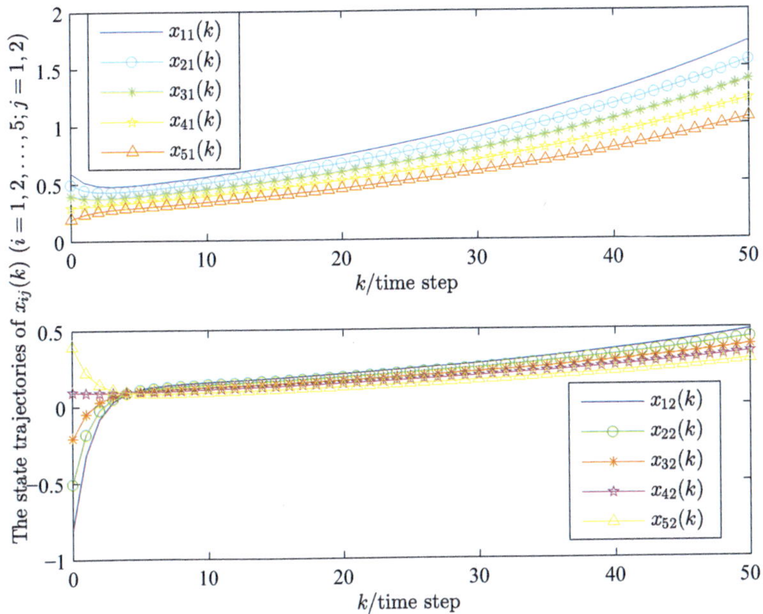

FIGURE 5.3
The decoding errors $\vec{\xi}_{ij}(k)$ $(i = 1, 2 \ldots, 5; j = 1, 2)$.

5.4 Summary

In this chapter, the observer-based consensus control problem has been dealt with for a class of discrete-time networked MASs with CDCP. The signal transmission of each agent between the observer and the controller has been implemented via the digital communication channel. By means of the uniform quantization technique, the CDCP has been employed to encode the relative measurements between adjacent agents to finite symbolic values before transmission and then decode the received symbolic data at the controller side. In virtue of the decoded signal, an observer-based control scheme has been put forward. Then, consensus criteria have been derived in terms of the solutions to certain matrix inequality constraints such that the closed-loop system achieves the desired performance requirement.

6

Recursive Filtering with Measurement Fading: A Multiple Description Coding Scheme

In recent years, in response to the ever-growing complexity of dynamical systems, the filtering or state estimation problem has been gaining increasing popularity from both control and signal processing communities. According to the categories of system noises and the specifications of estimation performance, a variety of filtering strategies have been developed in the literature with successful applications in real-world plants, see for example [45,99,148]. Among others, the notable extended Kalman filtering (EKF) algorithm has been recognized as an effective tool to deal with the state estimation problems for nonlinear stochastic systems. So far, a rich body of research results regarding the EKF algorithm has been reported either on the variants of the EKF algorithm accommodating more general systems or on its engineering applications in more general practice, see for example [97,170].

In practical engineering, the signal transmission via wireless channels might suffer from certain physical interferences such as reflection, diffraction, and scattering, and this gives rise to the signal fading issue. From the mathematical viewpoint, the fading phenomenon is usually modeled by a stochastic process to reflect the random fluctuations in both the amplitude and the phase of the transmitted signals. Till now, a great deal of research effort has been made toward the network-induced channel fading problems. Nevertheless, when it comes to the nonlinear recursive filtering issues with fading measurements, the corresponding results have been scattered in the literature.

In reality, the bandwidth of a communication network is usually limited and only finite bits of data are allowed to be transmitted in a certain time. As such, it is of both theoretical importance and practical significance to make full use of the limited network resource in order to achieve the desired system performance. In this case, the *data-coding* scheme appears to be particularly efficient in resource saving as this scheme maps the original data to certain codewords with *fewer* bits.

Among the existing coding algorithms, the multiple description coding (MDC) scheme stands out as an effective one for its distinguished advantages

DOI: 10.1201/9781003534853-6

in error resilience and rate-adaptive streaming. The MDC is based on the diversity principle [134] that possesses three features: 1) the information source is encoded into multiple descriptions with identical importance and then sent to the decoder via parallel independent channels; 2) the original data is decoded with an acceptable quality as long as at least one of the descriptions is successfully transmitted to the decoder; and 3) the more descriptions received by the decoder the higher accuracy of the decoded data will be attained. Up to date, the MDC scheme has found a wide range of applications including distributed storage systems, diversity communication system and image/audio/video coding. Unfortunately, despite their clear engineering insights, the MDC-based control/filtering issues for NSs have received very little research attention mainly because of the essential difficulties in quantifying the influences from the implementation of the MDC on the performance indices.

Motivated by the above discussions, in this chapter, we aim to investigate the recursive filtering problem for a class of discrete-time stochastic nonlinear systems subject to fading channels where the multiple description coding scheme is employed during the data exchange between the devices. In order to facilitate the data transmission in a resource-constrained communication network, the multiple description coding scheme is adopted to encode the fading measurements into two descriptions with the identical importance. Two independent Bernoulli distributed random variables are introduced to govern the occurrences of the packet dropouts in two channels from the encoders to the decoders. The channel fading phenomenon is characterized by the Mth-order Rice fading model whose coefficients are mutually independent random variables obeying certain probability distributions. The purpose of the problem addressed is to design a recursive filter such that, in the simultaneous presence of the stochastic noises, the channel fading and the data coding-decoding mechanism, an upper bound of the filtering error variance is obtained and then minimized at each time step. In virtue of the Riccati difference equation technique and the stochastic analysis approach, the explicit form of the desired filter parameters is derived by solving a sequence of coupled algebraic Riccati-like difference equations. Finally, a simulation experiment is provided to show the applicability of the developed filtering scheme.

6.1 Problem Formulation

Consider a discrete time-varying nonlinear stochastic system described by

$$\begin{cases} x_{k+1} = h(x_k, \zeta_k) + B_{1k}\omega_k \\ \tilde{y}_k = C_k x_k + B_{2k} v_k \end{cases} \tag{6.1}$$

where $x_k \in \mathbb{R}^{n_x}$ represents the state vector, $\zeta_k \in \mathbb{R}^{n_u}$ is the system input, and $\tilde{y}_k \in \mathbb{R}^{n_y}$ denotes the measurement output. $h(\cdot,\cdot) : \mathbb{R}^{n_x} \times \mathbb{R}^{n_\zeta} \mapsto \mathbb{R}^{n_x}$ is a deterministic and continuously differential nonlinear function. $\omega_k \in \mathbb{R}^{n_\omega}$ and $v_k \in \mathbb{R}^{n_v}$ are two families of mutually independent *truncated* Gaussian white noise sequences taking values over the interval $[-\bar{\omega},\bar{\omega}]$ and $[-\bar{v},\bar{v}]$, respectively, where $\bar{\omega}$ and $\bar{v}$ are given positive scalars. The probability density functions of the truncated Gaussian white sequences ω_k and v_k are, respectively, defined as

$$
f_\omega(\omega_k; \mu_\omega, \sqrt{R_{0k}}, -\bar{\omega}, \bar{\omega}) \triangleq \frac{\frac{1}{\sqrt{R_{0k}}}\phi\left(\frac{\omega_k-\mu_\omega}{\sqrt{R_{0k}}}\right)}{\Phi\left(\frac{\bar{\omega}-\mu_\omega}{\sqrt{R_{0k}}}\right) - \Phi\left(\frac{-\bar{\omega}-\mu_\omega}{\sqrt{R_{0k}}}\right)}
$$

and

$$
f_v(v_k; \mu_v, \sqrt{Q_{0k}}, -\bar{v}, \bar{v}) \triangleq \frac{\frac{1}{\sqrt{Q_{0k}}}\phi\left(\frac{v_k-\mu_v}{\sqrt{Q_{0k}}}\right)}{\Phi\left(\frac{\bar{v}-\mu_v}{\sqrt{Q_{0k}}}\right) - \Phi\left(\frac{-\bar{v}-\mu_v}{\sqrt{Q_{0k}}}\right)}
$$

where $\phi(x) \triangleq \frac{1}{\sqrt{2\pi}}\exp\left(-\frac{1}{2}x^2\right)$, $\Phi(x) \triangleq \int_{-\infty}^{x}\frac{1}{\sqrt{2\pi}}\exp\left(-\frac{s^2}{2}\right)ds$, μ_ω and R_{0k} are, respectively, the mean and variance of ω_k without truncation. Similarly, μ_v and Q_{0k} are, respectively, the mean and variance of v_k without truncation. Without loss of generality, letting $\mu_\omega = \mu_v = 0$, it is not difficult to calculate that the truncated ω_k and v_k are zero-mean random sequences with the variances

$$
R_k = R_{0k}\left[1 - \frac{\bar{\omega}\left(\phi\left(\frac{-\bar{\omega}}{\sqrt{R_{0k}}}\right) + \phi\left(\frac{\bar{\omega}}{\sqrt{R_{0k}}}\right)\right)}{\sqrt{R_{0k}}\left(\Phi\left(\frac{-\bar{\omega}}{\sqrt{R_{0k}}}\right) - \Phi\left(\frac{\bar{\omega}}{\sqrt{R_{0k}}}\right)\right)}\right]
$$

and

$$
Q_k = Q_{0k}\left[1 - \frac{\bar{v}\left(\phi\left(\frac{-\bar{v}}{\sqrt{Q_{0k}}}\right) + \phi\left(\frac{\bar{v}}{\sqrt{Q_{0k}}}\right)\right)}{\sqrt{Q_{0k}}\left(\Phi\left(\frac{-\bar{v}}{\sqrt{Q_{0k}}}\right) - \Phi\left(\frac{\bar{v}}{\sqrt{Q_{0k}}}\right)\right)}\right],
$$

respectively.

6.1.1 Channel Fading

In this chapter, the phenomenon of channel fading is considered between the sensor and the encoder due to the unreliable wireless communication. The fading signal y_k is described by the following Mth-order Rice fading model:

$$y_k = \sum_{m=0}^{M} \alpha_{mk}\tilde{y}_{k-m} + B_{3k}\varpi_k \tag{6.2}$$

where M stands for the given number of the paths where the signals get through. α_{mk} $(m=0,1,\ldots,M)$ are the channel coefficients which are mutually independent random variables in m and k. In addition, α_{mk} have the probability density functions $p_m(\alpha_{mk})$ on the interval $[0,\ 1]$ with mathematical expectations $\bar{\alpha}_m$ and variances ς_m. $\varpi_k \in \mathbb{R}^{n_\varpi}$ is a zero-mean Gaussian random vector that characterizes the channel noise with variance S_k. All the noise signals and random variables α_{mk} $(m=0,1,\ldots,M)$ are mutually uncorrelated. B_{3k} is a known time-varying matrix with appropriate dimension.

To facilitate the description, we set $\mathscr{M}_k \triangleq \min\{M,k\}$ and rewrite the model (6.2) as follows:

$$y_k = \sum_{m=0}^{\mathscr{M}_k} \alpha_{mk}\tilde{y}_{k-m} + B_{3k}\varpi_k. \tag{6.3}$$

6.1.2 Multiple Description Coding Scheme

In a networked environment, the phenomenon of packet dropouts is frequently encountered during data transmissions owing primarily to the limited capacity of the communication network. The MDC scheme is known for its error resilience and has been widely used to improve the communication quality. For example, with the use of the MDC scheme, the negative impacts from the packet dropouts are mitigated in the sense that the data stream is not interrupted at the cost of temporary quality degradation. As such, in this chapter, we aim to examine how the MDC scheme influences the state estimation performance. Without loss of generality, the *two-description* case is considered. As shown in Figure 6.1, the

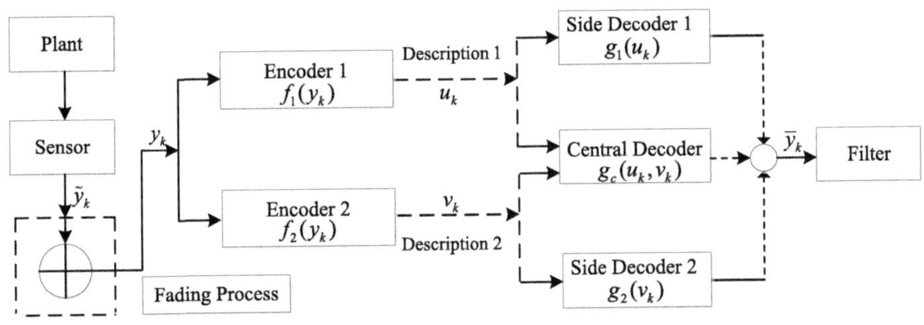

FIGURE 6.1
Structure of two-description coding-decoding for NSs with unreliable channels.

fading measurement is first encoded into two descriptions, which are then simultaneously transmitted to the decoders via two mutually independent channels.

According to Figure 6.1, the general form of the MDC is mathematically characterized as follows:

Encoder:

$$\begin{cases} u_{lk} = f_{1l}(y_{lk}) \\ v_{lk} = f_{2l}(y_{lk}) \end{cases} \tag{6.4}$$

Decoder:

$$\bar{y}_{lk} = \begin{cases} g_{1l}(u_{lk}), & \text{when } \lambda_k^1 = 1, \lambda_k^2 = 0 \\ g_{2l}(v_{lk}), & \text{when } \lambda_k^1 = 0, \lambda_k^2 = 1 \\ g_{cl}(u_{lk}, v_{lk}), & \text{when } \lambda_k^1 = 1, \lambda_k^2 = 1 \end{cases} \tag{6.5}$$

for $l = 1, \ldots, n_y$, where $f_{1l}(\,\cdot\,)$ and $f_{2l}(\,\cdot\,)$ are two coding functions whose outputs u_{lk} and v_{lk} can be regarded as the individual descriptions of the source y_{lk} with y_{lk} being the lth component of y_k. $\bar{y}_{lk}$ is the decoding value corresponding to y_{lk}. $g_{1l}(\,\cdot\,)$ and $g_{2l}(\,\cdot\,)$ are two side decoding functions and $g_{cl}(\cdot,\cdot)$ is the central decoding function. λ_k^i $(i = 1, 2)$ are two independent random variables which govern the packet dropout phenomena during the transmissions of the descriptions and satisfy the Bernoulli binary distribution taking values on either 1 or 0 with mathematical expectations $\bar{\lambda}_i$ and variances σ_i, respectively. Here, $\lambda_k^i = 1$ indicates that the ith channel works well, and $\lambda_k^i = 0$ means that the ith channel suffers from the packet dropouts at time instant k.

For notational convenience, we denote

$$u_k \triangleq [u_{1k} \quad u_{2k} \quad \cdots \quad u_{n_yk}]^T$$
$$v_k \triangleq [v_{1k} \quad v_{2k} \quad \cdots \quad v_{n_yk}]^T$$
$$\bar{y}_k \triangleq [\bar{y}_{1k} \quad \bar{y}_{2k} \quad \cdots \quad \bar{y}_{n_yk}]^T$$
$$f_1(y_k) \triangleq [f_{11}(y_{1k}) \quad f_{12}(y_{2k}) \quad \cdots \quad f_{1n_y}(y_{n_yk})]^T$$
$$f_2(y_k) \triangleq [f_{21}(y_{1k}) \quad f_{22}(y_{2k}) \quad \cdots \quad f_{2n_y}(y_{n_yk})]^T$$
$$g_1(y_k) \triangleq [g_{11}(y_{1k}) \quad g_{12}(y_{2k}) \quad \cdots \quad g_{1n_y}(y_{n_yk})]^T$$
$$g_2(y_k) \triangleq [g_{21}(y_{1k}) \quad g_{22}(y_{2k}) \quad \cdots \quad g_{2n_y}(y_{n_yk})]^T$$
$$g_c(y_k) \triangleq [g_{c1}(y_{1k}) \quad g_{c2}(y_{2k}) \quad \cdots \quad g_{cn_y}(y_{n_yk})]^T$$

$$\tag{6.6}$$

where u_k and v_k are two description vectors to be transmitted as two packets via different channels.

It is not difficult to see that there are a total of three cases at the decoder side.

1) Only one packet, either u_k or v_k, is available to the corresponding decoder $g_1(\,\cdot\,)$ or $g_2(\,\cdot\,)$.

2) Both the packets u_k and v_k are received by the decoder $g_c(\,\cdot\,)$.

3) Neither of the packets is received by the decoders. In this case, both the packets are lost and thus the zero-order holder (ZOH) strategy is implemented by using the latest decoded measurement $\bar{y}_{k-1}$ to compensate the current decoded value $\bar{y}_k$, that is, $\bar{y}_k = \bar{y}_{k-1}$ when $\lambda_{1k} = \lambda_{2k} = 0$.

6.1.3 The Filter Structure

After obtaining the decoded measurement $\bar{y}_k$ according to the decoding scheme (6.5) and the zero-order holder strategy, we are now in a position to design the recursive filter with the following form:

$$
\begin{cases}
\hat{x}_{k+1|k} = h(\hat{x}_{k|k}, \zeta_k) & \text{(6.7a)} \\
\hat{x}_{k+1|k+1} = \hat{x}_{k+1|k} + K_{k+1}\vartheta\left(\bar{y}_{k+1}\right) & \text{(6.7b)}
\end{cases}
$$

where $\hat{x}_{k|k} \in \mathbb{R}^{n_x}$ is the estimate of the system state x_k at time instant k, $\hat{x}_{k+1|k} \in \mathbb{R}^{n_x}$ is the one-step prediction of x_{k+1} at k, $K_{k+1} \in \mathbb{R}^{n_x \times n_y}$ is the time-varying filter parameter to be designed, and $\vartheta(\,\cdot\,) : \mathbb{R}^{n_y} \mapsto \mathbb{R}^{n_y}$ is the innovation function of y_{k+1} whose explicit expression is to be given later.

The objective of this chapter is twofold.

1) Provide a rigorous mathematical analysis on the MDC scheme, prove that the decoding error is uniformly bounded, and then establish a unified model for the decoded measurement $\bar{y}_k$.

2) Design a time-varying recursive filter of the form (6.7a)-(6.7b) for the stochastic nonlinear system (6.1) subject to fading measurements and the MDC such that the filtering error variance satisfies the following requirement:

$$
\mathbb{E}\{(x_{k+1} - \hat{x}_{k+1|k+1})(x_{k+1} - \hat{x}_{k+1|k+1})^T\} \leq \Pi_{k+1|k+1} \qquad \text{(6.8)}
$$

where $\Pi_{k+1|k+1}$ is certain upper bound of the filtering error variance, which is further minimized by solving an optimization problem with respect to the filter parameter K_{k+1}.

6.2 Design of the MDC and Analysis of the Decoding Error

In this section, we first endeavor to design the MDC via the scalar quantization approach. Then, for all different scenarios of the packet delivery, the corresponding decoding errors are analyzed. In addition, a comprehensive mathematical model of the decoded measurement $\bar{y}_k$ is established.

6.2.1 Design of the MDC

From the technical point of view, the design problem of the MDC procedure can be generally divided into two steps: 1) the *index generation* and 2) the *index assignment*. More specifically, in this chapter, we shall use a set of scalar uniform quantizers to convert the source into a set of specific indices. Then, by employing the so-called *nested* index assignment principle [156], the generated indices are mapped into the corresponding cells of an index mapping matrix. In what follows, we elaborate the multiple description coding procedure in accordance to the above steps.

Index Generation

First, we define the scalar uniform quantizer $q(\,\cdot\,)$ with the following structure:

$$
q(\tau) = \begin{cases} L, & \tau_k \geq L \\ -L, & \tau_k < -L \\ -L + \frac{(2s-1)L}{\rho}, & -L + v_1 \leq \tau < -L + v_2 \end{cases}
\tag{6.9}
$$

where $\tau \in \mathbb{R}$ is the signal to be quantized and L is the saturation value; $v_1 \triangleq \frac{2(s-1)L}{\rho}$ and $v_2 \triangleq \frac{2sL}{\rho}$ for $s \in \{1, 2, \ldots, \rho\}$. It can be seen from (6.9) that the interval $[-L, L]$ is partitioned into ρ regions $I_i(L)(i = 1, 2, \ldots, \rho)$, where $I_i(L) = \left[-L + \frac{2(i-1)L}{\rho}, -L + \frac{2iL}{\rho}\right)$ and $I_i(L) \cap I_j(L) = \emptyset$ for any $i, j \in \{1, 2, \ldots, \rho\}$ and $i \neq j$.

For the measurement component y_{lk}, define $q_l(y_{lk}) \triangleq \mu_{lk} q(\frac{y_{lk}}{\mu_{lk}})$, where $\mu_{lk} > 0$ is an adjustable parameter. In order to ensure that the quantizer $q_l(\,\cdot\,)$ is not saturated, at each time instant, we select a proper μ_{lk} to guarantee that the value y_{lk}/μ_{lk} is located within the interval $[-L, L]$ when $|y_{lk}| > L$. Hence, it is easily calculated that the quantization error $\varrho_{lk} \triangleq y_{lk} - q_l(y_{lk})$ satisfies

$$
|\varrho_{lk}| \leq \frac{\mu_{lk}L}{\rho}.
\tag{6.10}
$$

1	3						
2	4	5					
	6	7	9				
		8	10	11			
			12	13	15		
				14	16	17	
					18	19	21
						20	22

FIGURE 6.2
Nested index assignment for $d = 8$ and $p = 1$ in [156].

Remark 6.1: *It is observed from (6.10) that a larger μ_{lk} leads to a larger quantization error, which means that the scaling parameter μ_{lk} should be chosen as small as possible for the sake of decreasing the quantization error. On the other hand, μ_{lk} needs to be large enough to avoid the quantizer saturation. As such, μ_{lk} should be selected appropriately to balance the trade-off between the quantization error and the quantizer saturation.*

Next, for the lth quantized measurement component $q_l(y_{lk})$, we define the index generator function $\psi_l(\,\cdot\,) : \mathbb{R} \to \mathbb{N}^+$ satisfying

$$\psi_l(q_l(y_{lk})) = i_{lk}, \text{ when } q_l(y_{lk}) \in I_{i_{lk}}(L) \tag{6.11}$$

with $i_{lk} \in \{1, 2, \dots, \rho\}$ and $l = 1, 2, \dots, n_y$.

Index Assignment
After generating the index i_{lk}, we need to assign it into the corresponding cell of the mapping matrix $\mathcal{J}$ according to the nested assignment principle. Then, in terms of the location of i_{lk} in $\mathcal{J}$, we obtain a pair of descriptions (u_{lk}, v_{lk}) accordingly. Without loss of generality, we suppose that the mapping matrix $\mathcal{J}$ is a d−dimensional matrix with d being an even number.

Before proceeding, let us illustrate the nested assignment principle. It is inferred from [156] that the indices are allocated in those cells that lie on the main diagonal and on the $2p$ diagonals closest to the main diagonal. To simplify the exposition, in this chapter, we discuss the index assignment problem only for the case $p = 1$. As shown in Figure 6.2, all the indices are allocated in the cells that lie on $2p + 1 = 3$ diagonals including the main diagonal and other 2 diagonals which are closest to the main diagonal.

Remark 6.2: *It is apparent from (6.10) that the quantization level ρ has a significant impact on the quantization accuracy. That is, a large quantization level ρ results in a high quantization accuracy (or small quantization error) while occupying a large amount of bandwidth. It is noted that a real network channel is often subjected to the bandwidth constraint. Here, it is assumed that only χ-bits of data are allowed* *to be transmitted at each time instant. In this case, for the traditional uniform-quantization-based single description coding approach, the quantization level is determined by $\rho = 2^\chi$. However, it is seen from Figure 6.2 that the proposed two-description-based coding approach enlarges the quantization level from 2^χ to $3 \times 2^\chi - 2$, thereby improving the quantization accuracy.*

Now, we are in a position to conduct a rather detailed mathematical analysis on the implementation of the proposed index assignment scheme. Denote by $\varphi_l(\,\cdot\,) : \mathbb{N}^+ \to \mathbb{N}^+ \times \mathbb{N}^+$ the index assignment function, which is defined as $\varphi_l(i_{lk}) \triangleq (\varphi_{1l}(i_{lk}), \varphi_{2l}(i_{lk})) =$

$$
\begin{cases}
(r_{lk} + 1, r_{lk} + 1), & \text{if } m_{lk} = 1 \\
(r_{lk} + 1, r_{lk}), & \text{if } m_{lk} = 0 \text{ and } r_{lk} \text{ is even} \\
(r_{lk}, r_{lk} + 1), & \text{if } m_{lk} = 0 \text{ and } r_{lk} \text{ is odd} \\
(r_{lk} + 2, r_{lk} + 1), & \text{if } m_{lk} = 2 \text{ and } r_{lk} \text{ is even} \\
(r_{lk} + 1, r_{lk} + 2), & \text{if } m_{lk} = 2 \text{ and } r_{lk} \text{ is odd}
\end{cases}
\tag{6.12}
$$

where $\varphi_{1l}(i_{lk})$ and $\varphi_{2l}(i_{lk})$ are, respectively, the row assignment function and column assignment function; $r_{lk} = \lfloor \frac{i_{lk}}{2p+1} \rfloor = \lfloor \frac{i_{lk}}{3} \rfloor$ and $m_{lk} = \langle \frac{i_{lk}}{2p+1} \rangle = \langle \frac{i_{lk}}{3} \rangle$. It is easy to see that the index assignment function $\varphi_l(i_{lk})$ maps the single description i_{lk} into the corresponding description pair (u_{lk}, v_{lk}), where $u_{lk} = \varphi_{1l}(i_{lk})$ and $v_{lk} = \varphi_{2l}(i_{lk})$.

Based on the above discussion, the coding functions $f_{1l}(\,\cdot\,)$ and $f_{2l}(\,\cdot\,)$ in (6.4) can be finally expressed as

$$
\begin{cases}
f_{1l}(y_{lk}) = \varphi_{1l}(\psi_l(q_l(y_{lk}))) \\
f_{2l}(y_{lk}) = \varphi_{2l}(\psi_l(q_l(y_{lk}))).
\end{cases}
\tag{6.13}
$$

Up to now, by employing the index generator function $\psi_l(\,\cdot\,)$ and the index assignment function $\varphi_l(\,\cdot\,)$, the component-based encoder design problem has been dealt with. In the next stage, we shall address the decoder design problem based on the established coding procedure.

6.2.2 Analysis of the Decoding Error

In light of the established index generation and assignment schemes, in this subsection, we are to investigate the decoding accuracy of the proposed coding-decoding strategy. For this purpose, the first step is to construct an index estimation function to estimate each index i_{lk} according to the received descriptions. Then, based on the estimated index $\hat{i}_{lk}$, we will design a decoding function to obtain the decoded value $\bar{y}_{lk}$.

In the following, we discuss the possible situations of the decoding process on a one-by-one basis in terms of the individual case of the received descriptions at the decoder side.

- **Case 1**: both the descriptions u_{lk} and v_{lk} are successfully received at the decoder side. Then, the central decoder is applied to uniquely determine the index i_{lk} and obtain the decoded measurement component $\bar{y}_{lk}$.

- **Case 2**: one of the descriptions (u_{lk} or v_{lk}) is lost during the packet transmissions. The explicit location of the index i_{lk} is unknown and only its row number (or column number) can be acquired. Hence, in order to generate the decoded measurement $\bar{y}_{lk}$ by the side decoder, we let the diagonal element in the same row (or column) with i_{lk} be an estimation of i_{lk}.

- **Case 3**: neither of the description packets is available to the decoders. In this case, $\hat{i}_{lk}$ is set as zero and the latest decoded measurement $\bar{y}_{l(k-1)}$ is utilized to compensate the value of $\bar{y}_{lk}$ by using the ZOH strategy.

Based on the above analysis, the key point in the design of the decoding procedure is to determine the estimated index value $\hat{i}_{lk}$ for each measurement component y_{lk}. Obviously, with more descriptions received, we would expect a less error between $\hat{i}_{lk}$ and i_{lk}, and hence a more accurate decoded value $\bar{y}_{lk}$. As such, corresponding to the above three cases, we need to examine the relationships between the received description(s) and the estimated index $\hat{i}_{lk}$.

Let us start with the analysis for **Case 1**. In this case, it follows directly from (6.12) that $\hat{i}_{lk} \triangleq \phi_{cl}(u_{lk}, v_{lk}) =$

$$
\begin{cases}
3u_{lk} - 2, & \text{if } u_{lk} = v_{lk} \\[2mm]
3u_{lk} - 3, & \text{if } u_{lk} = v_{lk} + 1 \text{ and } u_{lk} \text{ is odd} \\[2mm]
3u_{lk}, & \text{if } u_{lk} = v_{lk} - 1 \text{ and } u_{lk} \text{ is odd} \\[2mm]
3u_{lk} - 4, & \text{if } u_{lk} = v_{lk} + 1 \text{ and } u_{lk} \text{ is even} \\[2mm]
3u_{lk} - 1, & \text{if } u_{lk} = v_{lk} - 1 \text{ and } u_{lk} \text{ is even}
\end{cases}
\tag{6.14}
$$

where $\phi_{cl}(\cdot, \cdot)$ is the estimation function of the central decoder. Furthermore, (6.14) means that the estimated index $\hat{i}_{lk}$ can be explicitly calculated and the index estimation error is $\tilde{i}_{lk} = \hat{i}_{lk} - i_{lk} = 0$.

For **Case 2**, suppose that only the description u_{lk} (the row number) is received while the description v_{lk} (the column number) is lost. In such a case, we let the index, which is located in the diagonal cell of the u_{lk}th row, be the estimation of i_{lk}. So, we have

$$\hat{i}_{lk} \triangleq \phi_{1l}(u_{lk}) = 3u_{lk} - 2 \tag{6.15}$$

where $\phi_{1l}(\,\cdot\,)$ is the index estimation function of the side decoder 1.

Similarly, if only the description v_{lk} is available, the index estimation function for i_{lk} is determined by

$$\hat{i}_{lk} \triangleq \phi_{2l}(v_{lk}) = 3v_{lk} - 2 \tag{6.16}$$

where $\phi_{2l}(\,\cdot\,)$ is the index estimation function of the side decoder 2.

For **Case 3**, there is no description packet received at the decoder side. Therefore, instead of estimating i_{lk}, the latest decoded measurement $\bar{y}_{l(k-1)}$ is directly employed to compensate the value of $\bar{y}_{lk}$.

Based on the above analysis, we introduce the following theorem that gives a quantitative evaluation on the decoding error for different cases of the descriptions received at the decoder. This theorem plays a crucial role for the subsequent filter design issue.

Theorem 6.1: *Under the framework of the multiple description coding scheme (6.4)–(6.5), for each measurement component y_{lk} ($l = 1, 2, \ldots, n_y$), the corresponding decoding error $\Delta_{lk} \triangleq \bar{y}_{lk} - y_{lk}$ satisfies*

$$\Delta_{lk} = \begin{cases} \Delta_{lk}^s, \ |\Delta_{lk}^s| \leq 5\varrho_{lk} & \textit{for the side decoder} \\ \Delta_{lk}^c, \ |\Delta_{lk}^c| \leq \varrho_{lk} & \textit{for the central decoder} \end{cases} \tag{6.17}$$

where ϱ_{lk} is defined previously as $\varrho_{lk} \triangleq y_{lk} - q_l(y_{lk})$.

Proof: *First, recalling that i_{lk} can be calculated as $i_{lk} = 3r_{lk} + m_{lk}$, we confirm from (6.12), (6.15)–(6.16) that the index estimation error satisfies*

$$|\tilde{i}_{lk}| \leq \begin{cases} 0, & \textit{both descriptions are received} \\ 2, & \textit{only one description is received.} \end{cases} \tag{6.18}$$

Then, based on the obtained index estimation $\hat{i}_{lk}$, we define the following inverse quantization function $\hat{q}_l(\,\cdot\,)$ as follows:

$$\hat{q}_l(\hat{i}_{lk}) \triangleq -L + \frac{(2\hat{i}_{lk} - 1)L}{\rho} \tag{6.19}$$

for $\hat{i}_{lk} \in \{1, 2, \ldots, \rho\}$.

Accordingly, the decoder functions are given by

$$
\begin{cases}
g_{1l}(u_{lk}) = \hat{q}_l(\phi_{1l}(u_{lk})), & \text{when } \lambda_{1k} = 1, \lambda_{2k} = 0 \\
g_{2l}(v_{lk}) = \hat{q}_l(\phi_{2l}(v_{lk})), & \text{when } \lambda_{1k} = 0, \lambda_{2k} = 1 \\
g_{cl}(u_{lk}, v_{lk}) = \hat{q}_l(\phi_{cl}(u_{lk}, v_{lk})), & \text{when } \lambda_{1k} = \lambda_{2k} = 1.
\end{cases}
\tag{6.20}
$$

Then, it follows directly from (6.20) that the decoding error Δ_{lk} of y_{lk} satisfies (6.17), and the proof of this theorem is complete.

6.2.3 A Unified Measurement Model

Having analyzed the decoding error for each measurement component y_{lk}, we are now ready to establish a unified measurement model accounting for the MDC as well as the channel fading. For notational simplicity, define

$$
\begin{aligned}
\Delta_k^s &\triangleq [\Delta_{1k}^s \quad \Delta_{2k}^s \quad \cdots \quad \Delta_{n_y k}^s]^T \\
\Delta_k^c &\triangleq [\Delta_{1k}^c \quad \Delta_{2k}^c \quad \cdots \quad \Delta_{n_y k}^c]^T \\
\beta_{ik} &\triangleq \delta(\tilde{\lambda}_k, i), \quad \tilde{\lambda}_k \triangleq \lambda_{1k} + \lambda_{2k}.
\end{aligned}
\tag{6.21}
$$

Based on the results in Theorem 6.1, the decoded measurement output $\bar{y}_k$ is described with the following form:

$$
\bar{y}_k = \beta_{0k}\bar{y}_{k-1} + \beta_{1k}(y_k + \Delta_k^s) + \beta_{2k}(y_k + \Delta_k^c)
\tag{6.22}
$$

where $|\Delta_k^s| \leq \kappa_k^s$ and $|\Delta_k^c| \leq \kappa_k^c$ with $\kappa_k^s = \sqrt{\sum_{l=1}^{n_y} (5\varrho_{lk})^2}$ and $\kappa_k^c = \sqrt{\sum_{l=1}^{n_y} \varrho_{lk}^2}$. In addition, it is derived from the definition of β_{ik} that $\sum_{i=0}^{2} \beta_{ik} = 1$ and

$$
\begin{aligned}
\bar{\beta}_0 &\triangleq \mathbb{E}\{\beta_{0k}\} = (1 - \bar{\lambda}_1)(1 - \bar{\lambda}_2) \\
\bar{\beta}_1 &\triangleq \mathbb{E}\{\beta_{1k}\} = \bar{\lambda}_1(1 - \bar{\lambda}_2) + \bar{\lambda}_2(1 - \bar{\lambda}_1) \\
\bar{\beta}_2 &\triangleq \mathbb{E}\{\beta_{2k}\} = \bar{\lambda}_1\bar{\lambda}_2.
\end{aligned}
\tag{6.23}
$$

Consequently, the decoded measurement output (6.22) is further converted into

$$
\bar{y}_k = \beta_{0k}\bar{y}_{k-1} + (1 - \beta_{0k})y_k + \beta_{1k}\Delta_k^s + \beta_{2k}\Delta_k^c.
\tag{6.24}
$$

which serves as a unified measurement model reflecting both the MDC scheme and the channel fading phenomenon.

Remark 6.3: *It follows from (6.17) and (6.21) that the decoding error can be expressed as $\Delta_k \triangleq \delta(\tilde{\lambda}_k, 0)|\Delta_k^d| + \delta(\tilde{\lambda}_k, 1)|\Delta_k^s| + \delta(\tilde{\lambda}_k, 2)|\Delta_k^c|$, where Δ_k^s and*

Δ_k^c, *which have been defined in (6.21), satisfy* $|\Delta_k^s| \leq \kappa_k^s \triangleq \sqrt{\sum_{l=1}^{n_y} (5\varrho_{lk})^2}$ *and* $|\Delta_k^c| \leq \kappa_k^c \triangleq \sqrt{\sum_{l=1}^{n_y} \varrho_{lk}^2}$, *respectively.* $|\Delta_k^d|$ *denotes the decoding error when both of the descriptions are missing, and therefore it is reasonable to assume that its upper bound is greater than* κ_k^s. *Then, in terms of the condition* $\kappa_k^s = 5\kappa_k^c$, *there exists a positive scalar* $p > 5$ *such that* $|\Delta_k^d| \leq p\kappa_k^c$. *Next, introducing an average decoding error which is defined as* $\bar{\Delta}_k \triangleq \mathbb{E}\{\Delta_k\}$, *one calculates* $\bar{\Delta}_k = \mathbb{E}\{\delta(\tilde{\lambda}_k, 0)\}|\Delta_k^d| + \mathbb{E}\{\delta(\tilde{\lambda}_k, 1)\}|\Delta_k^s| + \mathbb{E}\{\delta(\tilde{\lambda}_k, 2)\}|\Delta_k^c| = \bar{\beta}_0|\Delta_k^d| + \bar{\beta}_1|\Delta_k^s| + \bar{\beta}_2|\Delta_k^c| \leq \bar{\beta}_0 p\kappa_k^c + \bar{\beta}_1\kappa_k^s + \bar{\beta}_2\kappa_k^c = (p\bar{\beta}_0 + 5\bar{\beta}_1 + \bar{\beta}_2)\kappa_k^c \triangleq \aleph(\bar{\lambda}_1, \bar{\lambda}_2, \kappa_k^c)$, *where* $\bar{\beta}_0 = (1 - \bar{\lambda}_1)(1 - \bar{\lambda}_2)$, $\bar{\beta}_1 = \bar{\lambda}_1(1 - \bar{\lambda}_2) + \bar{\lambda}_2(1 - \bar{\lambda}_1)$ *and* $\bar{\beta}_2 = \bar{\lambda}_1\bar{\lambda}_2$. *It is not difficult to derive that* $\aleph(\bar{\lambda}_1, \bar{\lambda}_2, \kappa_k^c)$ *is monotonically decreasing with respect to* $\bar{\lambda}_i$ $(i = 1, 2)$. *Therefore, one can conclude that a larger packet dropout rate (i.e. a smaller* $\bar{\lambda}_i$ $(i = 1, 2)$*) would lead to a larger upper bound of the average decoding error, which complies with practical engineering.*

6.3 The Recursive Filtering Strategy

So far, we have discussed the design of the MDC procedure for the fading measurement y_k. In this section, we are interested in the filter design issue. For convenience of later analysis, for $i \in [-M, -1]$, we let $y_i = 0$. Based on the obtained decoded measurement $\bar{y}_k$ in (6.24), the innovation function $\vartheta(\bar{y}_{k+1})$ in (6.7b) is given as follows:

$$\vartheta(\bar{y}_{k+1}) \triangleq \bar{y}_{k+1} - \bar{\beta}_0\bar{y}_k - (1 - \bar{\beta}_0) \sum_{m=0}^{M} \bar{\alpha}_m C_{k-m+1}\hat{x}_{k-m+1|k-m}. \tag{6.25}$$

According to the proposed filter structure (6.7a)-(6.7b), we define the one-step prediction error and filtering error by $\tilde{x}_{k+1|k} \triangleq x_{k+1} - \hat{x}_{k+1|k}$ and $\tilde{x}_{k+1|k+1} \triangleq x_{k+1} - \hat{x}_{k+1|k+1}$, respectively. Then, we denote by $P_{k+1|k} \triangleq \mathbb{E}\{\tilde{x}_{k+1|k}\tilde{x}_{k+1|k}^T\}$ and $P_{k+1|k+1} \triangleq \mathbb{E}\{\tilde{x}_{k+1|k+1}\tilde{x}_{k+1|k+1}^T\}$ the prediction error variance and the filtering error variance, respectively.

By considering (6.1) and (6.7a), we have

$$\tilde{x}_{k+1|k} = h(x_k, \zeta_k) - h(\hat{x}_{k|k}, \zeta_k) + B_{1k}\omega_k. \tag{6.26}$$

In order to linearize the nonlinear function $h(x_k, \zeta_k)$, the Taylor series expansion method is employed around $\hat{x}_{k|k}$, which leads to

$$h(x_k, \zeta_k) = h(\hat{x}_{k|k}, \zeta_k) + A_k\tilde{x}_{k|k} + o(|\tilde{x}_{k|k}|) \tag{6.27}$$

where $A_k = \left.\frac{\partial h(x_k,\zeta_k)}{\partial x_k}\right|_{x_k=\hat{x}_{k|k}}$ and $o(|\tilde{x}_{k|k}|)$ means the linearization error caused by discarding the high-order terms. As discussed in [174], such a linearization error is further transformed into the following tractable form:

$$o(|\tilde{x}_{k|k}|) = M_k F_k N_k \tilde{x}_{k|k} \tag{6.28}$$

where M_k is a problem-dependent scaling matrix, the matrix N_k is used to provide extra design freedom, and the matrix F_k $(F_k F_k^T \leq I)$ describes the errors of the linear approximation manipulation.

Substituting (6.27) into (6.26) yields

$$\tilde{x}_{k+1|k} = (A_k + M_k F_k N_k)\tilde{x}_{k|k} + B_{1k}w_k. \tag{6.29}$$

Then, we derive the following dynamics of the prediction error variance:

$$P_{k+1|k} = (A_k + M_k F_k N_k)P_{k|k}(A_k + M_k F_k N_k)^T + B_{1k}Q_k B_{1k}^T. \tag{6.30}$$

Denoting $\gamma_{m,k} \triangleq (1-\beta_{0k})\alpha_{m,k}$, $\bar{\gamma}_m \triangleq \mathbb{E}\{\gamma_{m,k}\} = (1-\bar{\beta}_0)\bar{\alpha}_m$, $\tilde{\beta}_{i,k} \triangleq \beta_{i,k} - \bar{\beta}_i$ and $\tilde{\gamma}_{m,k} \triangleq \gamma_{m,k} - \bar{\gamma}_m$, we calculate the innovation of the filter as follows:

$$\vartheta(\bar{y}_{k+1}) = \bar{y}_{k+1} - \bar{\beta}_0 \bar{y}_k - (1-\bar{\beta}_0)\sum_{m=0}^{M} \bar{\alpha}_m C_{k-m+1}\hat{x}_{k-m+1|k-m}$$

$$= \beta_{0,k+1}\bar{y}_k + (1-\beta_{0,k+1})\sum_{m=0}^{M} \alpha_{m,k+1}\tilde{y}_{k-m+1}$$

$$+ (1-\beta_{0,k+1})B_{3,k+1}\varpi_{k+1} + \beta_{1,k+1}\Delta_{k+1}^s + \beta_{2,k+1}\Delta_{k+1}^c$$

$$- \bar{\beta}_0 \bar{y}_k - (1-\bar{\beta}_0)\sum_{m=0}^{M} \bar{\alpha}_m C_{k-m+1}\hat{x}_{k-m+1|k-m} \tag{6.31}$$

$$= \tilde{\beta}_{0,k+1}\bar{y}_k + \sum_{m=0}^{M} \tilde{\gamma}_{m,k+1}C_{k-m+1}x_{k-m+1} + \sum_{m=0}^{M} \gamma_{m,k+1}$$

$$\times B_{2,k-m+1}v_{k-m+1} + \sum_{m=0}^{M} \bar{\gamma}_m C_{k-m+1}\tilde{x}_{k-m+1|k-m}$$

$$+ (1-\beta_{0,k+1})B_{3,k+1}\varpi_{k+1} + \beta_{1,k+1}\Delta_{k+1}^s + \beta_{2,k+1}\Delta_{k+1}^c.$$

Hence, it follows from (6.1), (6.7b) and (6.31) that the filtering error dynamics is governed by

$$\tilde{x}_{k+1|k+1} = (I - \bar{\gamma}_0 K_{k+1} C_{k+1})\tilde{x}_{k+1|k} - K_{k+1}\sum_{m=1}^{M}\tilde{\gamma}_m C_{k-m+1}\tilde{x}_{k-m+1|k-m}$$

$$- K_{k+1}\sum_{m=0}^{M}\tilde{\gamma}_{m,k+1}C_{k-m+1}x_{k-m+1} - \sum_{m=0}^{M}\gamma_{m,k+1}K_{k+1}B_{2,k-m+1}v_{k-m+1}$$

$$- \tilde{\beta}_{0,k+1}K_{k+1}\bar{y}_k - (1-\beta_{0,k+1})K_{k+1}B_{3,k+1}\varpi_{k+1}$$

$$- \beta_{1,k+1}K_{k+1}\Delta_{k+1}^{s} - \beta_{2,k+1}K_{k+1}\Delta_{k+1}^{c}. \tag{6.32}$$

In the following, three useful lemmas are presented to facilitate the subsequent filter analysis and design issues.

Lemma 6.1: *Assume that A, M, N and F are real matrices of compatible dimensions with $FF^T \leq I$. For any positive scalar $\pi > 0$, if there exists a positive definite matrix $X > 0$ such that $\pi^{-1}I - NXN^T > 0$, then the following inequality is true:*

$$(A + MFN)X(A + MFN)^T \leq A(X^{-1} - \pi N^T N)^{-1}A^T + \pi^{-1}MM^T. \tag{6.33}$$

Lemma 6.2: *For the positive definite matrix $X > 0$, real matrix functions $\mathcal{U}_k(X) = \mathcal{U}_k^T(X) \in \mathbb{R}^{n\times n}$ and $\mathcal{V}_k(X) = \mathcal{V}_k^T(X) \in \mathbb{R}^{n\times n}$, if*

$$\mathcal{U}_k(X) \leq \mathcal{U}_k(Y), \forall X \leq Y = Y^T \tag{6.34}$$

and

$$\mathcal{U}_k(Y) \leq \mathcal{V}_k(Y) \tag{6.35}$$

over the time horizon [0 N], then the solutions to the following difference equations:

$$S_{k+1} = \mathcal{U}_k(S_k), \quad T_{k+1} = \mathcal{V}_k(T_k), \quad S_0 = T_0 \tag{6.36}$$

satisfy

$$S_k \leq T_k. \tag{6.37}$$

The proof of the following lemma is easily accessible from (6.32) and is therefore omitted.

Lemma 6.3: *The filtering error variance $P_{k+1|k+1}$ is determined by*

$$P_{k+1|k+1} = (I - \bar{\gamma}_0 K_{k+1}C_{k+1})P_{k+1|k}(I - \bar{\gamma}_0 K_{k+1}C_{k+1})^T$$

$$+ \mathbb{E}\{\mathscr{F}_k\mathscr{F}_k^T\} + \mathbb{E}\{\mathscr{G}_{k+1}\mathscr{G}_{k+1}^T\} + \bar{\beta}_0(1-\bar{\beta}_0)K_{k+1}\bar{y}_k\bar{y}_k^T K_{k+1}^T$$

$$+ (1-\bar{\beta}_0)K_{k+1}B_{3,k+1}K_{k+1}^T + \sum_{m=0}^{M}\bar{\gamma}_m K_{k+1}\mathcal{B}_{2,k-m+1}K_{k+1}^T$$

$$
\begin{aligned}
&+ \bar{\beta}_2 K_{k+1} \Delta^c_{k+1} \Delta^{cT}_{k+1} K^T_{k+1} + \bar{\beta}_1 K_{k+1} \Delta^s_{k+1} \Delta^{sT}_{k+1} K^T_{k+1} \\
&- \mathscr{L}_{k+1} - \mathscr{L}^T_{k+1} - \mathscr{M}_{k+1} - \mathscr{M}^T_{k+1} - \mathscr{N}_{k+1} - \mathscr{N}^T_{k+1} \\
&+ \mathscr{O}_{k+1} + \mathscr{O}^T_{k+1} + \mathscr{P}_{k+1} + \mathscr{P}^T_{k+1} + \mathscr{Q}_{k+1} + \mathscr{Q}^T_{k+1} \\
&+ \mathscr{R}_{k+1} + \mathscr{R}^T_{k+1} + \mathscr{S}_{k+1} + \mathscr{S}^T_{k+1} + \mathscr{T}_{k+1} + \mathscr{T}^T_{k+1} \\
&+ \mathscr{U}_{k+1} + \mathscr{U}^T_{k+1}
\end{aligned} \tag{6.38}
$$

where

$$
\mathscr{F}_k = K_{k+1} \left(\sum_{m=1}^{M} \bar{\gamma}_m C_{k-m+1} \tilde{x}_{k-m+1|k-m} \right)
$$

$$
\mathscr{G}_{k+1} = K_{k+1} \left(\sum_{m=0}^{M} \tilde{\gamma}_{m,k+1} C_{k-m+1} x_{k-m+1} \right)
$$

$$
\mathscr{L}_{k+1} = (I - \bar{\gamma}_0 K_{k+1} C_{k+1}) \mathbb{E} \left\{ \tilde{x}_{k+1|k} \left(\sum_{m=1}^{M} \bar{\gamma}_m C_{k-m+1} \tilde{x}_{k-m+1|k-m} \right)^T \right\} K^T_{k+1}
$$

$$
\mathscr{M}_{k+1} = \bar{\beta}_1 (I - \bar{\gamma}_0 K_{k+1} C_{k+1}) \mathbb{E} \left\{ \tilde{x}_{k+1|k} \right\} \Delta^{sT}_{k+1} K^T_{k+1}
$$

$$
\mathscr{N}_{k+1} = \bar{\beta}_2 (I - \bar{\gamma}_0 K_{k+1} C_{k+1}) \mathbb{E} \left\{ \tilde{x}_{k+1|k} \right\} \Delta^{cT}_{k+1} K^T_{k+1}
$$

$$
\mathscr{O}_{k+1} = \bar{\beta}_1 K_{k+1} \mathbb{E} \left\{ \left(\sum_{m=1}^{M} \bar{\gamma}_m C_{k-m+1} \tilde{x}_{k-m+1|k-m} \right) \right\} Delta^{sT}_{k+1} K^T_{k+1}
$$

$$
\mathscr{P}_{k+1} = \bar{\beta}_2 K_{k+1} \mathbb{E} \left\{ \left(\sum_{m=1}^{M} \bar{\gamma}_m C_{k-m+1} \tilde{x}_{k-m+1|k-m} \right) \right\} \Delta^{cT}_{k+1} K^T_{k+1}
$$

$$
\mathscr{Q}_{k+1} = K_{k+1} \mathbb{E} \left\{ \tilde{\beta}_{0,k+1} \sum_{m=0}^{M} \tilde{\gamma}_{m,k+1} C_{k-m+1} x_{k-m+1} \right\} \bar{y}^T_k K^T_{k+1}
$$

$$
\mathscr{R}_{k+1} = K_{k+1} \mathbb{E} \left\{ \beta_{1,k+1} \sum_{m=0}^{M} \tilde{\gamma}_{m,k+1} C_{k-m+1} x_{k-m+1} \right\} \Delta^{sT}_{k+1} K^T_{k+1}
$$

$$
\mathscr{S}_{k+1} = K_{k+1} \mathbb{E} \left\{ \beta_{2,k+1} \sum_{m=0}^{M} \tilde{\gamma}_{m,k+1} C_{k-m+1} x_{k-m+1} \right\} \Delta^{cT}_{k+1} K^T_{k+1}
$$

$$
\mathscr{T}_{k+1} = K_{k+1} \bar{y}_k \mathbb{E} \left\{ \tilde{\beta}_{0,k+1} \beta_{1,k+1} \right\} \Delta^{sT}_{k+1} K^T_{k+1}
$$

$$
\mathscr{U}_{k+1} = K_{k+1} \bar{y}_k \mathbb{E} \left\{ \tilde{\beta}_{0,k+1} \beta_{2,k+1} \right\} \Delta^{cT}_{k+1} K^T_{k+1}
$$

$$
\mathcal{B}_{2,k-m+1} = B_{2,k-m+1} R_{k-m+1} B^T_{2,k-m+1}
$$

$$
\mathcal{B}_{3,k+1} = B_{3,k+1} S_{k+1} B^T_{3,k+1}.
$$

On the basis of Lemmas 6.1–6.3, we shall now handle the design issue of the filter parameter K_k. Due to uncertainty terms (e.g. F_k, Δ_k^s and Δ_k^c) in the filtering error variance matrix $P_{k|k}$, an upper bound of $P_{k|k}$ is first determined, and then the filter parameter K_k is calculated by minimizing the derived upper bound at each sampling instant k.

Theorem 6.2: *Consider the prediction error variance (6.30) and the filtering error variance (6.38). Let the positive scalars ϵ_i, ε_{hi} ($h = l, o, p, q, r, s$ and $i = 1, 2, \ldots, M$) and ε_j ($j = m, n, t, u$) be given. Assume that the following discrete-time coupled algebraic Riccati-like difference equations:*

$$\Pi_{k+1|k} = \mathcal{V}_{1k}(\Pi_{k|k})$$
$$\triangleq A_k(\Pi_{k|k}^{-1} - \pi N_k N_k^T)^{-1} A_k^T + \pi^{-1} M_k M_k^T + B_{1k} Q_k B_{1k}^T \tag{6.39}$$

and

$$\Pi_{k+1|k+1} = \mathcal{V}_{2k}(\Pi_{k+1|k})$$
$$\triangleq \xi_1 (I - \bar{\gamma}_0 K_{k+1} C_{k+1}) \Pi_{k+1|k} (I - \bar{\gamma}_0 K_{k+1} C_{k+1})^T$$
$$+ \xi_{20}^{(2)} K_{k+1} C_{k+1} \Pi_{k+1|k} C_{k+1}^T K_{k+1}^T$$
$$+ \sum_{m=1}^{M} \xi_{2m} K_{k+1} C_{k-m+1} \Pi_{k-m+1|k-m} C_{k-m+1}^T K_{k+1}^T$$
$$+ K_{k+1} \left(\xi_3 \theta_k^2 I + \xi_4 \bar{y}_k \bar{y}_k^T + \sum_{m=0}^{M} \bar{\gamma}_m \mathcal{B}_{2,k-m+1} + (1 - \bar{\beta}_0) \mathcal{B}_{3,k+1} \right) K_{k+1}^T$$
$$+ \sum_{m=1}^{M} \xi_{5m} K_{k+1} C_{k-m+1} \hat{X}_{k-m+1|k-m} C_{k-m+1}^T K_{k+1}^T \tag{6.40}$$

subject to

$$\begin{cases} \Pi_{i|i} = P_{i|i} > 0; \ i = -M, -M+1, \ldots, 0 & \text{(6.41a)} \\ \pi^{-1} I - N_k \Pi_{k|k} N_k^T > 0 & \text{(6.41b)} \end{cases}$$

have the positive definite solutions $\Pi_{k+1|k}$ and $\Pi_{k+1|k+1}$. Then, the desired filter parameter K_{k+1} is determined by

$$K_{k+1} = \xi_1 \bar{\gamma}_0 \Pi_{k+1|k} C_{k+1}^T \left(\bar{\xi}_1 C_{k+1} \Pi_{k+1|k} C_{k+1}^T \right.$$
$$\left. + \sum_{m=1}^{M} \xi_{2m} C_{k-m+1} \Pi_{k-m+1|k-m} C_{k-m+1}^T + \xi_3 \theta_k^2 I \right.$$

$$+ \xi_4 \bar{y}_k \bar{y}_k^T + \sum_{m=0}^{M} \mathcal{B}_{2,k-m+1} + (1 - \bar{\beta}_0)\mathcal{B}_{3,k+1}$$

$$+ \sum_{m=0}^{M} \xi_{5m} C_{k-m+1} \hat{X}_{k-m+1|k-m} C_{k-m+1}^T \Bigg)^{-1} \tag{6.42}$$

where

$$\xi_1 = 1 + \bar{\varepsilon}_l + \varepsilon_m + \varepsilon_n, \ \xi_{2m} = \xi_{2m}^{(1)} + \xi_{2m}^{(2)}$$

$$\xi_3 = 25\xi_3^{(1)} + \xi_3^{(2)}, \ \sigma_{\gamma_{mn}} = \mathbb{E}\{\tilde{\gamma}_{m,k}\tilde{\gamma}_{n,k}\}$$

$$\xi_4 = \bar{\beta}_0(\bar{\varepsilon}_q \bar{\beta}_0 + \varepsilon_t^{-1}\bar{\beta}_0 + \varepsilon_u^{-1}\bar{\beta}_0 + 1 - \bar{\beta}_0)$$

$$\xi_{5m} = (1 + \epsilon_m^{-1})\Im_m, \ \bar{\xi}_1 = \xi_1 \bar{\gamma}_0^2 + \xi_{20}^{(2)}$$

$$\xi_{2m}^{(1)} = \varepsilon_{lm}^{-1}\bar{\gamma}_m^2 + \varepsilon_{om}^{-1}\bar{\gamma}_m^2 + \varepsilon_{pm}^{-1}\bar{\gamma}_m^2 + \bar{\gamma}\bar{\gamma}_m$$

$$\xi_3^{(1)} = \bar{\beta}_1^2(\varepsilon_m^{-1} + \bar{\varepsilon}_o + \bar{\varepsilon}_r + \varepsilon_t + \bar{\beta}_1^{-1})$$

$$\xi_3^{(2)} = \bar{\beta}_2^2(\varepsilon_n^{-1} + \bar{\varepsilon}_p + \bar{\varepsilon}_s + \varepsilon_u + \bar{\beta}_2^{-1})$$

$$\Im_m = \varepsilon_{qm}^{-1}\bar{\gamma}_m^2 + \bar{\varepsilon}_{rm}^{-1}\bar{\beta}_0^2\bar{\alpha}_m^2 + \bar{\varepsilon}_{sm}^{-1}\bar{\beta}_0^2\bar{\alpha}_m^2 + \sigma_{\gamma_m}$$

$$\xi_{2m}^{(2)} = (1 + \epsilon_m)\Im_m, \ \bar{\gamma} = \sum_{m=1}^{M} \bar{\gamma}_m$$

$$\hat{X}_{k-m+1|k-m} = \hat{x}_{k-m+1|k-m}\hat{x}_{k-m+1|k-m}^T$$

$$\bar{\varepsilon}_\hbar = \sum_{m=1}^{M} \varepsilon_{\hbar m}, \hbar = l, o, p, q, r, s$$

$$\theta_k = \left(\sum_{l=1}^{n_y} \mu_{lk}^2\right)^{\frac{1}{2}} \frac{L}{\rho}, \ \sigma_{\gamma_m} = \sum_{n=0}^{M} \sigma_{\gamma_{mn}}$$

$\mathcal{V}_{1k}(\cdot) : \mathbb{R}^{n_x \times n_x} \to \mathbb{R}^{n_x \times n_x}$ and $\mathcal{V}_{2k}(\cdot) : \mathbb{R}^{n_x \times n_x} \to \mathbb{R}^{n_x \times n_x}$ stand for two matrix-valued functions. In addition, the matrix $\Pi_{k+1|k+1}$ is an upper bound of $P_{k+1|k+1}$, namely, $P_{k+1|k+1} \leq \Pi_{k+1|k+1}$. Also, such an upper bound is minimized by the filter parameter K_{k+1} at each sampling instant.

Proof: *The proof is carried out by applying the mathematical induction. First, according to (6.30), the prediction error variance $P_{k+1|k}$ is actually a function of the filtering error variance $P_{k|k}$, namely, $P_{k+1|k} \triangleq \mathcal{U}_{1k}(P_{k|k})$, and then one easily obtains that $\mathcal{U}_{1k}(X_k) \leq \mathcal{U}_{1k}(Y_k), \forall \ 0 \leq X_k \leq Y_k$. Noting that $P_{i|i} = \Pi_{i|i}$ $(i = -M, -M+1, \ldots, 0)$ and assuming that $P_{i|i} \leq \Pi_{i|i}$ $(i = 1, 2, \ldots, k)$, we have*

$\mathcal{U}_{1k}(P_{k|k}) \leq \mathcal{U}_{1k}(\Pi_{k|k})$. Subsequently, it is readily observed from Lemma 6.1, (6.30), (6.39) and (6.41b) that $P_{k+1|k} = \mathcal{U}_{1k}(P_{k|k}) \leq \mathcal{U}_{1k}(\Pi_{k|k}) \leq \mathcal{V}_{1k}(\Pi_{k|k}) = \Pi_{k+1|k}$. In the following, we only need to show that $P_{k+1|k+1} \leq \Pi_{k+1|k+1}$. For this purpose, let us deal with some terms of the right-hand side of (6.38).

By resorting to the elementary inequality $(\varepsilon^{\frac{1}{2}} x - \varepsilon^{-\frac{1}{2}} y)(\varepsilon^{\frac{1}{2}} x - \varepsilon^{-\frac{1}{2}} y)^T \geq 0$ for any $x, y \in \mathbb{R}^n$ and $\varepsilon > 0$, one has

$$-\mathcal{L}_{k+1}^T - \mathcal{L}_{k+1} \leq \bar{\varepsilon}_l (I - \bar{\gamma}_0 K_{k+1} C_{k+1}) P_{k+1|k} (I - \bar{\gamma}_0 K_{k+1} C_{k+1})^T$$
$$+ \sum_{m=1}^{M} \varepsilon_{lm}^{-1} \bar{\gamma}_m^2 K_{k+1} C_{k-m+1} P_{k-m+1|k-m} C_{k-m+1}^T K_{k+1}^T, \quad (6.43)$$

$$\mathcal{M}_{k+1}^T - \mathcal{M}_{k+1} \leq \varepsilon_m (I - \bar{\gamma}_0 K_{k+1} C_{k+1}) P_{k+1|k} (I - \bar{\gamma}_0 K_{k+1} C_{k+1})^T$$
$$+ \varepsilon_m^{-1} \bar{\beta}_1^2 K_{k+1} \Delta_{k+1}^s \Delta_{k+1}^{sT} K_{k+1}^T \quad (6.44)$$

$$-\mathcal{N}_{k+1}^T - \mathcal{N}_{k+1} \leq \varepsilon_n (I - \bar{\gamma}_0 K_{k+1} C_{k+1}) P_{k+1|k} (I - \bar{\gamma}_0 K_{k+1} C_{k+1})^T$$
$$+ \varepsilon_n^{-1} \bar{\beta}_2^2 K_{k+1} \Delta_{k+1}^c \Delta_{k+1}^{cT} K_{k+1}^T \quad (6.45)$$

$$\mathcal{O}_{k+1}^T + \mathcal{O}_{k+1} \leq \sum_{m=1}^{M} \varepsilon_{om}^{-1} \bar{\gamma}_m^2 K_{k+1} C_{k-m+1} P_{k-m+1|k-m} C_{k-m+1}^T K_{k+1}^T$$
$$+ \bar{\varepsilon}_o \bar{\beta}_1^2 K_{k+1} \Delta_{k+1}^s \Delta_{k+1}^{sT} K_{k+1}^T \quad (6.46)$$

$$\mathcal{P}_{k+1}^T + \mathcal{P}_{k+1} \leq \sum_{m=1}^{M} \varepsilon_{pm}^{-1} \bar{\gamma}_m^2 K_{k+1} C_{k-m+1} P_{k-m+1|k-m} C_{k-m+1}^T K_{k+1}^T$$
$$+ \bar{\varepsilon}_p \bar{\beta}_2^2 K_{k+1} \Delta_{k+1}^c \Delta_{k+1}^{cT} K_{k+1}^T \quad (6.47)$$

$$\mathcal{Q}_{k+1}^T + \mathcal{Q}_{k+1} \leq \sum_{m=0}^{M} \varepsilon_{qm}^{-1} \bar{\gamma}_m^2 K_{k+1} C_{k-m+1} \mathbb{E}\{x_{k-m+1} x_{k-m+1}^T\} C_{k-m+1}^T K_{k+1}^T$$
$$+ \bar{\varepsilon}_q \bar{\beta}_0^2 K_{k+1} \bar{y}_k \bar{y}_k^T K_{k+1}^T \quad (6.48)$$

$$\mathcal{R}_{k+1}^T + \mathcal{R}_{k+1} \leq \sum_{m=0}^{M} \varepsilon_{rm}^{-1} \bar{\beta}_0^2 \bar{\alpha}_m^2 K_{k+1} C_{k-m+1} \mathbb{E}\{x_{k-m+1} x_{k-m+1}^T\} C_{k-m+1}^T K_{k+1}^T$$

$$(6.49)$$

$$+ \bar{\varepsilon}_r \bar{\beta}_1^2 K_{k+1} \Delta_{k+1}^s \Delta_{k+1}^{sT} K_{k+1}^T \quad (6.50)$$

$$\mathcal{S}_{k+1}^T + \mathcal{S}_{k+1} \leq \sum_{m=0}^{M} \varepsilon_{sm}^{-1} \bar{\beta}_0^2 \bar{\alpha}_m^2 K_{k+1} C_{k-m+1} \mathbb{E}\{x_{k-m+1} x_{k-m+1}^T\} C_{k-m+1}^T K_{k+1}^T$$
$$+ \bar{\varepsilon}_s \bar{\beta}_2^2 K_{k+1} \Delta_{k+1}^c \Delta_{k+1}^{cT} K_{k+1}^T \quad (6.51)$$

$$\mathcal{T}_{k+1}^T + \mathcal{T}_{k+1} \leq \varepsilon_t^{-1} \bar{\beta}_0^2 K_{k+1} \bar{y}_k \bar{y}_k^T K_{k+1}^T + \varepsilon_t \bar{\beta}_1^2 K_{k+1} \Delta_{k+1}^s \Delta_{k+1}^{sT} K_{k+1}^T \quad (6.52)$$

and

$$\mathscr{U}_{k+1}^T + \mathscr{U}_{k+1} \leq \varepsilon_u^{-1} \bar{\beta}_0^2 K_{k+1} \bar{y}_k \bar{y}_k^T K_{k+1}^T + \varepsilon_u \bar{\beta}_2^2 K_{k+1} \Delta_{k+1}^c \Delta_{k+1}^{cT} K_{k+1}^T. \tag{6.53}$$

Moreover, invoking the Jensen inequality yields

$$\mathbb{E}\{\mathscr{F}_k \mathscr{F}_k^T\} \leq \bar{\gamma} \sum_{m=1}^{M} \bar{\gamma}_m K_{k+1} C_{k-m+1} P_{k-m+1|k-m} C_{k-m+1}^T K_{k+1}^T \tag{6.54}$$

$$\mathbb{E}\{\mathscr{G}_k \mathscr{G}_k^T\} \leq \sum_{m=0}^{M} \sigma_{r_m} K_{k+1} C_{k-m+1} \mathbb{E}\{x_{k-m+1} x_{k-m+1}^T\} C_{k-m+1}^T K_{k+1}^T. \tag{6.55}$$

By utilizing the elementary inequality again, it is shown that the term $\mathbb{E}\{x_{k-m+1} x_{k-m+1}^T\}$ *on the right-hand sides of (6.48)–(6.51) and (6.55) further satisfies*

$$\mathbb{E}\{x_{k-m+1} x_{k-m+1}^T\} \leq \bar{\epsilon}_{1m} P_{k-m+1|k-m} + \bar{\epsilon}_{2m} \hat{x}_{k-m+1k-m} \hat{x}_{k-m+1|k-m}^T \tag{6.56}$$

where $\bar{\epsilon}_{1m} = 1 + \epsilon_m$ *and* $\bar{\epsilon}_{2m} = 1 + \epsilon_m^{-1}$. *On the other hand, based on (6.41a) and the assumption* $P_{i|i} \leq \Pi_{i|i}$ $(i = 1, 2, \ldots, k)$, *it is obvious that* $P_{i+1|i} \leq \Pi_{i+1|i}$ $(i = -M, -M+1, \ldots, k)$. *Hence, according to the above discussion, it is not difficult to verify that*

$$P_{k+1|k+1} \leq \mathcal{V}_2(P_{k+1|k}) \leq \mathcal{V}_2(\Pi_{k+1|k}) = \Pi_{k+1|k+1}. \tag{6.57}$$

In the rest of the proof, it remains to show that the given filter parameter (6.42) minimizes the variance upper bound $\Pi_{k+1|k+1}$ *at each sampling instant. Applying the partial derivative operation to the trace of* $\Pi_{k+1|k+1}$ *with respect to* K_{k+1}, *we have*

$$\frac{\partial tr\{\Pi_{k+1|k+1}\}}{\partial K_{k+1}}$$

$$= -2\xi_1 \bar{\gamma}_0 (I - \bar{\gamma}_0 K_{k+1} C_{k+1}) \Pi_{k+1|k} C_{k+1}^T + 2\xi_{20}^{(2)} K_{k+1} C_{k+1} \Pi_{k+1} C_{k+1}^T$$

$$+ 2 \sum_{m=1}^{M} \xi_{2m} K_{k+1} C_{k-m+1} \Pi_{k-m+1|k-m} C_{k-m+1}^T + 2 K_{k+1} \left(\xi_3 \theta_k^2 I + \xi_4 \bar{y}_k \bar{y}_k^T \right.$$

$$\tag{6.58}$$

$$\left. + \sum_{m=0}^{M} \bar{\gamma}_m \mathcal{B}_{2,k-m+1} + (1 - \bar{\beta}_0) \mathcal{B}_{3k} \right)$$

$$+ 2 \sum_{m=1}^{M} \xi_{5m} K_{k+1} C_{k-m+1} \hat{X}_{k-m+1|k-m} C_{k-m+1}^T.$$

Setting $\frac{\partial tr\{\Pi_{k+1|k+1}\}}{\partial K_{k+1}} = 0$, the filter parameter K_{k+1} is immediately obtained with the form in (6.42), and the proof of this theorem is thus complete.

Remark 6.4: *The computational burden is a crucial issue which should be taken into account during the filtering algorithm design. For the considered system (1) and (2), the dimensions of system parameters are given as follows: $x_k \in \mathbb{R}^{n_x}$, $y_k \in \mathbb{R}^{n_y}$, $B_{1k} \in \mathbb{R}^{n_x \times n_\omega}$, $B_{2k} \in \mathbb{R}^{n_y \times n_v}$, $B_{3k} \in \mathbb{R}^{n_y \times n_\varpi}$ and $C_k \in \mathbb{R}^{n_y \times n_x}$. By carrying out some matrix manipulations (e.g. addition, multiplication, transpose, inversion, etc.), it is not difficult to calculate that the proposed filtering algorithm mainly contains a nonlinear operation, $(7M + 18)$ addition operations, $(18M + 35)$ multiplication operations, $(7M + 12)$ transpose operations and $(3M + 7)$ inversion operations, respectively. Based on the well-known computation complexity of the matrix manipulations, the total computational complexity of the proposed filtering algorithm for each iteration is $\mathcal{O} = \max_{i=1,2...,6}\{\mathcal{O}_i\}$, where $\mathcal{O}_1 = \mathcal{O}(h(\hat{x}_{k|k}, \zeta_k))$, $\mathcal{O}_2 \triangleq \mathcal{O}(9n_x^{2.376} + 3n_x^2 + n_x n_\omega(n_x + n_\omega))$, $\mathcal{O}_3 \triangleq \mathcal{O}_{a3} + \mathcal{O}_{m3} + \mathcal{O}_{t3} + \mathcal{O}_{i3}$, $\mathcal{O}_4 \triangleq \mathcal{O}((M+1)n_x n_y + (M+2)n_y)$, $\mathcal{O}_5 \triangleq \mathcal{O}((1+n_x)n_y)$, and $\mathcal{O}_6 = \mathcal{O}_{a6} + \mathcal{O}_{m6} + \mathcal{O}_{t6}$ with $\mathcal{O}_{a3} = \mathcal{O}((3M+5)n_y^2)$, $\mathcal{O}_{m3} = \mathcal{O}(2(1+M)n_x n_y(n_x+n_y) + n_y^2 + (M+1)n_y n_v(n_y + n_v) + n_y n_\varpi(n_y + n_\varpi))$, $\mathcal{O}_{t3} = \mathcal{O}((3M+4)n_x n_y + n_y)$ and $\mathcal{O}_{i3} = \mathcal{O}((3M+5)n_y^{2.376})$ and $\mathcal{O}_{a6} = \mathcal{O}((7 + 3M)n_x^2)$, $\mathcal{O}_{m6} = \mathcal{O}((M + 3)n_x^{2.376} + (6M + 5)n_x^2 n_y + (M + 1)(n_y n_v^2 + n_y^2 n_v) + n_y n_\varpi^2 + n_y^2 n_\varpi)$, $\mathcal{O}_{t6} = \mathcal{O}(n_x^2 + (4M+3)n_y n_x)$. It is noted that the computational complexity increases polynomially with the growth of the dimensions of system parameters. Fortunately, research on matrix computations is a quite active field in applied mathematics and operations research community, and substantial speedups can be expected in the near future.*

Remark 6.5: *The main results of this chapter are dependent on some parameters such as ϵ_i, ε_{hi} ($\hbar = l, o, p, q, r, s$ and $i = 1, 2, \ldots, M$) and ε_j ($j = m, n, t, u$). In fact, these parameters are involved in the upper bound of the filtering error variance (i.e. $\Pi_{k|k}$). In general, a basic principle for selecting these parameters is to minimize the obtained upper bound of the filtering error variance (i.e. $\Pi_{k|k}$) at time instant k, that is, it is desirable to determine a set of parameters that would (locally) minimize $\Pi_{k|k}$ in the sense of matrix trace. Unfortunately, the upper bound of the filtering error variance (i.e. $\Pi_{k|k}$) is actually a non-convex function of these parameters, and this makes it extremely difficult (if not impossible) to develop an analytic algorithm for selecting the best parameters in order to achieve the local minimum of the obtained upper bound. In this case, a rather practical way would be to select these parameters for optimal performance by using the evolutionary computation algorithms (e.g. particle swarm optimization algorithm and genetic algorithm), and this would be one of the future research topics.*

Remark 6.6: *In this chapter, we make tremendous efforts to examine the impact from the MDC scheme on the filtering performance for a class of stochastic nonlinear systems subject to measurement fading. It can be seen from Theorem 6.2 that all* *the factors contributing to the system complexity are reflected in the filter design procedure, which include: 1) the stochasticity and nonlinearity of the system; 2) the channel coefficient of the measurement fading phenomenon; 3) the coding accuracy of the proposed MDC scheme; and 4) the occurrence probabilities of the packet dropout during the transmission of the individual descriptions. In comparison with the existing literature, there are three distinct features with our main results: 1) this is the first attempt to consider the recursive filtering issue for the nonlinear stochastic system with the MDC scheme; 2) a concrete design procedure is, for the first time, proposed in a mathematically rigorous way on the MDC scheme; and 3) the developed filtering algorithm is of a recursive nature and is therefore suitable for the online application.*

6.4 An Illustrative Example

The tank systems are widely applied in engineering practice such as process control for petrochemical and metallurgical industries. As a simplified tank system, the three-tank system is commonly adopted in the laboratory to simulate the dynamics of the industrial tank systems.

In this section, in order to validate the effectiveness and the applicability of the proposed nonlinear filtering strategy in a networked environment, an experimental simulation is conducted on the internet-based three-tank system [189]. Figure 6.3 shows the diagram of the three-tank system. It can be seen from Figure 6.3 that the three tanks constitute the main body of the system. The three tanks are of the equivalent cross section S_a and interconnected by two cylindrical pipes with the cross section S_n. For Tank 1 and Tank 2, two pumps are equipped to pump water to each one with the

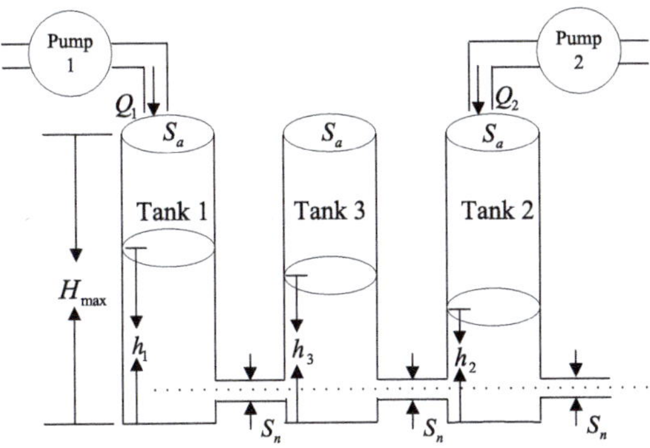

FIGURE 6.3
The diagram of three-tank system.

flow rate $Q_1(t)$ and $Q_2(t)$. Also, two sensors are installed to measure their level heights, respectively.

In the simulation, it is assumed that the levels of three tanks satisfy the condition $h_1(t) > h_3(t) > h_2(t)$, that is, water always flows from Tank 1 to Tank 2 via Tank 3. In terms of the "mass balance principle", the dynamical behavior of the level heights of the tanks can be described with the following differential equations:

$$\begin{cases} \dot{h}_1(t) = \frac{1}{S_a}(Q_1(t) - Q_{13}(t)) + b_{11}\omega_1(t) \\ \dot{h}_2(t) = \frac{1}{S_a}(Q_2(t) + Q_{32}(t) - Q_{20}(t)) + b_{12}\omega_2(t) \\ \dot{h}_3(t) = \frac{1}{S_a}(Q_{13}(t) - Q_{32}(t)) + b_{13}\omega_3(t) \end{cases} \tag{6.59}$$

where $Q_{ij}(t)$ stands for the water flow rate from the ith tank to the jth tank and can be calculated as $Q_{ij}(t) = az_i S_n \text{sign}(h_i(t) - h_j(t))\sqrt{2g|h_i(t) - h_j(t)|}$, in which az_i is the real value of the outflow coefficient of pipe i, S_n is the cross section area of the connection pipe, and $g = 9.8\text{m/s}^2$ is the acceleration of gravity. $Q_{20}(t)$ is the rate of outflow of the liquid from tank 2 and can be determined by $Q_{20}(t) = az_2 S_n\sqrt{2gh_2(t)}$. $\omega_i(t)$ is the process noise with the intensive coefficient b_{1i}. Consequently, substituting the detailed forms of the above parameters to (6.59) yields

$$\begin{cases} \dot{h}_1(t) = \frac{1}{S_a}\Big(- az_1 S_n\sqrt{2g(h_1(t) - h_3(t))} + Q_1(t)\Big) \\ \qquad + b_{11}\omega_1(t) \\ \dot{h}_2(t) = \frac{1}{S_a}\Big(az_3 S_n\sqrt{2g(h_3(t) - h_2(t))} - az_2 S_n\sqrt{2gh_2(t)} \\ \qquad + Q_2(t)\Big) + b_{12}\omega_2(t) \\ \dot{h}_3(t) = \frac{1}{S_a}\Big(az_1 S_n\sqrt{2g(h_1(t) - h_3(t))} \\ \qquad - az_3 S_n\sqrt{2g(h_3(t) - h_2(t))}\Big) + b_{13}\omega_3(t). \end{cases} \tag{6.60}$$

By applying the Euler discretization approach [43] to the above continuous-time state equation at each sampling instant t_k, the corresponding approximated discrete-time counterpart is obtained as follows:

$$\begin{cases} h_1(t_{k+1}) = h_1(t_k) - \frac{TS_n}{S_a}az_1\sqrt{2g(h_1(t_k) - h_3(t_k))} \\ \qquad + \frac{T}{S_a}Q_1(t_k) + Tb_{11}\omega_1(t_k) \\ h_2(t_{k+1}) = h_2(t_k) + \frac{TS_n}{S_a}az_3\sqrt{2g(h_3(t_k) - h_2(t_k))} \\ \qquad - \frac{TS_n}{S_a}az_2\sqrt{2gh_2(t_k)} + \frac{T}{S_a}Q_2(t_k) \\ \qquad + Tb_{12}\omega_2(t_k) \\ h_3(t_{k+1}) = h_3(t_k) + \frac{TS_n}{S_a}az_1\sqrt{2g(h_1(t_k) - h_3(t_k))} \\ \qquad - \frac{TS_n}{S_a}az_3\sqrt{2g(h_3(t_k) - h_2(t_k))} \\ \qquad + Tb_{13}\omega_3(t_k). \end{cases} \tag{6.61}$$

Here, T is the constant sampling period, that is, $T = t_{k+1} - t_k$. Similarly, the measurement equation is obtained by the following formula:

$$\begin{cases} y_1(t_k) = (1 + \sin(t_k))h_1(t_k) + b_{21}v_1(t_k) \\ y_2(t_k) = h_2(t_k) + b_{22}v_2(t_k) \end{cases} \tag{6.62}$$

where $\sin(t_k)$ reflects the measurement error and $v_i(t_k)$ is the measurement noise with the intensive coefficient b_{2i}.

For convenience, in the sequel, we denote k as the shorthand of t_k. Also, the process noises ω_{ik} ($i = 1, 2, 3$) are assumed to be the same one and obey the truncated Gaussian distribution, that is, ω_{ik} ($i = 1, 2, 3$) obey the Gaussian distribution with zero mean and variance Q_k and lie within the interval $[r_1, r_2]$. Moreover, the similar assumption is made on the measurement noises v_{ik} ($i = 1, 2$), that is, $v_{ik} = v_k \sim \mathcal{N}(0, R_k)$ ($i = 1, 2$). From (6.60)–(6.62), system parameters are given as follows:

$$B_{1k} = \begin{bmatrix} 0.05 & 0.05 & 0.05 \end{bmatrix}^T, B_{2k} = \begin{bmatrix} 0.01 & 0.01 \end{bmatrix}^T,$$

$$B_{3k} = \begin{bmatrix} 0.01 & 0.01 \end{bmatrix}^T, r_1 = -0.5, r_2 = 0.5, \tag{6.63}$$

$$C_k = \begin{bmatrix} 1 + 0.1\sin(k) & 0 & 0 \\ 0 & 1 & 0 \end{bmatrix}, Q_k = 0.05, R_k = S_k = 0.01$$

and other system parameters are listed in Table 6.1.

The order of the fading measurement model (6.2) is set to be $M = 2$ and the probability density functions of channel coefficients are chosen as

TABLE 6.1

System Parameters

Parameters	Symbol	Value
Cross section area of tanks	S_a	154 cm^2
Cross section area of pipes	S_n	0.5 cm^2
Sampling period	T	8 s
Max. height of tanks	H_{max}	620 cm
Max. flow rate of tank 1	$Q_{1\,max}$	93 cm^3/s
Max. flow rate of tank 2	$Q_{2\,max}$	91 cm^3/s
The control law of tank 1	Q_1	20 cm^3/s
The control law of tank 2	Q_2	21 cm^3/s
The outflow coefficient of pipe 1	az_1	0.48
The outflow coefficient of pipe 2	az_2	0.58
The outflow coefficient of pipe 3	az_3	0.48

$$\begin{cases} p_0(\alpha_{0k}) = 0.0005(e^{9.8933\alpha_{0k}} - 1), \quad 0 \le \alpha_{0k} \le 1 \\[2mm] p_1(\alpha_{1k}) = \begin{cases} 10\alpha_{1k}, & 0 \le \alpha_{1k} \le 0.2 \\ -2.5(\alpha_{1k} - 1), & 0.2 < \alpha_{1k} \le 1 \end{cases} \\[2mm] p_2(\alpha_{2k}) = 8.5017e^{-8.5\alpha_{2k}}, \quad 0 \le \alpha_{2k} \le 1, \end{cases} \tag{6.64}$$

from which it can be verified that mathematical expectations $\bar{\alpha}_m$ of α_{mk} ($m = 0, 1, 2$) are 0.8991, 0.4000 and 0.1174, and the variances ς_m are 0.0133, 0.0467 and 0.01364, respectively. In addition, the successful arrival rates of the two description packets from the encoder to the decoder are set as $\mathrm{Prob}\{\lambda_k^i = 1\} = 0.9$ ($i = 1, 2$). Thus, the corresponding mathematical expectations and variances are easily obtained as $\bar{\lambda}_i = 0.9$ and $\sigma_i = 0.09$ ($i = 1, 2$), respectively. For the uniform quantizer (6.9), let the saturation value be $L = 1$, the number of the quantization level be $\rho = 100$ and the adjustable parameter be $\mu_{lk} = 1$. The sampling period T is taken as $T = 13$.

Letting $x_k = [x_{1k} \quad x_{2k} \quad x_{3k}]^T = [h_{1k} \quad h_{2k} \quad h_{3k}]^T$, in this simulation, the initial heights of the tanks and their estimates are, respectively, set as $x_0 = [0.45 \quad 0.14 \quad 0.18]^T$ and $\hat{x}_0 = [0.15 \quad 0.04 \quad 0.08]^T$. At each time step k, for the ith component ($i = 1, 2, 3$), define the filtering error (FE) as $\mathrm{FE}_i \triangleq \sqrt{(x_{ik} - \hat{x}_{ik})^2}$.

In order to illustrate the superiority of the proposed filtering scheme, the performances (in terms of the upper bound of the filtering error (UBFE)) of the MDC-based filter and the single-description-coding-based filter are compared in the simulation. Moreover, for the sake of revealing the effects of the system complexities (e.g. the channel fading, the process noises and the probability of the packet dropout) on the filtering performance, some comparison simulations are also constructed.

The simulation results are shown in Figures 6.4–6.11. Figure 6.4 describes the measurements received by the encoder and the filter. Figure 6.5 plots the actual level heights of the tanks and their estimates. Figures 6.6–6.8 illustrate the FE_i ($i = 1, 2, 3$) and their upper bounds. In Figure 6.9, the upper bounds of the filtering error for each component with single description coding scheme and two description coding scheme are, respectively, plotted, which shows that the filtering performance with the two description coding is superior to the single description coding case. Figure 6.10 gives the effect of different channel coefficients on the UBFE, which indicates that, as the fading parameter M grows from 0 (without channel fading) to 2, the UBFE increases accordingly. The influence of the process noise is reflected in Figure 6.11, where the UBFE increases with the growth of the noise intensity. In Figure 6.12, the impact of the packet-arrival probability is exhibited, which verifies that a higher arrival probability leads to a better filtering performance. It is shown from the simulation results that the proposed

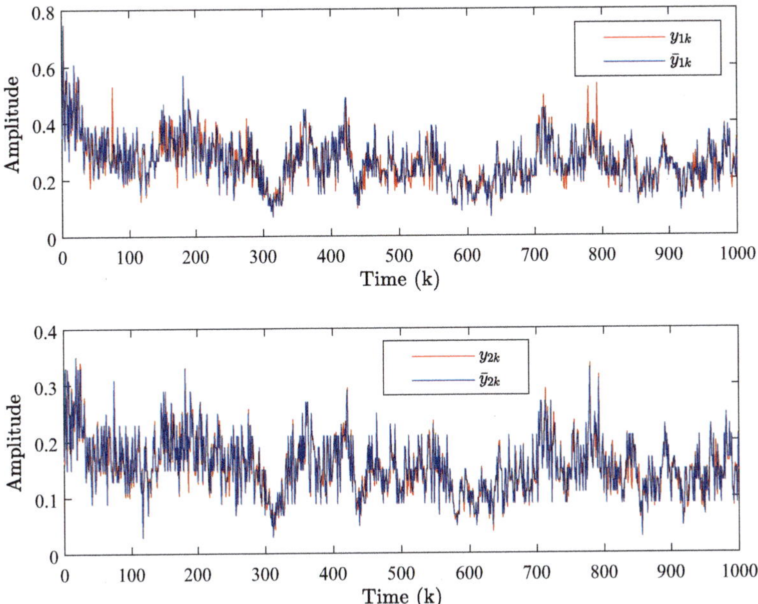

FIGURE 6.4

The fading measurement components y_{ik} $(i = 1, 2)$ and their decoded values $\bar{y}_{ik}$.

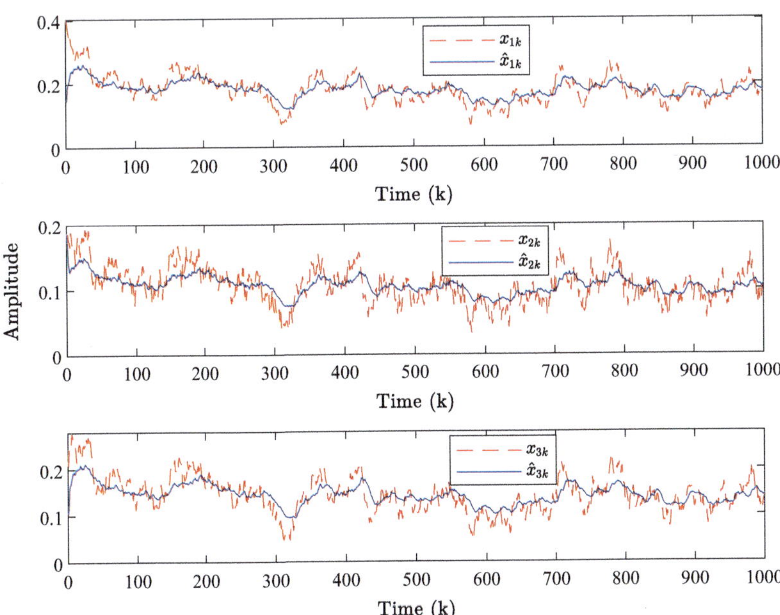

FIGURE 6.5

The actual state components x_{ik} $(i = 1, 2, 3)$ and their estimates $\hat{x}_{ik}$.

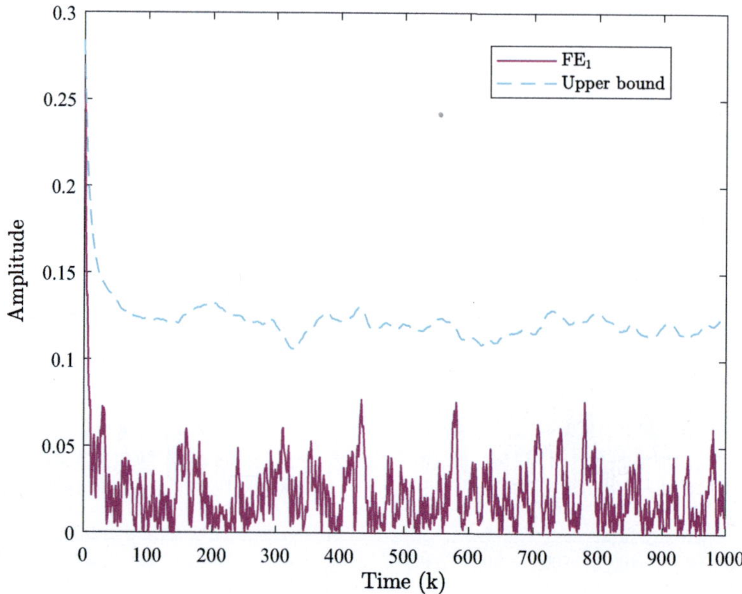

FIGURE 6.6

The FE of x_{1k} and its upper bound.

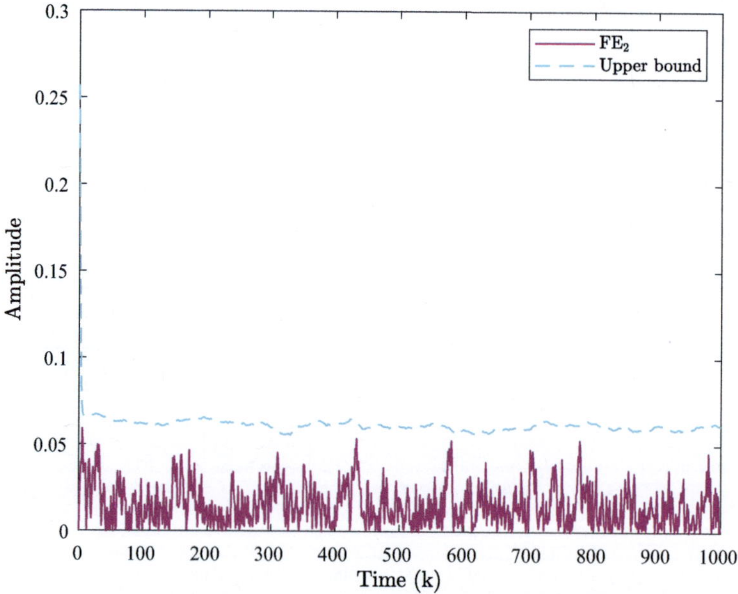

FIGURE 6.7

The FE of x_{2k} and its upper bound.

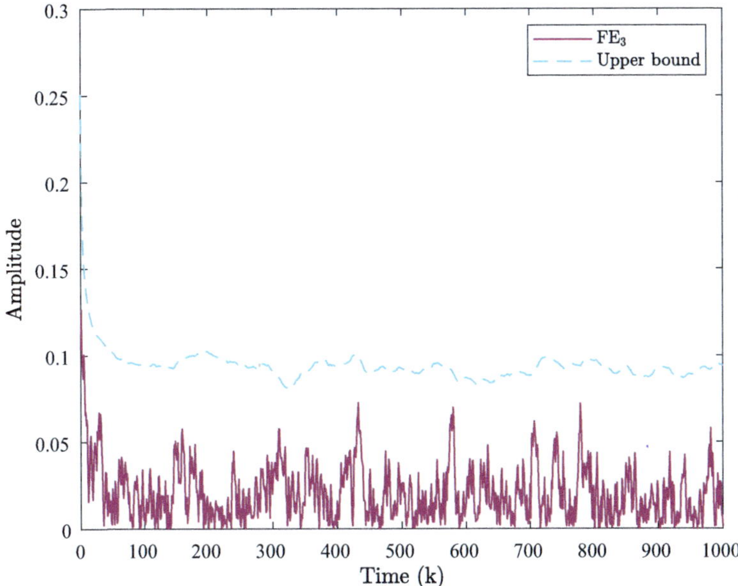

FIGURE 6.8

The FE of x_{3k} and its upper bound.

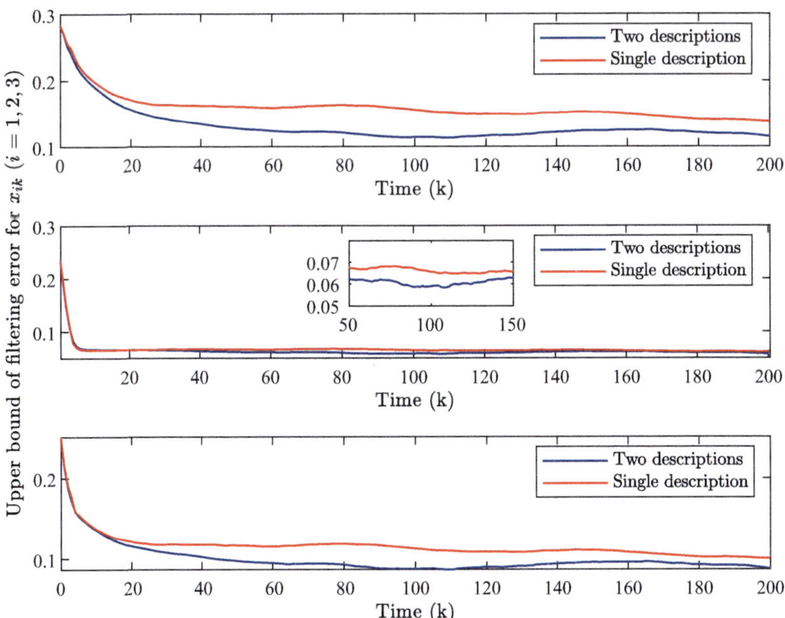

FIGURE 6.9

The upper bounds of filtering errors of x_{ik} ($i = 1, 2, 3$) with single description and two descriptions.

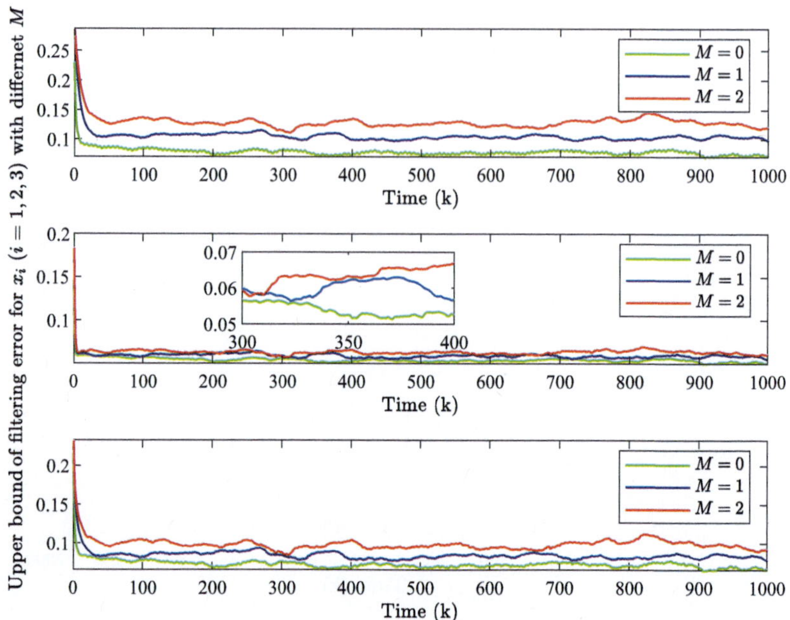

FIGURE 6.10

The effect of fading channel parameter M on filtering performance.

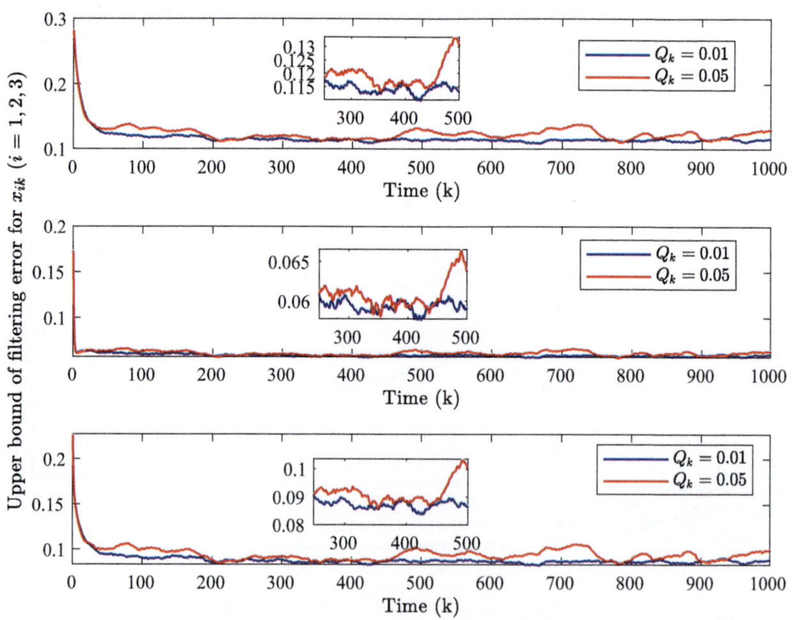

FIGURE 6.11

The filtering errors of x_{ik} $(i = 1, 2, 3)$ and their upper bounds with different noise intensities.

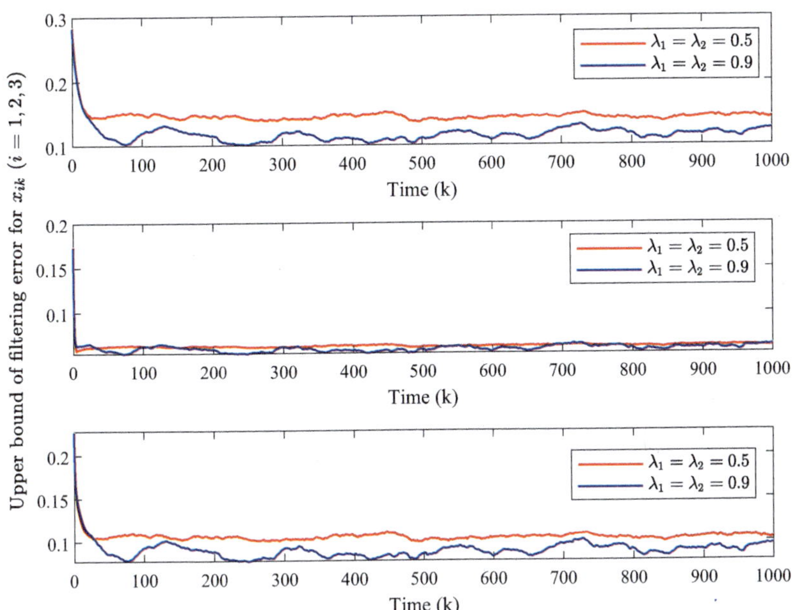

FIGURE 6.12

The filtering errors of x_{ik} ($i = 1, 2, 3$) and their upper bounds with different packet-arrival probabilities.

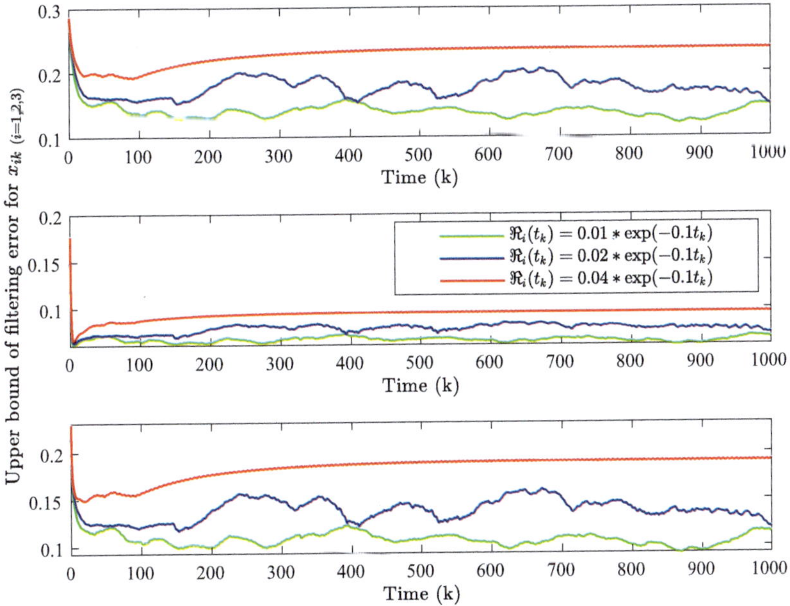

FIGURE 6.13

The effect of the model uncertainty on the filtering performance.

MDC-based filtering scheme performs very well and the considered system complexities do have a major effect on the filtering performance.

In order to examine the sensitivity of the developed filter to the model uncertainty, a simulation test has been conducted. In this case, the terms $h_i(t_k)$ ($i = 1, 2, 3$) in (6.61) have been replaced by $(1 + \Re_i(t_k))h_i(t_k)$ where $\Re_i(t_k)$ ($i = 1, 2, 3$) represent the parameter uncertainties and are taken as $\Re_i(t_k) = \bar{\Re}_i \exp(-0.1t_k)$ with $\bar{\Re}_i$ being positive scalars. It is easy to see that $\Re_i$ satisfies $|\Re_i(t_k)| \leq \bar{\Re}_i$. It is observed from Figure 6.13 that as the degree of the model uncertainties increases (i.e. the increase of $\bar{\Re}_i$), the upper bound of the filtering error variance grows accordingly, which indicates that the proposed filter is sensitive to the model uncertainties. One of our future investigations would be focused on the improvement of the robustness for the developed filtering scheme.

6.5 Summary

In this chapter, a multiple description coding scheme has been employed to deal with the recursive filtering issue for a class of discrete-time nonlinear stochastic networked systems. The *M*th-order Rice fading model has been adopted to describe the measurement fading from the sensor to the encoder. By utilizing the distorted data resulting from the channel fading, coding process and measurement noises, a recursive sub-optimal filter has been designed to estimate the actual system state. In addition, a minimized upper bound of filtering error variance has been derived and the corresponding filter parameter has been explicitly determined by solving the coupled algebraic Riccati-like difference equations. Finally, a simulation example has been given to demonstrate the effectiveness of the proposed filtering strategy.

7

Stabilization of Linear Discrete-Time Systems over Resource-Constrained Networks under Dynamical Multiple Description Coding Scheme

In networked control systems (NCSs), the communication among components is realized via a shared network, thereby merits decreased system wiring, easy installation and low maintenance cost. In NCS-related research, the scarcity of communication resources is one of the main concerns that lead to imperfect network communication and subsequently the degradation of the overall system performance. In fact, NCSs have been acting as a research frontier for almost two decades and a large amount of literature has been available on the control/filtering problems in networked environments, see for example [6,11,187].

Network-induced phenomena (NIP), as a consequence of limited communication capacity of NCSs, are likely to occur during data transmission and pose a major impact on the system stability. An effective way of mitigating the occurrence frequency of the undesirable NIPs is to improve the communication quality of the shared channel by deploying the transmission scheduling protocols/mechanisms. Up to now, much research effort has been devoted to the analysis and/or design problems for NCSs subject to scheduling protocols with some pioneering work appearing in the literature. Roughly speaking, the aforementioned protocols/mechanisms adopt the following two ways to improve the efficiency of utilizing network resources: 1) reducing communication frequency (e.g. static, dynamic or adaptive event-triggering mechanisms) and 2) reducing the communication traffic (e.g. Round-Robin, random access and Try-Once-Discard protocols). These protocols/mechanisms are, however, based on an implicit assumption that the signals scheduled by the communication protocols are *successfully* transmitted without any distortions, and this assumption is often unrealistic in some engineering practices. To ensure distortion-free transmissions, a huge communication resource would be occupied (which is especially true for high precision/amplitude signals), and such a resource overuse

DOI: 10.1201/9781003534853-7

is simply unaffordable in a typically bit-rate-constrained communication environment.

A recently popular strategy for data transmission is the data coding scheme that aims at converting the raw data into specific codewords with less bit occupancy according to a specific rule. It should be noted that most of the existing coding schemes are only valid on the premise of perfect codeword transmissions, that is, *no* packet loss occurs during the codeword transmission. This premise is, unfortunately, sometimes incorrect in high-throughput signal transmissions due primarily to the unreliable network circumstance. As such, it appears both theoretically significant and practically important to further improve existing coding schemes with the aim to cope with the unreliability of the codeword transmission.

With the aspiration to handle the unreliability induced by the packet dropout phenomenon during data transmission, the so-called *multiple description coding* (MDC) scheme has been developed which comprises the following three steps: 1) the encoder maps the raw data into several descriptions with fewer bit occupancies and identical importance; 2) the descriptions are transmitted to the decoder side through individual channels; and 3) the decoder restores the decoded data in terms of the received descriptions. Therefore, the primary motivation of current chapter is to generalize the MDC strategy to examine the trade-off between communication capacity and system stability.

Following the discussions made thus far, we conclude that there appears to be a lack of systematic investigation on the stabilization problem for a class of linear discrete-time NCSs subject to the randomly occurring packet dropouts and bit-rate constraints. As such, the main objective of this paper is to bridge such a gap by developing an MDC strategy capable of alleviating the negative impact from randomly occurring packet dropouts and limited network resources on system stability. This task is fairly demanding as we are confronted with the following technical challenges: 1) how to construct an appropriate coding-decoding framework to ensure that the decoding error converges to 0 asymptotically? 2) how to develop a control scheme capable of effectively handling the distorted signal while achieving the desired stability condition? and 3) how to quantitatively analyze the influences from the packet-dropout probabilities (of the channels and the encoder parameters) on the system stability?

7.1 Problem Formulation

Consider the following linear discrete-time system:

$$x(k+1) = Ax(k) + Bu(k) \tag{7.1}$$

where $x(k) \in \mathbb{R}^n$ is the system state, $u(k) \in \mathbb{R}^r$ is the control input signal, and A and B are known matrices with appropriate dimensions.

In an NCS framework, the system states are transmitted to the controller via a resource-constrained communication media. Accordingly, the MDC scheme is introduced with a view to enhancing the reliability of signal transmission. Without loss of generality, we consider the case of two descriptions in this paper. To be more specific, the data is first encoded into two descriptions in terms of certain coding criterion, and the obtained descriptions are subsequently transmitted to the decoding devices via the two independent channels (labeled as "C1" and "C2"), respectively. In addition, the component-based MDC is adopted in this paper in order to meet the requirement of encoding each component $x_i(k)$ $(i \in \mathfrak{N} \triangleq \{1, 2, \ldots, n\})$ of the system state $x(k)$.

For the ith state component, a scalar-valued encoder is constructed as follows:

$$
\begin{cases}
\varepsilon_i(\tau h) = f_{1i}\big(x_i(\tau h)\big) \\
\sigma_i(\tau h) = f_{2i}\big(x_i(\tau h)\big)
\end{cases}
\tag{7.2}
$$

for $i \in \mathfrak{N}$, where $f_{1i}(\cdot) : \mathbb{R} \mapsto \mathbb{R}$ and $f_{2i}(\cdot) : \mathbb{R} \mapsto \mathbb{R}$ are two coding functions, $\varepsilon_i(\tau h) \in \mathbb{R}$ and $\sigma_i(\tau h) \in \mathbb{R}$ are two scalar-valued descriptions of $x_i(\tau h)$ $(\tau = 0, 1, 2, \ldots)$, and $h \in \mathbb{N}^+$ is the coding period.

The acquired descriptions $\varepsilon_i(\tau h)$ and $\sigma_i(\tau h)$ are, respectively, packed as two data packets $\varepsilon(\tau h)$ and $\sigma(\tau h)$ with

$$
\varepsilon(\tau h) \triangleq \begin{bmatrix} \varepsilon_1(\tau h) & \varepsilon_2(\tau h) & \cdots & \varepsilon_n(\tau h) \end{bmatrix}^T
$$
$$
\sigma(\tau h) \triangleq \begin{bmatrix} \sigma_1(\tau h) & \sigma_2(\tau h) & \cdots & \sigma_n(\tau h) \end{bmatrix}^T,
$$

which are correspondingly transmitted to the decoder via channels "C1" and "C2".

Due to the inevitable unreliability of the communication media, the packet dropout phenomenon is frequently encountered during the packet transmission. In this situation, for $i \in \mathfrak{N}$, the corresponding scalar-valued decoder is characterized as follows:

$$
\hat{x}_i(\tau h) = \begin{cases}
g_{1i}\big(\varepsilon_i(\tau h)\big), & \text{if } \gamma_1(\tau h) = 1,\ \gamma_2(\tau h) = 0 \\
g_{2i}\big(\sigma_i(\tau h)\big), & \text{if } \gamma_1(\tau h) = 0,\ \gamma_2(\tau h) = 1 \\
g_{ci}\big(\varepsilon_i(\tau h), \sigma_i(\tau h)\big), & \text{if } \gamma_1(\tau h) = \gamma_2(\tau h) = 1
\end{cases}
\tag{7.3}
$$

where $g_{1i}(\cdot)$ and $g_{2i}(\cdot)$ are two side decoding functions, and $g_{ci}(\cdot, \cdot)$ is the central decoding function. $\hat{x}_i(\tau h)$ is the decoded value of $x_i(\tau h)$. The random variables $\gamma_i(\tau h)$ $(i = 1, 2)$ are two Bernoulli-distributed sequences with the following probability distributions:

$$\text{Prob}\{\gamma_i(\tau h) = 1\} = \bar{\gamma}_i, \ \text{Prob}\{\gamma_i(\tau h) = 0\} = 1 - \bar{\gamma}_i \tag{7.4}$$

where $\bar{\gamma}_i \in [0,1]$ ($i = 1,2$) are given constants.

For presentation clarity, we set

$$\hat{x}(\cdot) \triangleq \begin{bmatrix} \hat{x}_1(\cdot) & \hat{x}_2(\cdot) & \cdots & \hat{x}_n(\cdot) \end{bmatrix}^T$$

$$f_1(x(\cdot)) \triangleq \begin{bmatrix} f_{11}(x_1(\cdot)) & f_{12}(x_2(\cdot)) & \cdots & f_{1n}(x_n(\cdot)) \end{bmatrix}^T$$

$$f_2(x(\cdot)) \triangleq \begin{bmatrix} f_{21}(x_1(\cdot)) & f_{22}(x_2(\cdot)) & \cdots & f_{2n}(x_n(\cdot)) \end{bmatrix}^T$$

$$g_1(\varepsilon(\cdot)) \triangleq \begin{bmatrix} g_{11}(\varepsilon_1(\cdot)) & g_{12}(\varepsilon_2(\cdot)) & \cdots & g_{1n}(\varepsilon_n(\cdot)) \end{bmatrix}^T \tag{7.5}$$

$$g_2(\sigma(\cdot)) \triangleq \begin{bmatrix} g_{21}(\sigma_1(\cdot)) & g_{22}(\sigma_2(\cdot)) & \cdots & g_{2n}(\sigma_n(\cdot)) \end{bmatrix}^T$$

$$g_c(\varepsilon(\cdot),\sigma(\cdot)) \triangleq \begin{bmatrix} g_{c1}(\varepsilon_1(\cdot),\sigma_1(\cdot)) & \cdots & g_{cn}(\varepsilon_n(\cdot),\sigma_n(\cdot)) \end{bmatrix}^T.$$

The scalar-valued encoder-decoder pair (7.2)–(7.3) can be compacted into the following form:

$$\begin{cases} \varepsilon(\tau h) = f_1(x(\tau h)) \\ \sigma(\tau h) = f_2(x(\tau h)) \end{cases} \tag{7.6}$$

and

$$\hat{x}(\tau h) = \begin{cases} g_1(\varepsilon(\tau h)), & \text{if } \gamma_1(\tau h) = 1, \ \gamma_2(\tau h) = 0 \\ g_2(\sigma(\tau h)), & \text{if } \gamma_1(\tau h) = 0, \ \gamma_2(\tau h) = 1 \\ g_c(\varepsilon(\tau h),\sigma(\tau h)), & \text{if } \gamma_1(\tau h) = \gamma_2(\tau h) = 1 \\ g(\hat{x}(\tau h - 1)), & \text{if } \gamma_1(\tau h) = \gamma_2(\tau h) = 0 \end{cases} \tag{7.7}$$

where $\hat{x}(\tau h)$ is the vector-valued decoded data and $g(\cdot)$ is a function of $\hat{x}(\tau h - 1)$ which will be given later.

Remark 7.1: *It is observed from (7.4) that the random variables $\gamma_i(\tau h)$ ($i = 1,2$) are employed to regulate the randomly occurring packet-dropout phenomenon in the communication channels "C1" and "C2". In accordance with the MDC scheme (7.6)–(7.7), at each coding time step τh, the raw signal $x(\tau h)$ is encoded into two descriptions $\varepsilon(\tau h)$ and $\sigma(\tau h)$ by the coding functions $f_1(\cdot)$ and $f_2(\cdot)$, respectively. Then, $\varepsilon(\tau h)$ and $\sigma(\tau h)$ are individually transmitted to the decoder side via communication channels "C1" and "C2". As shown in Figure 7.1, if no packet-dropout occurs, that is, $\gamma_1(\tau h) = \gamma_2(\tau h) = 1$, which means that the description packets $\varepsilon(\tau h)$ and $\sigma(\tau h)$ are successfully transmitted and the central decoder (labeled as "CD") is triggered to generate the decoded value. If the packet-dropout occurs in channel "C1" only (i.e. $\gamma_1(\tau h) = 0$, $\gamma_2(\tau h) = 1$), the description packet $\sigma(\tau h)$ is available and the side decoder 2 (labeled as "SD2") is activated to execute the decoding procedure. If the packet-dropout occurs in channel "C2" only (i.e. $\gamma_1(\tau h) = 1$,*

1	3							
2	4	5						
	6	7	9					
		8	10	11				
			12	13	15			
				14	16	17		
					18	19	21	
						20	22	

FIGURE 7.1
The structure of the closed-loop control system with the MDC scheme.

$\gamma_2(\tau h) = 0$), *the description packet $\varepsilon(\tau h)$ is available and the side decoder 1 (labeled as "SD1") is enabled to perform the decoding operation. Furthermore, if the packet-dropout occurs in both channels "C1" and "C2" (i.e. $\gamma_1(\tau h) = \gamma_2(\tau h) = 0$), none of the description packets is available and all the decoders fail to work. In this case, without loss of generality, we assume that the decoded value $\hat{x}(\tau h - 1)$ is used to generate the control signal.*

In this paper, the remote control scenario is considered that the encoder and controller are deployed at different places. The schematic structure is shown in Figure 7.1, where the network-based communication scheme is employed and descriptions generated by two encoders are transmitted to the remote controller via bandwidth-constrained network channels. The main purpose of this paper is to investigate the remote control issue under the dynamical MDC scheme. In particular, we aim to

- design the coding and decoding procedures based on the MDC scheme;
- guarantee the convergence of the decoding error in the statistical sense;
- provide sufficient conditions on the stochastic stability of the closed-loop system; and
- reveal the influences of the major factors (e.g. system parameters, encoding-decoding scheme and the probabilities of the channel packet dropouts) on the stochastic stability of the closed-loop system.

7.2 Main Results

Coding Scheme
In this subsection, we are devoted to formalizing the coding procedure based on the MDC scheme. The coding procedure developed in this paper includes

two steps, namely, the index generation step and the index assignment step. For the step of index generation, the uniform quantization method is employed to generate the indices and, for the step of index assignment, the *nested* index assignment principle [156] is adopted to assign the generated indices to a certain mapping matrix.

In the sequel, we slightly abuse the notation by using x, ε, σ, ε_i, σ_i and x_i to denote $x(\tau h)$, $\varepsilon(\tau h)$, $\sigma(\tau h)$, $\varepsilon_i(\tau h)$, $\sigma_i(\tau h)$ and $x_i(\tau h)$, respectively.

Index Generation Scheme

For a scalar quantizer $\mathcal{Q}(\cdot) : \mathbb{R} \mapsto \mathbb{R}$, the scaling parameter s and the positive integer N correspond to the quantization range and the quantization level, respectively.

Let us divide the hyperrectangles $\mathcal{H}_s = \{x \in \mathbb{R}^n : |x|_\infty \leq s\}$ into N^n parts $H_{1,p_1}(s) \times H_{2,p_2}(s) \times \cdots \times H_{n,p_n}(s)$ with equal volume, where $p_1, p_2, \ldots, p_n \in \{1, 2, \ldots, N\}$ and

$$H_{i,1}(s) \triangleq \left\{ x_i \middle| -s \leq x_i < -s + \frac{2s}{N} \right\}$$

$$H_{i,2}(s) \triangleq \left\{ x_i \middle| -s + \frac{2s}{N} \leq x_i < -s + \frac{4s}{N} \right\}$$

$$\vdots$$

$$H_{i,N}(s) \triangleq \left\{ x_i \middle| s - \frac{2s}{N} \leq x_i \leq s \right\}$$

(7.8)

with $i = 1, 2, \ldots, n$. Based on the above partition, we can see that, for an arbitrary $x \in \mathcal{H}_s$, there exists a sequence of positive scalars $p_1, p_2, \ldots, p_n$ such that x belongs to the sub-hyperrectangle $H_{1,p_1}(s) \times H_{2,p_2}(s) \times \cdots \times H_{n,p_n}(s)$. Thus, a vector-valued index $p \triangleq [p_1 \quad p_2 \quad \cdots \quad p_n]^T$ is exclusively to determine the sub-hyperrectangle $H_{1,p_1}(s) \times H_{2,p_2}(s) \times \cdots \times H_{n,p_n}(s)$.

Having established the relationship between the sub-hyperrectangle $H_{1,p_1}(s) \times H_{2,p_2}(s) \times \cdots \times H_{n,p_n}(s)$ and the vector p, we now construct the index generator functions $\phi_i(\cdot)$ $(i = 1, 2, \ldots, n) : \mathbb{R} \mapsto \{1, 2, \ldots, N\}$ as follows:

$$\phi_i(x_i) = p_i \tag{7.9}$$

for $x \in H_{1,p_1}(s) \times H_{2,p_2}(s) \times \cdots \times H_{n,p_n}(s)$.

In light of (7.8), it is easy to verify that the center of the sub-hyperrectangle $H_{1,p_1}(s) \times H_{2,p_2}(s) \times \cdots \times H_{n,p_n}(s)$ is

$$\xi(p, s) \triangleq \left[-s + \frac{(2p_1-1)s}{N} \quad -s + \frac{(2p_2-1)s}{N} \quad \cdots \quad -s + \frac{(2p_n-1)s}{N} \right]^T. \tag{7.10}$$

Noticing the fact that $x \in H_{1,p_1}(s) \times H_{2,p_2}(s) \times \cdots \times H_{n,p_n}(s)$, one directly obtains the following relationship:

$$|x - \xi(p,s)|_\infty \leq \frac{s}{N}. \tag{7.11}$$

Index Assignment Principle

Inspired by the *nested* assignment principle, we shall assign the generated index p_i associated with the component x_i into the certain cell of the matrix $\mathcal{M}_i$. Without loss of generality, the mapping matrix $\mathcal{M}_i$ is assumed to be an l-dimensional matrix with l being an even integer. Subsequently, in terms of the locations of the row and the column of the cell containing p_i, a pair of descriptions ε_i and σ_i is thus obtained, where ε_i denotes the row location of p_i and σ_i stands for its column location.

The index assignment function is given as follows:

$$\Phi_i(p_i) = (\varphi_{1i}(p_i), \ \varphi_{2i}(p_i)) \triangleq (\varepsilon_i, \sigma_i) \tag{7.12}$$

where $\varphi_{1i}(\cdot)$ and $\varphi_{2i}(\cdot)$ are the row assignment function and the column assignment function, respectively, and ε_i and σ_i are two descriptions of x_i.

Remark 7.2: *As stated in [156], the nested assignment principle stipulates that the indices are placed in the cells that lie on the main diagonal and its nearest 2d diagonals of the mapping matrix. For the convenience of the exposition, the scenario $d = 1$ is discussed in this paper, which means that the indices are placed in the cells lying on the main diagonal and its nearest 2 diagonals of the mapping matrix. Figure 7.2 gives an example of the nested assignment principle with $l = 8$ and $d = 1$.*

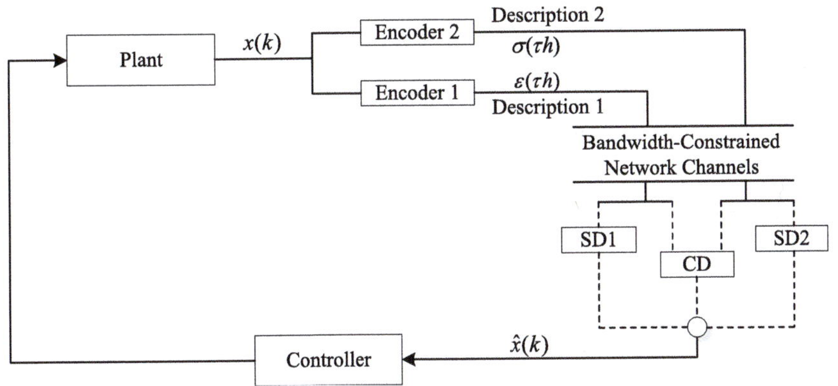

FIGURE 7.2
Nested index assignment for $l = 8$, $d = 1$.

The specific form of the index assignment function $\Phi_i(\cdot) : \mathbb{N}^+ \mapsto \mathbb{N}^+ \times \mathbb{N}^+$ is proposed as follows:

$$\Phi_i(p_i) \triangleq (\varepsilon_i, \sigma_i)$$

$$= \begin{cases} (r_i + 1, r_i + 1), & \text{if } q_i = 1 \\ (r_i + 1, r_i), & \text{if } q_i = 0 \text{ and } r_i \text{ is even} \\ (r_i, r_i + 1), & \text{if } q_i = 0 \text{ and } r_i \text{ is odd} \\ (r_i + 2, r_i + 1), & \text{if } q_i = 2 \text{ and } r_i \text{ is even} \\ (r_i + 1, r_i + 2), & \text{if } q_i = 2 \text{ and } r_i \text{ is odd} \end{cases} \tag{7.13}$$

where $r_i = \lfloor \frac{p_i}{2d+1} \rfloor$ and $q_i = \langle \frac{p_i}{2d+1} \rangle$. As such, it is not difficult to verify that all the indices p_i are mapped to the corresponding description pairs $(\varepsilon_i, \sigma_i)$.

7.2.1 Decoding Scheme

Due to the possible dropout of the description packets (i.e. ε and σ) during the data transmission, it is difficult to obtain an accurate value of the vector-valued index p. In view of this, an index estimation strategy, which is dependent on the reception of the descriptions, is first put forward to estimate the index of the encoded signal.

Index Estimation Strategy

The index p_i, which carries essential information of x_i, plays an important role in the design of the decoding procedure. Moreover, it is inferred from the index assignment function (7.13) that an index p_i exclusively determines by two descriptions ε_i and σ_i, and vice versa. Unfortunately, the exact value of index p_i cannot be always accessible at the decoder side due to the probabilistic packet dropout during the transmission of the descriptions. In this sense, it is necessary to develop an index estimation strategy with the hope to obtain an estimation of the index p_i in terms of the reception of the descriptions.

Case I: both the description packets ε and σ are successfully transmitted to the central decoder. In this case, the index estimation function $\psi(\cdot, \cdot) : \mathbb{Z}^+ \times \mathbb{Z}^+ \mapsto \mathbb{Z}^+$ for each index p_i $(i = 1, 2, \ldots, n)$ is given as

$$\psi(\varepsilon_i, \sigma_i) = \begin{cases} 3\varepsilon_i - 2, & \text{if } \varepsilon_i = \sigma_i \\ 3\varepsilon_i - 3, & \text{if } \varepsilon_i = \sigma_i + 1 \text{ and } \varepsilon_i \text{ is odd} \\ 3\varepsilon_i, & \text{if } \varepsilon_i = \sigma_i - 1 \text{ and } \varepsilon_i \text{ is odd} \\ 3\varepsilon_i - 4, & \text{if } \varepsilon_i = \sigma_i + 1 \text{ and } \varepsilon_i \text{ is even} \\ 3\varepsilon_i - 1, & \text{if } \varepsilon_i = \sigma_i - 1 \text{ and } \varepsilon_i \text{ is even} \end{cases} \tag{7.14}$$

and thus, the value of $\psi(\varepsilon_i, \sigma_i)$ (denoted as $\hat{p}_{ci}$) is the estimation of the index p_i. It is obvious that the estimation accuracy is 100%, namely, $\hat{p}_{ci} = p_i$.

Case II: only one of the description packet (either ε or σ) is received by the side decoder. In this situation, the index estimation function $\psi(\cdot) : \mathbb{Z}^+ \mapsto \mathbb{Z}^+$ for the index component p_i is defined by

$$\psi(z) = 3z - 2, \forall z \in \{\varepsilon_i, \sigma_i\}, \tag{7.15}$$

and the value of $\psi(z)$ is denoted as $\hat{p}_{si}$.

Case III: neither description packet is transmitted to the decoder. In this scenario, the estimation for the index p_i is set as 0.

Bearing $p_i = 3r_i + q_i$ in mind, it follows from (7.13)–(7.15) that the index estimation errors $\tilde{p}_{ci} \triangleq p_i - \hat{p}_{ci}$ and $\tilde{p}_{si} \triangleq p_i - \hat{p}_{si}$ satisfy $\tilde{p}_{ci} = 0$ for **Case I** and $|\tilde{p}_{si}| \leq 2$ for **Case II**, respectively.

7.2.2 Design of the Encoder-Decoder Pair

After discussing the generation, assignment and estimation of the index p_i ($i = 1, 2, \ldots, n$), we are in a position to specifically design the coding-decoding procedure via a dynamical quantization approach.

Before presenting the encoder-decoder structure, let us introduce a random sequence $\theta(\tau h) \triangleq \gamma_1(\tau h) + \gamma_2(\tau h)$ satisfying the following probability distribution:

$$\begin{aligned}
\text{Prob}\{\theta(\tau h) = 0\} &= \bar{\theta}_0 \\
\text{Prob}\{\theta(\tau h) = 1\} &= \bar{\theta}_1 \\
\text{Prob}\{\theta(\tau h) = 2\} &= \bar{\theta}_2
\end{aligned} \tag{7.16}$$

where

$$\begin{aligned}
\bar{\theta}_0 &\triangleq (1 - \bar{\gamma}_1)(1 - \bar{\gamma}_2) \\
\bar{\theta}_1 &\triangleq \bar{\gamma}_1(1 - \bar{\gamma}_2) + (1 - \bar{\gamma}_1)\bar{\gamma}_2 \\
\bar{\theta}_2 &\triangleq 1 - \bar{\theta}_0 - \bar{\theta}_1.
\end{aligned}$$

Encoder: at each coding instant τh ($\tau = 0, 1, 2, \ldots$), for the system state $x(\tau h)$, the coding function is designed as follows:

$$\begin{cases}
\varepsilon_i(\tau h) \triangleq \varphi_{1i}(\phi_i(\tilde{x}_i(\tau h))) \\
\sigma_i(\tau h) \triangleq \varphi_{2i}(\phi_i(\tilde{x}_i(\tau h))), \quad i = 1, 2, \ldots n
\end{cases} \tag{7.17}$$

where $\tilde{x}_i(\tau h)$ is the ith component of $\tilde{x}(\tau h)$ and $\tilde{x}(\tau h) \triangleq x(\tau h) - \bar{x}(\tau h) \in H_{1,p_1}(s(\tau h)) \times H_{2,p_2}(s(\tau h)) \times \cdots \times H_{n,p_n}(s(\tau h))$. Moreover, the dynamics of $\bar{x}(k)$ satisfies

$$\begin{cases} \bar{x}(k) = A\check{x}(k-1) + B\check{u}(k-1), \; \forall\, k > 0 \\ \bar{x}(0) = 0 \end{cases} \tag{7.18}$$

and the dynamics of $\check{x}(k)$ is governed by the following auxiliary system:

$$\begin{cases} \check{x}(k+1) = A\check{x}(k) + B\check{u}(k), \; k \neq \tau h - 1 \\ \check{x}(\tau h) = A\check{x}(\tau h - 1) + B\check{u}(\tau h - 1) \\ \qquad\qquad + \xi(p(\tau h), s(\tau h)) \\ \check{u}(k) = K\check{x}(k) \\ \check{x}(0) = 0 \end{cases} \tag{7.19}$$

where $\xi(\cdot,\cdot)$ is defined in (7.10) and K is the controller gain matrix to be designed later.

It follows from (7.2), (7.5) and (7.17) that

$$f_{1i}(x_i(\tau h)) = \varphi_{1i}(\phi_i(\tilde{x}_i(\tau h))), \; f_1(x(\tau h)) = \varphi_1(\phi(\tilde{x}(\tau h)))$$
$$f_{2i}(x(\tau h)) = \varphi_{2i}(\phi_i(\tilde{x}_i(\tau h))), \; f_2(x(\tau h)) = \varphi_2(\phi(\tilde{x}(\tau h)))$$

where

$$\varphi_1(\phi(\tilde{x}(\tau h))) \triangleq [\varphi_{11}(\phi_1(\tilde{x}_1(\tau h))) \;\cdots\; \varphi_{1n}(\phi_n(\tilde{x}_n(\tau h)))]^T$$
$$\varphi_2(\phi(\tilde{x}(\tau h))) \triangleq [\varphi_{21}(\phi_1(\tilde{x}_1(\tau h))) \;\cdots\; \varphi_{2n}(\phi_n(\tilde{x}_n(\tau h)))]^T.$$

According to the reception of the description packets $\varepsilon(\tau h)$ and $\sigma(\tau h)$ at the decoder side, the estimation model of the index vector $p(\tau h)$ is described by

$$\hat{p}(\tau h) = \delta\{\theta(\tau h), 2\}\hat{p}_c(\tau h) + \delta\{\theta(\tau h), 1\}\hat{p}_s(\tau h) \tag{7.20}$$

where

$$\hat{p}_c(\tau h) \triangleq [\hat{p}_{c1}(\tau h) \quad \hat{p}_{c2}(\tau h) \quad \cdots \quad \hat{p}_{cn}(\tau h)]^T$$
$$\hat{p}_s(\tau h) \triangleq [\hat{p}_{s1}(\tau h) \quad \hat{p}_{s2}(\tau h) \quad \cdots \quad \hat{p}_{sn}(\tau h)]^T$$

and $\delta\{\cdot,\cdot\}$ is the Kronecker delta function defined as

$$\delta\{s, t\} = \begin{cases} 1, s = t \\ 0, s \neq t, \end{cases}$$

Remark 7.3: *It is observed from (7.16) and (7.20) that: 1) $\theta(\tau h) = 2$ indicates the successful transmissions of both $\varepsilon(\tau h)$ and $\sigma(\tau h)$, under which the estimated index $\hat{p}_c(\tau h)$ is obtained; 2) $\theta(\tau h) = 1$ implies that one of the packet (either $\varepsilon(\tau h)$ or*

$\sigma(\tau h))$ *is lost during the transmission, and hence the index* $\hat{p}_s(\tau h)$ *is estimated; and* 3) $\theta(\tau h) = 0$ *means the simultaneous transmission failures of both* $\varepsilon(\tau h)$ *and* $\sigma(\tau h)$, *under which the estimation of the index vector* $p(\tau h)$ *is set as 0, namely,* $\hat{p}(\tau h) = 0$.

In terms of the estimation index $\hat{p}(\tau h)$ obtained at the decoder side, one derives the center of the sub-hyperrectangle $H_{1,p_1(\tau h)}(s(\tau h)) \times H_{2,p_2(\tau h)}(s(\tau h)) \times \cdots \times H_{n,p_n(\tau h)}(s(\tau h))$ as $\xi(\hat{p}(\tau h), s(\tau h))$. Moreover, recalling the definitions of $\xi(\cdot, \cdot)$ in (7.10) and $\hat{p}(\tau h)$ in (7.20), $\xi(\hat{p}(\tau h), s(\tau h))$ can be further rewritten as

$$\begin{aligned}
\xi(\hat{p}(\tau h), s(\tau h)) = {} & \delta\{\theta(\tau h), 2\}\xi(\hat{p}_c(\tau h), s(\tau h)) \\
& + \delta\{\theta(\tau h), 1\}\xi(\hat{p}_s(\tau h), s(\tau h)),
\end{aligned} \tag{7.21}$$

which infers that $\xi(\hat{p}(\tau h), s(\tau h))$ is set to be 0 when both of the vector-valued descriptions (i.e. $\varepsilon(\tau h)$ and $\sigma(\tau h)$) are lost (i.e. $\theta(\tau h) = 0$).

It should be emphasized that the model (7.21) reflects all the cases of the reception of $\varepsilon(\tau h)$ and $\sigma(\tau h)$ at the decoder side, based on which we intend to tackle with the decoder design issue.

Decoder: at the decoder side, the decoding function is designed with the following form:

$$\begin{cases}
\hat{x}(k+1) = A\hat{x}(k) + Bu(k), \ k \neq \tau h - 1 \\
\hat{x}(\tau h) = A\hat{x}(\tau h - 1) + Bu(\tau h - 1) \\
\qquad\quad + \xi(\hat{p}(\tau h), s(\tau h)) \\
u(k) = K\hat{x}(k) \\
\hat{x}(0) = 0
\end{cases} \tag{7.22}$$

where $u(k)$ is the decoded-value-based control input of the plant (7.1).

Denoting the decoding error as $e(k) \triangleq x(k) - \hat{x}(k)$, the closed-loop system is obtained as follows:

$$x(k+1) = \mathcal{A}x(k) - BKe(k) \tag{7.23}$$

where $\mathcal{A} \triangleq A + BK$.

Definition 7.1: *The discrete-time closed-loop system (7.23) is said to be stochastically stable if, for any initial condition* $x(0) \in \mathbb{R}^n$, *the system dynamics satisfies*

$$\lim_{k \to \infty} \mathbb{E}\{\|x(k)\|\} = 0, \ k = 1, 2, \dots. \tag{7.24}$$

Next, we shall pay our attention to developing a new dynamical multiple description coding scheme such that the closed-loop system (7.23) is stochastically stable under the decoded-value-based control input $\hat{u}(k)$.

For convenience, some notations are defined as follows:

$$\vec{x}_1(k) \triangleq x(k) - \check{x}(k), \quad \vec{x}_2(k) \triangleq \hat{x}(k) - \check{x}(k)$$

$$s_0 \triangleq N\|x(0)\|_\infty, \quad \mu_h \triangleq \|\mathcal{A}^h\|_\infty, \quad \bar{\mu}_h \triangleq \|A^h\|_\infty$$

$$\zeta_h \triangleq \left\| \sum_{i=1}^{h} A^{h-i} BK\mathcal{A}^{i-1} \right\|_\infty.$$

Before proceeding, we introduce the following lemmas which will be utilized in later discussions.

Lemma 7.1: *Let the coding procedure (7.17)–(7.19) be given. The following relationship*

$$\mathbb{E}\{\|x(\tau h) - \bar{x}(\tau h)\|_\infty\} \leq s(\tau h)$$

is guaranteed for all $\tau = 1, 2, \ldots$, where $s(\tau h)$ is governed by the iteration

$$s(\tau h) = \frac{\bar{\mu}_h}{N} s((\tau - 1)h) + \zeta_h \bar{b}((\tau - 1)h) \tag{7.25}$$

with $s(0) \triangleq s_0$ and $\bar{b}(\tau h) \triangleq \mathbb{E}\{\|\vec{x}_2(\tau h)\|_\infty\}$.

Proof: *The proof is conducted by mathematical induction.*
  For $\tau = 1$, one has

$$x(h) - \bar{x}(h) = A\vec{x}_1(h - 1) + BK\vec{x}_2(h - 1)$$

$$= A^2\vec{x}_1(h - 2) + ABK\vec{x}_2(h - 2) + BK\vec{x}_2(h - 1)$$

$$= A^h\vec{x}_1(0) + A^{h-1}BK\vec{x}_2(0) + A^{h-2}BK\vec{x}_2(1)$$

$$+ A^{h-3}BK\vec{x}_2(2) + \cdots + ABK\vec{x}_2(h - 2)$$

$$+ BK\vec{x}_2(h - 1).$$

Noting the fact that

$$\vec{x}_2(i) = \mathcal{A}^i\vec{x}_2(0), \quad (i = 1, 2, \ldots, h - 1),$$

we obtain

$$x(h) - \bar{x}(h) = A^h\vec{x}_1(0) + \sum_{i=1}^{h} A^{h-i} BK\mathcal{A}^{i-1}\vec{x}_2(0),$$

which gives

$$\mathbb{E}\{\|x(h) - \bar{x}(h)\|_\infty\} \le \frac{\bar{\mu}_h}{N}s(0) + \zeta_h\bar{b}(0).$$

Assuming that the inequality holds for $\tau = 2, 3, \cdots, j$, we explore the case of $\tau = j + 1$. Similar to the situation of $\tau = 1$, one finds

$$\begin{aligned}
x((j+1)h) - \bar{x}((j+1)h) &= A\vec{x}_1((j+1)h - 1) + BK\vec{x}_2((j+1)h - 1) \\
&= A^2\vec{x}_1((j+1)h - 2) + ABK\vec{x}_2((j+1)h - 2) \\
&\quad + BK\vec{x}_2((j+1)h - 1) \\
&= A^h\vec{x}_1(jh) + A^{h-1}BK\vec{x}_2(jh) + A^{h-2}BK\vec{x}_2(jh + 1) \\
&\quad + A^{h-3}BK\vec{x}_2(jh + 2) + \cdots + ABK\vec{x}_2((j+1)h - 2) \\
&\quad + BK\vec{x}_2((j+1)h - 1).
\end{aligned}$$

By using the fact $\vec{x}_2(i) = \mathcal{A}^{i-jh}\vec{x}_2(jh)$ for $i = jh + 1, jh + 2, \cdots, (j+1)h - 1$ again, one has

$$x((j+1)h) - \bar{x}((j+1)h) = A^h\vec{x}_1(jh) + \sum_{i=1}^{h} A^{h-i}BK\mathcal{A}^{i-1}\vec{x}_2(jh),$$

which implies

$$\mathbb{E}\{\|x((j+1)h) - \bar{x}((j+1)h)\|_\infty\} \le \bar{\mu}_h\mathbb{E}\{\|\vec{x}_1(jh)\|_\infty\} + \zeta_h\bar{b}(jh).$$

In addition, it is inferred from (7.11) that

$$\begin{aligned}
\mathbb{E}\{\|\vec{x}_1(jh)\|_\infty\} &= \mathbb{E}\{\|x(jh) - \bar{x}(jh) - \xi(p(jh), s(jh))\|_\infty\} \\
&\le \frac{1}{N}s(jh).
\end{aligned}$$

Consequently, we have

$$\begin{aligned}
\mathbb{E}\{\|x((j+1)h) - \bar{x}((j+1)h)\|_\infty\} &\le \frac{\bar{\mu}_n}{N}s(jh) + \zeta_h\bar{b}(jh) \\
&= s((j+1)h),
\end{aligned}$$

which completes the proof.

It should be pointed out that since the iteration (7.25) with the given initial condition and system parameters A, B, K, h and N are known a prior to both the encoder and the decoder, $s(\tau h)$ can be calculated and kept synchronized at both the encoder side and the decoder side. This means that the synchronization of the quantization interval $s(\tau h)$ can be always guaranteed at the encoder side and the decoder side no matter whether the packet dropout occurs or not.

Lemma 7.2: *Let the coding procedure (7.17)–(7.19) and the decoding procedure (7.22) be given. The error between the auxiliary system state $\check{x}(k)$ and the decoded value $\hat{x}(k)$ satisfies*

$$\mathbb{E}\{\|\check{x}(\tau h) - \hat{x}(\tau h)\|_\infty\} \leq b(\tau h),$$

where $b(\tau h)$ is determined by the recursive rule

$$b(\tau h) = \mu_h b((\tau - 1)h) + vs(\tau h)$$

with $b(0) \triangleq 0$, $v \triangleq \frac{4}{N}\bar{\theta}_1 + \bar{\theta}_0$ and $\tau = 1, 2, \ldots$.

Proof: *The mathematical induction is applied again to the proof of this lemma. For $\tau = 1$, it follows from (7.19) and (7.22) that*

$$\begin{aligned}
\mathbb{E}\{\|\check{x}(h) - \hat{x}(h)\|_\infty\} &\leq \mu_h\|\check{x}(0) - \hat{x}(0)\|_\infty + \mathbb{E}\{\|\xi(p(h), s(h)) - \xi(\hat{p}(h), s(h))\|_\infty\} \\
&= \|\xi(p(h), s(h)) - \xi(\hat{p}_c(h), s(h))\|_\infty \times Prob\{\theta(h) = 2\} \\
&\quad + \|\xi(p(h), s(h)) - \xi(\hat{p}_s(h), s(h))\|_\infty \times Prob\{\theta(h) = 1\} \\
&\quad + \|\xi(p(h), s(h))\|_\infty \times Prob\{\theta(h) = 0\} + \mu_h\|\check{x}(0) - \hat{x}(0)\|_\infty \\
&\leq vs(h) + \mu_h b(0)
\end{aligned}$$

and

$$\begin{aligned}
\mathbb{E}\{\|\check{x}(h) - \hat{x}(h)\|_\infty\} &\leq \mu_h\|\check{x}(0) - \hat{x}(0)\|_\infty + \mathbb{E}\{\|\xi(p(h), s(h)) - \xi(\hat{p}(h), s(h))\|_\infty\} \\
&= \|\xi(p(h), s(h)) - \xi(\hat{p}_c(h), s(h))\|_\infty \times Prob\{\theta(h) = 2\} \\
&\quad + \|\xi(p(h), s(h)) - \xi(\hat{p}_s(h), s(h))\|_\infty \times Prob\{\theta(h) = 1\} \\
&\quad + \|\xi(p(h), s(h))\|_\infty \times Prob\{\theta(h) = 0\} \\
&\quad + \mu_h\|\check{x}(0) - \hat{x}(0)\|_\infty \\
&\leq \mu_h b(0) + vs(h).
\end{aligned}$$

Assume that the inequality holds when $\tau = 2, 3, \ldots, j$. Then, for $\tau = j + 1$, we have

$$\begin{aligned}
&\mathbb{E}\{\|\check{x}((j+1)h) - \hat{x}((j+1)h)\|_\infty\} \\
&\quad = \mathbb{E}\{\|A\check{x}((j+1)h - 1) + \xi(p((j+1)h), s((j+1)h)) \\
&\qquad - A\hat{x}((j+1)h - 1) \\
&\qquad - \delta\{\theta((j+1)h), 1\} \times \xi(\hat{p}_c((j+1)h), s((j+1)h))
\end{aligned}$$

$$- \delta\{\theta((j+1)h), 2\} \times \xi(\hat{p}_s((j+1)h), s((j+1)h))\|_\infty\}$$
$$\leq \mathbb{E}\{\mu_h\|\check{x}((jh) - \hat{x}(jh)\|_\infty + \|\xi(p((j+1)h), s((j+1)h))$$
$$- \delta\{\theta((j+1)h), 1\} \times \xi(\hat{p}_c((j+1)h), s((j+1)h))$$
$$- \delta\{\theta((j+1)h), 2\} \times \xi(\hat{p}_s((j+1)h), s((j+1)h))\|_\infty\}. \tag{7.26}$$

By tedious calculations, one arrives at

$$\mathbb{E}\{\|\xi(p((j+1)h), s((j+1)h)) - \delta\{\theta((j+1)h\}, 1)$$
$$\times \xi(\hat{p}_c((j+1)h), s((j+1)h)) - \delta\{\theta((j+1)h), 2\}$$
$$\times \xi(\hat{p}_s((j+1)h), s((j+1)h))\|_\infty\}$$
$$= \|\xi(p((j+1)h), s((j+1)h))$$
$$- \xi(\hat{p}_c((j+1)h), s((j+1)h))\|_\infty \times Prob\{\theta(\tau h) = 2\}$$
$$+ \|\xi(p((j+1)h), s((j+1)h))$$
$$- \xi(\hat{p}_s((j+1)h), s((j+1)h))\|_\infty \times Prob\{\theta(\tau h) = 1\} \tag{7.27}$$
$$+ \|\xi(p((j+1)h), s((j+1)h))\|_\infty \times Prob\{\theta(\tau h) = 0\}$$
$$= \bar{\theta}_1 \|\xi(p((j+1)h), s((j+1)h))$$
$$- \xi(\hat{p}_s((j+1)h), s((j+1)h))\|_\infty$$
$$+ \bar{\theta}_0 \|\xi(p((j+1)h), s((j+1)h))\|_\infty$$
$$= \left(\frac{4}{N}\bar{\theta}_1 + \bar{\theta}_0\right) s((j+1)h).$$

Combining (7.26) and (7.27), it is easy to obtain that $b(\tau h) \leq \mu_h b((\tau - 1)h) + vs(\tau h)$ for $\tau = j+1$, and the proof of this lemma is thus complete.

Lemma 7.3: *The sequences $s(\tau h)$ and $b(\tau h)$ $(\tau = 1, 2, ..)$ in Lemma 7.1 and Lemma 7.2 satisfy*

$$\lim_{\tau \to \infty} s(\tau h) = 0 \text{ and } \lim_{\tau \to \infty} b(\tau h) = 0$$

if $\rho(\Omega_h) < 1$, where

$$\Omega_h \triangleq \begin{bmatrix} \frac{\bar{\mu}_h}{N} & \zeta_h \\ \tilde{v} & \tilde{\mu} \end{bmatrix}, \quad \tilde{\mu} \triangleq \mu_h + v\zeta_h, \quad \tilde{v} \triangleq v\frac{\bar{\mu}_h}{N}$$

and other parameters are defined in Lemmas 7.1–7.2.

Proof: *On the basis of Lemmas 7.1–7.2, we obtain the recursive expressions*

$$s(\tau h) = \frac{\bar{\mu}_h}{N}s((\tau - 1)h) + \zeta_h b((\tau - 1)h)$$
$$b(\tau h) = \mu_h b((\tau - 1)h) + vs(\tau h).$$

Let us introduce two auxiliary sequences $\mathcal{S}(\tau h)$ and $\mathcal{B}(\tau h)$ satisfying

$$\begin{cases} \mathcal{S}(\tau h) = \dfrac{\bar{\mu}_h}{N}\mathcal{S}((\tau-1)h) + \zeta_h \mathcal{B}((\tau-1)h) \\ \mathcal{S}(0) = s(0) \end{cases}$$

and

$$\begin{cases} \mathcal{B}(\tau h) = \mu_h \mathcal{B}((\tau-1)h) + v\mathcal{S}(\tau h) \\ \mathcal{B}(0) = b(0). \end{cases}$$

It is obvious that

$$s(\tau h) \leq \mathcal{S}(\tau h) \text{ and } b(\tau h) \leq \mathcal{B}(\tau h), \ (\tau = 1, 2, \dots).$$

Substituting the dynamics $\mathcal{S}(\tau h)$ into $\mathcal{B}(\tau h)$, we further obtain that

$$\mathcal{B}(\tau h) = \tilde{\mu}\mathcal{B}((\tau-1)h) + \tilde{v}\mathcal{S}((\tau-1)h).$$

Denoting $\eta(\tau h) \triangleq [\mathcal{S}(\tau h) \quad \mathcal{B}(\tau h)]^T$, the dynamics of the augmented system is acquired as

$$\eta\big((\tau+1)h\big) = \Omega_h \eta(\tau h)$$

and it follows immediately from $\rho(\Omega_h) < 1$ that

$$\lim_{\tau \to \infty} \eta(\tau h) = 0,$$

which indicates that

$$\lim_{\tau \to \infty} s(\tau h) = 0 \text{ and } \lim_{\tau \to \infty} b(\tau h) = 0.$$

The proof of this lemma is complete.

Lemma 7.4: *The decoding error $e(k)$ $(k = 1, 2, \dots)$ satisfies*

$$\lim_{k \to \infty} \mathbb{E}\{\|e(k)\|\} = 0$$

if $\rho(\Omega_h) < 1$, where Ω_h is defined in Lemma 7.3.

Proof: *Rewriting $e(k)$ as $e(k) = \vec{x}_1(k) - \vec{x}_2(k)$, we have*

$$\mathbb{E}\{\|e(k)\|_\infty\} \leq \mathbb{E}\{\|\vec{x}_1(k)\|_\infty\} + \mathbb{E}\{\|\vec{x}_2(k)\|_\infty\}.$$

In order to prove that the decoding error $e(k)$ achieves the stochastic stability, in what follows, we will show that

$$\lim_{k \to \infty} \mathbb{E}\{\|e(k)\|\} = 0$$

for $k = \tau h$ and

$$\tau h \leq k < (\tau + 1)h \ (\tau = 1, 2, \dots).$$

It follows from the results obtained in Lemmas 7.1–7.3 that

$$\mathbb{E}\{\|e(\tau h)\|_\infty\} \leq \mathbb{E}\{\|\vec{x}_1(\tau h)\|_\infty\} + \mathbb{E}\{\|\vec{x}_2(\tau h)\|_\infty\}$$

$$\leq \frac{1}{N}s(\tau h) + b(\tau h) \ (\tau = 1, 2, \dots),$$

which gives rise to

$$\lim_{\tau \to \infty} \{\|e(\tau h)\|_\infty\} = 0.$$

For $\tau h \leq j < (\tau + 1)h$, it is easily seen from the proofs of Lemmas 7.1–7.2 that

$$\mathbb{E}\{\|\vec{x}_1(j)\|_\infty\} = \mathbb{E}\{\|x(j) - \bar{x}(j)\|_\infty\}$$

$$\leq \bar{\mu}_j\|\vec{x}_1(\tau h)\|_\infty + \zeta_{j-\tau h}\|\vec{x}_2(\tau h)\|_\infty$$

and

$$\mathbb{E}\{\|\vec{x}_2(j)\|_\infty\} \leq \|A\|_\infty^{j-\tau h}b(\tau h)$$

where

$$\|A^j\|_\infty \triangleq \bar{\mu}_j, \quad \zeta_{j-\tau h} \triangleq \|\sum_{i=1}^{j-\tau h} A^{j-\tau h-i}BKA^{i-1}\|_\infty.$$

Letting

$$\ell \triangleq \max\{\bar{\mu}_j, \zeta_{j-\tau h}\|A\|_\infty^{j-\tau h}\} \ (\tau h \leq j < (\tau + 1)h),$$

we have

$$\mathbb{E}\{\|\vec{x}_1(j)\|_\infty\} + \mathbb{E}\{\|\vec{x}_2(j)\|_\infty\} \leq \ell(s(\tau h) + b(\tau h)),$$

which shows that

$$\lim_{j \to \infty} \mathbb{E}\{\|e(j)\|_\infty\} = 0$$

and the proof of this lemma is ended.

Having obtained the stochastic convergence condition for the decoding error, it is time for us to deal with the stochastic stabilization problem for the system (7.1). The following theorem is provided to establish a sufficient condition for the stochastic stability of the closed-loop system (7.23).

Theorem 7.1: *The closed-loop system (7.23) is stochastically stable if the conditions $\rho(\mathcal{A}) < 1$ and $\rho(\Omega_h) < 1$ are simultaneously satisfied, where Ω_h is defined in Lemma 7.3.*

Proof: *It follows from the dynamics of the closed-loop system (7.23) that*

$$x(k) = \mathcal{A}^k x(0) - \sum_{i=1}^{k} \mathcal{A}^{k-i} BKe(i-1),$$

which leads to

$$\mathbb{E}\{\|x(k)\|\} \leq \|\mathcal{A}^k\| \|x(0)\| + \sum_{i=1}^{k} \|\mathcal{A}\|^{k-i} \|BK\| \mathbb{E}\{\|e(i-1)\|\}.$$

Since the modulus of the eigenvalues of Ω_h lies in the unit circle on the complex plane, it is known from Lemma 7.4 that $\mathbb{E}\{\|e(i-1)\|\}$ is bounded, whose upper bound is denoted by $\bar{e}$. Accordingly, we further derive that

$$\mathbb{E}\{\|x(k)\|\} \leq \|\mathcal{A}^k\| \|x(0)\| + \sum_{i=1}^{k} \|\mathcal{A}\|^{k-i} \|BK\| \bar{e}$$

$$= \|\mathcal{A}^k\| \|x(0)\| + \frac{\|BK\|\bar{e}(1 - \|\mathcal{A}\|^k)}{1 - \|\mathcal{A}\|}.$$

Moreover, by applying the property that

$$\lim_{k \to \infty} \mathcal{A}^k = 0 \iff \rho(\mathcal{A}) < 1,$$

we confirm that

$$\lim_{k \to \infty} \mathbb{E}\{\|x(k)\|\} \leq \frac{\|BK\|\bar{e}}{1 - \|\mathcal{A}\|}.$$

In fact, the term $\sum_{i=1}^{k} \|\mathcal{A}\|^{k-i}\|BK\|\mathbb{E}\{\|e(i-1)\|\}$ can be rewritten as $\frac{\hbar_k}{\ell_k}$, where

$$\hbar_k \triangleq \sum_{i=1}^{k} \|\mathcal{A}\|^{-i}\|BK\|\mathbb{E}\{\|e(i-1)\|\}, \quad \ell_k \triangleq \|\mathcal{A}\|^{-k}.$$

Noting that ℓ_k is a strictly monotonously increasing positive sequence with

$$\lim_{k\to\infty} \ell_k = \infty,$$

it follows immediately from the Stolz-Cesàro Theorem that

$$\lim_{k\to\infty} \frac{\hbar_k}{\ell_k} = \lim_{k\to\infty} \frac{\hbar_{k+1} - \hbar_k}{\ell_{k+1} - \ell_k} = 0,$$

which yields

$$\lim_{k\to\infty} \mathbb{E}\{\|x(k)\|\} = 0.$$

The proof is complete.

Remark 7.4: *As a matter of fact, the L'Hôpital's rule is recognized as a powerful means in obtaining the convergence of the closed-loop system in the continuous-time setting with additive convergence disturbances. Nevertheless, when it comes to the discrete-time system, where the state trajectory is a sequence and the derivative of system state is unavailable, the L'Hôpital's rule is no longer valid. In view of this, the Stolz-Cesàro Theorem (recognized as a discrete-time version of L'Hôpital's rule) is employed to derive the stochastic stability of closed-loop system (7.23) in the proof of Theorem 7.1.*

The following corollary is put forward to analyze the influences from the encoder parameters and the probability of packet dropouts on the feasibility of the addressed MDC-based control problem.

Corollary 7.1: *Suppose that the spectral radius of matrix $\mathcal{A}$ satisfies $\rho(\mathcal{A}) < 1$. The closed-loop system (7.23) is stochastically stable if the probability of the simultaneous packet dropout of both communication channels satisfies the following relationship:*

$$\bar{\theta}_0 \leq \frac{(1 - \frac{\bar{\mu}_h}{N})(1 - \mu_h)}{\zeta_h} - \frac{4}{N}\bar{\theta}_1. \tag{7.28}$$

Proof: *Denote λ_i ($i = 1, 2$) as the two eigenvalues of matrix Ω_h, which are, respectively, calculated as*

$$\lambda_1 = \frac{1}{2}\left(\frac{\bar{\mu}_h}{N} + v\zeta_h + \mu_h + \sqrt{(\frac{\bar{\mu}_h}{N} + v\zeta_h - \mu_h)^2 + 4v\zeta_h\mu_h}\right)$$

$$\lambda_2 = \frac{1}{2}\left(\frac{\bar{\mu}_h}{N} + v\zeta_h + \mu_h - \sqrt{(\frac{\bar{\mu}_h}{N} + v\zeta_h - \mu_h)^2 + 4v\zeta_h\mu_h}\right).$$

Note that $\lambda_1(v)$ $(\lambda_2(v))$ is a monotonically increasing (decreasing) function of v on the interval $[0, +\infty)$ with

$$\lambda_1(0) = \frac{\frac{\bar{\mu}_h}{N} + \mu_h + |\frac{\bar{\mu}_h}{N} - \mu_h|}{2}$$

$$\lambda_2(0) = \frac{\frac{\bar{\mu}_h}{N} + \mu_h - |\frac{\bar{\mu}_h}{N} - \mu_h|}{2}.$$

In addition, it is easy to verify that

$$\lambda_i(0) \in \left\{\frac{\bar{\mu}_h}{N}, \mu_h\right\} \subset (0, 1) \ (i = 1, 2),$$

which implies that the equation $\lambda_2(v) = -1$ is infeasible. In this sense, letting $\lambda_1(v) = 1$ yields

$$v = \frac{(1 - \frac{\bar{\mu}_h}{N})(1 - \mu_h)}{\zeta_h}$$

and the proof of this corollary is therefore complete.

Corollary 7.2: *Suppose that the spectral radius of matrix A satisfies $\rho(A) < 1$. The closed-loop system (7.23) is stochastically stable if the encoder parameter N (i.e. quantization level) satisfies the following relationship:*

$$N > \frac{4\zeta_h\bar{\theta}_1 + (1 - \mu_h)\bar{\mu}_h}{1 - \mu_h - \bar{\theta}_0\zeta_h} \tag{7.29}$$

where all the parameters on the right-hand side of the above inequality have been defined previously.

Proof: *It follows from the calculation of the eigenvalues of matrix Ω_h in Corollary 7.1 that*

$$\lambda_i > 0 \ (i = 1, 2).$$

Consider the fact that λ_2 is a monotonically increasing function with respect to $\frac{1}{N}$ on $[0, +\infty)$. Then, solving the equation $\lambda_2(\frac{1}{N}) = 1$ gives

$$\frac{1}{N} = \frac{1}{\bar{\mu}_h} - \frac{\nu \zeta_h}{\bar{\mu}_h(1 - \mu_h)}.$$

It is inferred from (7.29) that the requirement $\Omega_h < 1$ is met and, therefore, the stochastic stability of the closed-loop system (7.29) is guaranteed. The proof of this corollary is complete.

Remark 7.5: *Corollary 7.1 (Corollary 7.2) reveals the relationship between the occurrence probability packet dropouts (the quantization level of the coding function) and the system stability. To be more specific, in Corollary 7.1, an upper bound of the probability of the simultaneous packet dropouts of both channels (i.e. $\bar{\theta}_0$) is obtained, which poses a basic constraint on the communication quality of the channels. Then, the relationship between system parameters (e.g. $\frac{\bar{\mu}_h}{N}$, μ_h, N and ζ_h) and the bound upper is explicitly characterized. In Corollary 7.2, a lower bound of the quantization level N is determined, from which we can conclude that such a lower bound is dependent on the major system parameters (e.g. $\bar{\mu}_h$, $\bar{\theta}_0$, $\bar{\theta}_1$, ζ_h and μ_h). In particular, it is readily seen from the right-hand sides of (7.28) and (7.29) that 1) the derived upper bound of the probability of the simultaneous packet dropouts increases with the growth of the quantization level N, and 2) the acquired lower bound of the quantization level N decreases with the descent of the packet-dropout probabilities $\bar{\theta}_0$ and $\bar{\theta}_1$, and such observations are in agreement with the practice.*

Remark 7.6: *In comparison with the existing studies, the main results of this paper exhibit the following distinctive contributions: 1) the MDC strategy developed in this paper is new in the sense that the dynamical uniform quantization approach is, for the first time, adopted to cope with the unreliability of the communication channel; 2) the constructed control scheme is new that ensures the stochastic stability of the controlled system subject to distorted control signals; and 3) the obtained stability criterion is new that reveals the influence of both the system parameters and the encoder parameters on the stability.*

7.3 Illustrative Examples

In this section, a numerical example is provided to evaluate the effectiveness and applicability of the proposed MDC-based control scheme.

Let the discrete-time system (7.1) have following parameters:

$$A = \begin{bmatrix} 0.4 & 0.63 \\ 0.18 & 0.83 \end{bmatrix}, \quad B = \begin{bmatrix} 1.5 & 0.5 \\ 0.8 & -0.45 \end{bmatrix}.$$

It is easy to calculate that the eigenvalues of A are $\lambda_1 = 0.2155$ and $\lambda_2 = 1.0145$, respectively, which indicates that the open-loop system is *unstable*.

The controller gain matrix K is chosen as

$$K = \begin{bmatrix} -0.4 & -0.21 \\ -0.36 & -0.45 \end{bmatrix}$$

under which the spectral radius of $\mathcal{A}$ satisfies $\rho(\mathcal{A}) = 0.8661 < 1$.

The initial parameters are set as $x(0) = [0.85 \quad -1.14]^T$ and $\check{x}(0) = \hat{x}(0) = [0 \quad 0]^T$. The coding period is given by $h = 3$ and the quantization level is selected as $N = 20$. For the two communication channels, the successful-arrival probabilities of the packets are set as $\gamma_1 = 0.6$ and $\gamma_2 = 0.7$, respectively.

The matrix Ω_h is calculated as

$$\Omega_h = \begin{bmatrix} 0.0532 & 1.0680 \\ 0.0113 & 0.8876 \end{bmatrix},$$

whose eigenvalues are, respectively, 0.0390 and 0.9018, which satisfies $\rho(\mathcal{A}) < 1$.

The simulation results are illustrated in Figures 7.3–7.7. Figure 7.3 describes the state trajectories of the open-loop system. Figure 7.4 shows

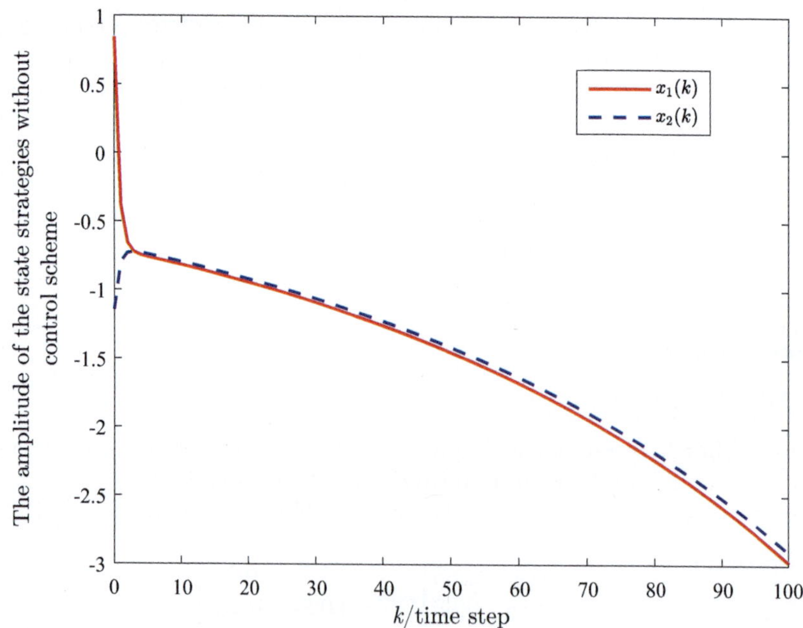

FIGURE 7.3

State trajectories of the open-loop system.

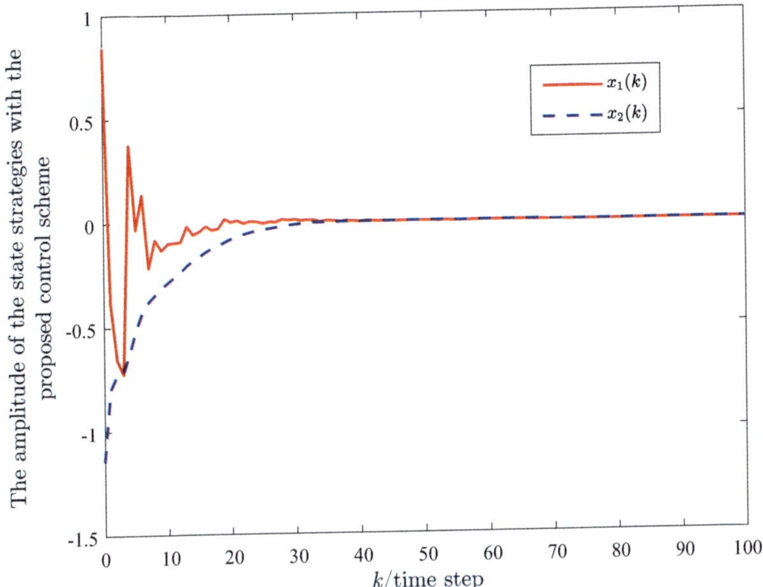

FIGURE 7.4

State trajectories of the closed-loop system.

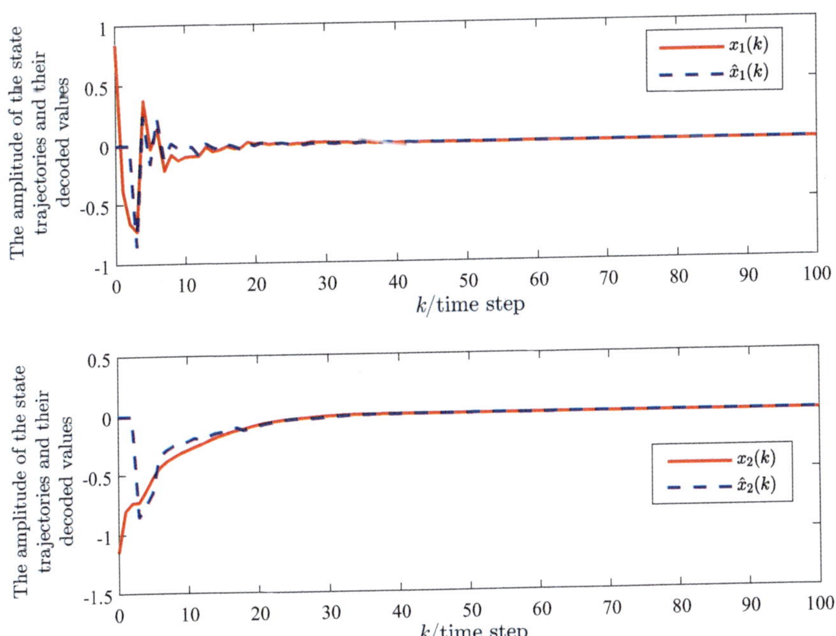

FIGURE 7.5

State trajectories and their decoded values.

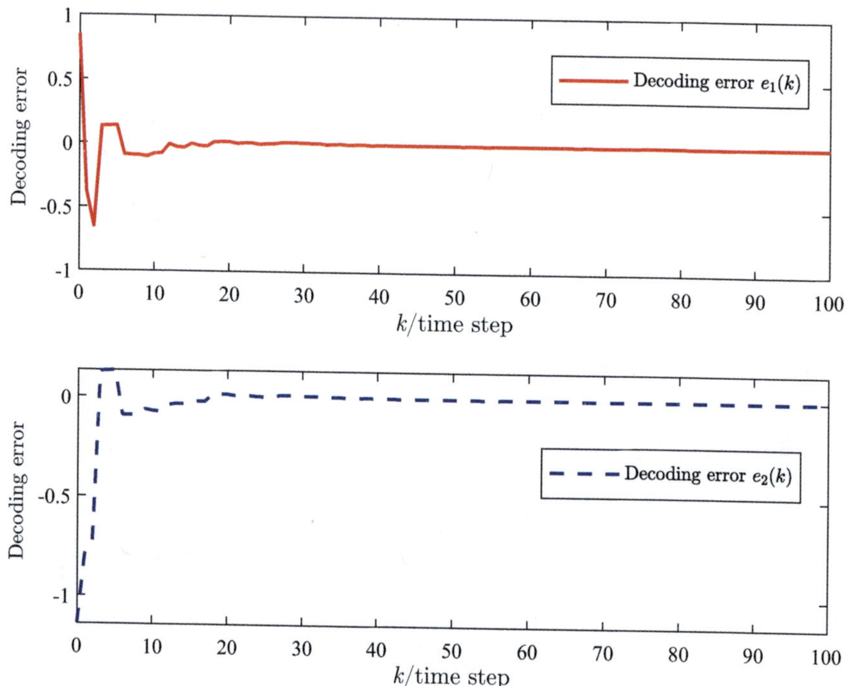

FIGURE 7.6

The trajectories of decoding errors.

the state trajectories of the corresponding closed-loop system with the proposed MDC-based control scheme. Figure 7.5 exhibits the controlled state trajectories and their decoded values. Figure 7.6 displays the decoding errors for different state components. Figure 7.7 plots the values of the Bernoulli sequences $\gamma_i(\tau h)$ $(i = 1, 2)$ that characterize the packet-dropout phenomenon of the communication channels. All the figures verify the applicability of the developed control strategy, which firmly confirms the obtained results.

7.4 Summary

This paper has dealt with the stabilization problem for a class of linear discrete-time systems under a resource-constrained channel. A new dynamical multiple description coding approach has been developed to cope with the unreliability of the communication channel. Two sequences of Bernoulli-distributed random variables with known probability

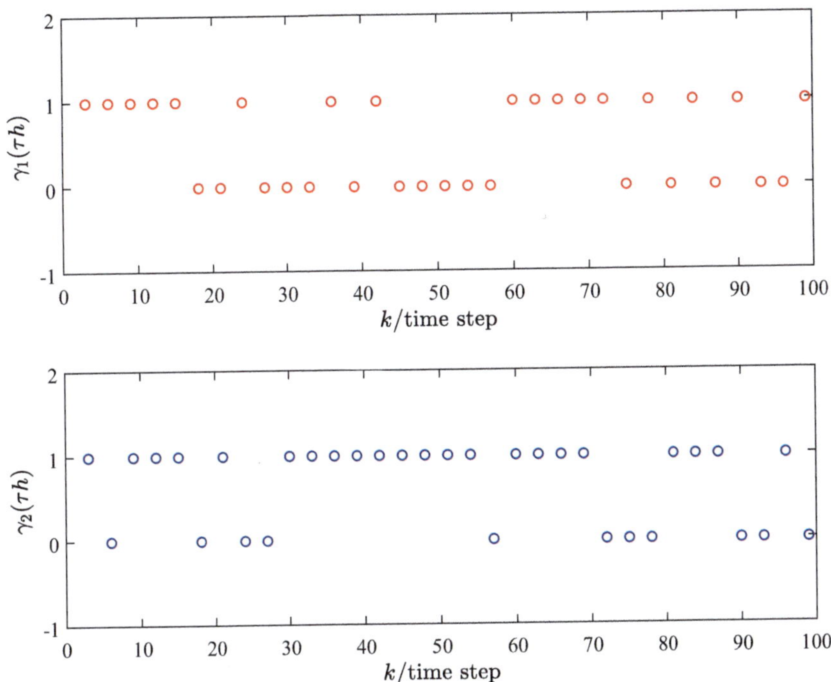

FIGURE 7.7

The Bernoulli sequences $\gamma_i(\tau h)$ $(i = 1, 2)$.

distributions have been employed to regulate the randomly occurring packet-dropout phenomenon in the sensor to controller channels. Based on the designed dynamical MDC strategy, the convergence criterion of the decoding error has been first obtained, and then the stochastic stability of the closed-loop system has been guaranteed with the help of the matrix theory and stochastic analysis technique. Moreover, a sufficient condition has been established to guarantee the stability of the closed-loop system, in which the packet-dropout probabilities and the encoder parameters have been involved. Finally, a simulation example has been provided to manifest the applicability of the derived main results.

8

An Event-Triggered Encoding Approach to Control of Linear Systems under Bit Rate Conditions

The past few decades have witnessed a surge of research interest on the dynamics analysis of networked control systems (NCSs) due primarily to their pervasive applications in real-world industrial systems. Because of the digital nature of the communication network, the original analog signal is required to be converted into the digitized one with finite number of bits, which inevitably results in the conversion error. Generally speaking, a high bandwidth (i.e. bit rate) of the communication channel gives rise to a small conversion error and little control performance degradation. However, in most of practical applications (such as micro-electro mechanical systems, sensor networks and industrial control networks), the network resource is scarce and the available bit rate of the network is often limited. Thus, it is of both theoretical significance and practical importance to look for a bit-rate condition as tight as possible under which the prescribed control objective is guaranteed.

In general, the network bandwidth occupancy of a data transmission task is determined by two factors: the bits of the data itself and its communication frequency. From this point of view, in order to comply with the bit-rate requirement of a digital communication channel, the signal usually needs to be "compressed" before being transmitted. One of the effective compression approaches is the utilization of certain coding algorithms to encode the raw data into specific codewords which occupy fewer bits. More specifically, the data encoding/decoding operation in closed-loop systems can be generally described as follows: 1) the sender encodes the state/measurement signals into the corresponding codewords by certain appropriate coding schemes; and 2) the generated codewords are transmitted to the decoder and suited decoding algorithms are applied to reconstruct the raw data by means of the received codewords. It should be pointed out that since the codewords carry only partial information of the raw data, how to develop an ingenious encoding-decoding algorithm to recover the raw data with an acceptable accuracy level appears to be quite crucial.

DOI: 10.1201/9781003534853-8

In recent years, considerable research effort has been directed toward the control problems for networked systems under certain bit-rate conditions, in most of which, the implementation of the encoding algorithms is governed by a time-scheduled manner due to its simplicity in design, that is, the encoder generates and sends the codewords periodically at those fixed time instants. However, under such a circumstance, the encoder could potentially cause unnecessary coding actions as well as a high codeword transmission frequency. This would further induce a severe occupancy of the network resource, which is actually a challenge for the practical networked systems subject to the limited network bandwidth. In order to mitigate the occupancy of the network bandwidth, apart from the adoption of certain coding strategies mentioned above, another effective way is to reduce the triggering ratio of the encoding operation.

Benefiting from the swift development of the digital control technology in the past decades, the data interaction in most engineering applications has been upgraded to the digital manner, and thus the event-triggered control strategy has been attracting substantial research attention. It is worth mentioning that, in engineering practice, the event-based control strategy is usually executed by a digital control device, which constitutes the so-called sampled-data control system. Among others, a prevalent approach is to utilize the time-based periodic sampling strategy to update the control signal by selecting a feasible sampling period. Instead of the time-triggered sampled-data control problems, the corresponding topics related to the event-triggered sampling mechanism have recently gained particular research attention. The most outstanding advantage of the event-triggered sampled-data control strategy is that the controller is updated only if some conditions are met, and therefore it might be more efficient in reducing the controller update frequency and prolonging the controller operating lifespan. However, to the best of the authors' knowledge, in most of the published results, only the event-based sampled-data control problem has been taken into account, when it comes to the simultaneous consideration of the bit-rate condition, the related results on this topic are quite few.

In response to the discussions aforesaid, in this paper, we aim to deal with the control problem for continuous-time linear systems via an event-triggered difference coding strategy, under which the corresponding bit rate conditions are derived. In fact, such an investigation is non-trivial with the following three major challenges: 1) how to design an event-triggered communication mechanism to schedule the updates of both the coder and the controller in a unified way? 2) how to reveal the relationship between the required bit rate and the system parameters? 3) how to develop appropriate methodologies to analyze the dynamics of the closed-loop system and synthesize the controller design? This motives our investigation in this chapter.

8.1 Problem Formulation

Consider a continuous-time linear system described by

$$\frac{d}{dt}x(t) = Ax(t) + Bu(t) \tag{8.1}$$

where $x(t) \in \mathbb{R}^n$ represents the state vector and $u(t) \in \mathbb{R}^p$ is the control input. The initial state $x(0)$ is bounded belonging to a given set $\mathcal{X}_0$. A and B are known matrices with appropriate dimensions. In this paper, it is assumed that at least one of the eigenvalues of A has positive real part, that is, the open-loop system is unstable.

From the resource-saving point of view, in a networked circumstance, it is preferable that the data transmission is executed in an event-driven manner. To this end, inspired by the idea in [73,89,90], an event-based coding scheme is developed in this paper, which is shown in Figure 8.1. It should be pointed out that the coder design is based on the idea of the difference coding scheme, which means that the input signal of the coder is the difference between the system state and the previous decoder (to be designed later) state. In order to access the decoder state at the encoder side and avoid the signal postback (i.e. signal flows from the decoder to the encoder), the following auxiliary system, which can be viewed as a "copy" of the decoder, is constructed:

$$\begin{cases} \dfrac{d}{dt}x_a(t) = Ax_a(t) + Bu_a(t) & \text{(8.2a)} \\[2mm] x_a(t_k^+) = x_a(t_k^-) + L\theta(s_k)\eta(t_k) & \text{(8.2b)} \end{cases}$$

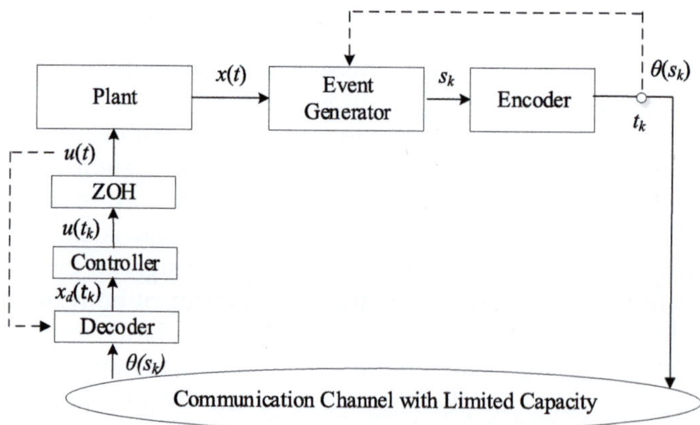

FIGURE 8.1

The schematic of the event-based control with encoding-decoding scheme.

where $x_a(t) \in \mathbb{R}^n$ and $u_a(t) \in \mathbb{R}^p$ are, respectively, the state and the control input of the auxiliary system. $x_a(t_k^+)$ and $x_a(t_k^-)$ are, respectively, defined as $x_a(t_k^-) \triangleq \lim_{t \to t_k^-} x_a(t)$ and $x_a(t_k^+) \triangleq \lim_{t \to t_k^+} x_a(t)$. s_k is the encoding instant and t_k is the time instant when both the decoder and the auxiliary system receive the data packet $\theta(s_k) \triangleq \mathrm{vec}_n\{\theta_i(s_k)\}$ with $\theta_i(s_k)$ $(i = 1, 2, \ldots, n)$ being the codewords encoded by the i-th encoder at time instant s_k. $\tau \triangleq t_k - s_k \geq 0$ is the time delay caused by the encoding process. $L \in \mathbb{R}^{n \times n}$ and $\eta(t_k) \in \mathbb{R}$ are the parameters to be designed later.

Denoting by $0 = s_0 < s_1 < \cdots s_k < s_{k+1} < \cdots$ the sequence of event occurring instants (encoding instants) and setting $\tilde{x}_a(s_k) \triangleq x(s_k) - x_a(s_k)$ as the encoder input. The encoding procedure is triggered when the following event generation condition is satisfied:

$$f(\tilde{x}_a(s_k), \sigma(s_k)) = 0 \tag{8.3}$$

where $f(\tilde{x}_a(s_k), \sigma(s_k)) \triangleq \|\tilde{x}_a(s_k)\| - \sigma(s_k)$ is called the triggering function and $\sigma(\cdot)$ is an exponentially decaying function defined as $\sigma(t) \triangleq \sigma_0 e^{-\lambda t}$ with σ_0 being a given positive scalar and λ being a parameter to be determined.

The triggering instants (or encoding instants) can be successively calculated by the following formula:

$$s_{k+1} = \inf\{t \mid t > s_k, \|\tilde{x}_a(t)\| = \sigma(t)\}. \tag{8.4}$$

Next, denote by $\tilde{x}_{ai}(t)$ the ith component of $\tilde{x}_a(t)$. As such, at the triggering time s_k, for each component $\tilde{x}_{ai}(t)$ $(i = 1, 2, \ldots, n)$, the individual codeword $\theta_i(s_k)$ is determined by the following rule:

$$\theta_i(s_k) = \mathrm{sign}(\tilde{x}_{ai}(s_k)) \tag{8.5}$$

where "sign$(\cdot)$" stands for the signum function.

As stated before, due to the encoding delay, the codeword $\theta_i(s_k)$ can only be available at time t_k. In view of this, for $t \in [t_k, t_{k+1})$, we construct the decoder with the following dynamics:

$$\begin{cases} \dfrac{d}{dt} x_d(t) = Ax_d(t) + Bu(t) & \text{(8.6a)} \\[2mm] x_d(t_k^+) = x_d(t_k^-) + L\theta(s_k)\eta(t_k) & \text{(8.6b)} \end{cases}$$

where $x_d(t) \in \mathbb{R}^n$ is the decoder state, $u(t)$ is the control input to be designed and other parameters are the same as the auxiliary system (8.2a)-(8.2b).

It is seen from (8.2b) and (8.6b) that, with the obtained data packet $\theta(s_k)$, both the auxiliary system state $x_a(t)$ and the decoding system state $x_d(t)$ are simultaneously updated at time t_k. Moreover, by employing the zero-order holder (ZOH) strategy, the control input $u(t)$ in (8.1) and (8.6a) is updated at t_k and will maintain constant until the next event time t_{k+1}, that is,

$$u(t) = Kx_d(t_k), \quad \forall t \in [t_k, t_{k+1}) \tag{8.7}$$

and, along the same line, the controller $u_a(t)$ in (8.2a) is designed as

$$u_a(t) = Kx_a(t_k), \quad \forall t \in [t_k, t_{k+1}) \tag{8.8}$$

where K is the controller parameter to be designed. Combining (8.2a)–(8.2b) and (8.6a)–(8.6b), it is not difficult to verify that $x_a(t) \equiv x_d(t)$ $\forall t > 0$ provided that $x_a(0) = x_d(0)$.

Remark 8.1: *In this paper, an event-based update mechanism is employed for the purpose of reducing the frequency of both the data transmission and the controller update. To be more specific, when the event generation condition (8.3) is satisfied at the time instant s_k, the encoder starts to implement the encoding procedure. In addition, owing to the existence of the time delay during the encoding process, the codeword $\theta_i(s_k)$ is only available to both the auxiliary system and the decoder system at t_k. Then, based on the renewed system states $x_a(t_k)$ and $x_d(t_k)$, the control input is updated at t_k. Therefore, not only the encoder executions but also the controller updates are dependent on the event function (8.3), and such an event-triggered mechanism achieves a great reduction of the update frequency of both the coder and the controller.*

The objective of this paper is summarized as the following threefold.

1. Develop an event-triggered encoding-decoding mechanism (8.3)–(8.6b), by which a sufficient condition for the bit-rate of the communication network is established such that the decoding error (i.e. $\tilde{x}_d(t) \triangleq x(t) - x_d(t)$) is convergent, that is $\lim_{t \to \infty} \|\tilde{x}_d(t)\| = 0$.
2. Derive a necessary condition to obtain a lower bound of the required bit rate, below which the decoding error diverges.
3. Design the event-based control strategy (8.7) for the system (8.1) to guarantee that the closed-loop system is exponentially stable.

8.2 Circumventing the "Zeno" Phenomenon

It is worth pointing out that, for the event-triggered continuous-time system (8.1), if the inter-event execution interval (namely, the length of two arbitrary successive event-triggering instants) trends to be infinitesimal, the so-called

"Zeno" phenomenon happens. In this case, the triggering frequency of the encoder would approach infinity and so does the update frequency of the controller. Obviously, the "Zeno" behavior will adversely affect the encoding/sampling process and even damage the relevant hardware devices, which should be strictly avoided. To this end, in the following, a useful lemma will be provided to eliminate the "Zeno" behavior. Also, an upper bound of the required bit rate to stabilize the addressed system (8.1) is derived.

Lemma 8.1: *If the encoding time-delay τ (i.e. $t_k - s_k$) satisfies $0 \le \tau < \kappa$, then there exists a positive scalar λ such that the inter-event execution interval satisfies $s_{k+1} - s_k \ge \kappa$, where $\kappa = \frac{\ln\frac{1}{\delta_1}}{\lambda_1}$ with $\delta_1 = \sqrt{1 - \frac{3}{4n}}$ and $\lambda_1 = \lambda + \|A\|$*

Proof: *In order to make our statement clearer, the exclusion of the "Zeno" behavior is conducted by the following two steps.*

Step I. In this step, our intention is to ensure that there is no any other triggering instant(s) over the interval (s_k, t_k), which is equivalent to show that $s_{k+1} > t_k$. To begin with, it follows from (8.1) and (8.2a)–(8.2b) that the dynamics of the $\tilde{x}_a(t)$ is governed by

$$\begin{cases} \dfrac{d}{dt}\tilde{x}_a(t) = A\tilde{x}_a(t) & \text{(8.9a)} \\ \tilde{x}_a(t_k) = \tilde{x}_a(t_k^-) - L\theta(s_k)\eta(t_k). & \text{(8.9b)} \end{cases}$$

For $t \in (s_k, t_k)$, the evolution of the state trajectory $\tilde{x}_a(t)$ can be determined by $\tilde{x}_a(t) = e^{A(t-s_k)}\tilde{x}_a(s_k)$, which indicates that $\|\tilde{x}_a(t)\| = \sqrt{\tilde{x}_a^T(s_k)\mathcal{A}(t - s_k)\tilde{x}_a(s_k)}$ with $\mathcal{A}(t - s_k) \triangleq e^{A^T(t-s_k)}e^{A(t-s_k)}$. Due to the inverse property of the matrix exponential, it is easy to verify that $\mathcal{A}(t - s_k)$ is a positive definite matrix and there exists a positive scalar function $\varphi(t - s_k) > 0$ such that $\mathcal{A}(t - s_k) \ge \varphi^2(t - s_k)I$ for $\forall t \in (s_k, t_k)$. Based on such a fact, one can further confirm that $\|\tilde{x}_a(t)\| \ge \varphi(t - s_k)\|\tilde{x}_a(s_k)\| = \varphi(t - s_k)\sigma(s_k)$ for $\forall t \in (s_k, t_k)$. Recalling that $\sigma(t)$ can be rewritten as $\sigma(t) = e^{-\lambda(t-s_k)}\sigma(s_k)$ for $\forall t \in (s_k, t_k)$, it is obvious that there must exist a suitable λ meeting the requirement $\varphi(t - s_k) > e^{-\lambda(t-s_k)}$, which leads to $\|\tilde{x}_a(t)\| > \sigma(t)$. Consequently, the triggering event is excluded over the interval (s_k, t_k) by properly choosing the parameter λ.

Step II. In this step, we are interested in investigating the length of $s_{k+1} - t_k$ and subsequently determining a lower bound of the inter-event execution interval $s_{k+1} - s_k$. To this end, our attention is focused on the dynamics evolution of $\|\tilde{x}_a(t)\|$ over the time interval $[t_k, s_{k+1})$. For $t \in [t_k, t_{k+1})$, the upper right-hand Dini derivative of $\|\tilde{x}_a(t)\|$ satisfies

$$D^+\|\tilde{x}_a(t)\| \le \|\dot{\tilde{x}}_a(t)\| \le \|A\|\|\tilde{x}_a(t)\|, \tag{8.10}$$

which further indicates

$$\|\tilde{x}_a(t)\| \leq e^{\|A\|(t-t_k)}\|\tilde{x}_a(t_k)\|. \tag{8.11}$$

Letting $L = \frac{1}{2n}e^{A\tau}e^{\lambda\tau}$ *and* $\eta(t) = \sigma(t)$, *it follows from (8.9b) that*

$$\begin{aligned}
\tilde{x}_a(t_k) &= \tilde{x}_a(t_k^-) - \frac{1}{2n}e^{A\tau}e^{\lambda\tau}\theta(s_k)\sigma(t_k) \\
&= e^{A(t_k-s_k)}\tilde{x}_a(s_k) - \frac{1}{2n}e^{A\tau}e^{\lambda\tau}\theta(s_k)\sigma(t_k).
\end{aligned} \tag{8.12}$$

Also, it is inferred from (8.3) and (8.5) that $\tilde{x}_a(s_k) = \Sigma(s_k)\theta(s_k)\sigma(s_k)$, *where* $\Sigma(s_k) = diag_n\left\{\frac{\sigma_i(s_k)}{\sigma(s_k)}\right\}$ *with* $\sigma_i(s_k) = |\tilde{x}_{ai}(s_k)|$. *Substituting* $\tilde{x}_a(s_k) = \Sigma(s_k)\theta(s_k)\sigma(s_k)$ *into (8.12) yields*

$$\begin{aligned}
\tilde{x}_a(t_k) &= e^{A(t_k-s_k)}\Sigma(s_k)\theta(s_k)\sigma(s_k) \\
&\quad - \frac{1}{2n}e^{A\tau}e^{\lambda\tau}\theta(s_k)\sigma(t_k) \\
&= e^{A\tau}\left(\Sigma(s_k)\sigma(s_k) - \frac{\sigma(t_k)}{2n}e^{\lambda\tau}I_n\right)\theta(s_k).
\end{aligned} \tag{8.13}$$

In addition, it is observed from (8.9a) that, over the interval $[t_k, s_{k+1})$, *the state trajectory of* $\tilde{x}_a(t)$ *can be determined by*

$$\begin{aligned}
\tilde{x}_a(t) &= e^{A(t-t_k)}\tilde{x}_a(t_k) \\
&= e^{A(t-t_k)}e^{A\tau}\left(\Sigma(s_k)\sigma(s_k) - \frac{\sigma(t_k)}{2n}e^{\lambda\tau}I_n\right)\theta(s_k) \\
&= e^{A(t-t_k)}e^{A\tau}\left(\Sigma(s_k) - \frac{1}{2n}I_n\right)\theta(s_k)e^{\lambda\tau}e^{\lambda(t-t_k)}\sigma(t)
\end{aligned} \tag{8.14}$$

and, by means of the matrix norm inequalities, one can further derive that

$$\begin{aligned}
\|\tilde{x}_a(t)\| &= \left\|e^{A(t-t_k)}e^{A\tau}\left(\Sigma(s_k) - \frac{1}{2n}I_n\right)\theta(s_k)e^{\lambda\tau}e^{\lambda(t-t_k)}\sigma(t)\right\| \\
&\leq \left\|e^{A(t-t_k)}\right\|\left\|e^{A\tau}\left(\Sigma(s_k) - \frac{1}{2n}I_n\right)\theta(s_k)\right\|e^{\lambda\tau}e^{\lambda(t-t_k)}\sigma(t) \\
&\leq e^{\|A\|(t-t_k)}e^{\|A\|\tau}\left\|(\Sigma(s_k) - \frac{1}{2n}I_n)\theta(s_k)\right\|e^{\lambda\tau}e^{\lambda(t-t_k)}\sigma(t)
\end{aligned}$$

$$= \left\| \left(\Sigma(s_k) - \frac{1}{2n}I_n \right) \theta(s_k) \right\| e^{(\|A\|+\lambda)(t+\tau-t_k)} \sigma(t)$$

$$= \left\| \left(\Sigma(s_k) - \frac{1}{2n}I_n \right) \theta(s_k) \right\| e^{(\|A\|+\lambda)(t-s_k)} \sigma(t). \tag{8.15}$$

Recalling the definition of $\Sigma(s_k)$, it is easy to verify that

$$\left\| (\Sigma(s_k) - \frac{1}{2n}I_n)\theta(s_k) \right\| = \bar{\Sigma}(s_k) \tag{8.16}$$

where

$$\bar{\Sigma}(s_k) = \sqrt{\sum_{i=1}^{n} \frac{1}{\sigma^2(s_k)} \left(\sigma_i^2(s_k) - \frac{\sigma_i(s_k)\sigma(s_k)}{n} + \frac{\sigma^2(s_k)\theta_i^2(s_k)}{4n^2} \right)}.$$

Noticing that $\sigma_i(s_k) \geq 0$ $(i = 1, 2, \ldots, n)$ and $\sum_{i=1}^{n} \sigma_i^2(s_k) = \sigma^2(s_k)$, we have

$$1 \leq \frac{\sigma_1(s_k) + \cdots + \sigma_n(s_k)}{\sigma(s_k)} \leq \sqrt{n}. \tag{8.17}$$

Moreover, it is readily seen that $1 \leq \sum_{i=1}^{n} \theta_i^2(s_k) \leq n$, which indicates

$$\delta_0 \leq \left\| \left(\Sigma(s_k) - \frac{1}{2n}I_n \right) \theta(s_k) \right\| \leq \delta_1 \tag{8.18}$$

where $\delta_0 = \sqrt{1 - \frac{\sqrt{n}}{n} + \frac{1}{4n^2}}$ and $\delta_1 = \sqrt{1 - \frac{3}{4n}}$.

Therefore, it follows from (8.15) that, for $t \in [t_k, t_{k+1})$, one has

$$\|\tilde{x}_a(t)\| \leq \delta_1 e^{\lambda_1(t-s_k)} \sigma(t) \tag{8.19}$$

where $\lambda_1 = \lambda + \|A\|$.

Letting $\delta_1 e^{\lambda_1(t-s_k)} = 1$, we have

$$t - s_k \triangleq \kappa = \frac{\ln\frac{1}{\delta_1}}{\lambda_1}, \tag{8.20}$$

namely,

$$t - t_k = \kappa - \tau = \Delta_{T_a}, \tag{8.21}$$

which further indicates that, for $t \in [t_k, t_k + \Delta_{T_a})$, $\|\tilde{x}_a(t)\| < \sigma(t)$. Consequently, it is seen from the encoding event generation condition (8.3) that $s_{k+1} - s_k \geq \kappa$, and the "Zeno" behavior is thus eliminated.

8.3 Convergence Analysis of $\tilde{x}_a(t)$

Lemma 8.2: *For the proposed event-based encoding-decoding scheme (8.3)–(8.6b), the dynamics $\tilde{x}_a(t)$ is exponentially convergent, that is, $\|\tilde{x}_a(t)\| \leq \bar{\sigma}_0 e^{-\lambda t}$ for $\forall t \geq 0$, where $\bar{\sigma}_0 = \sigma_0 e^{\lambda_1 \tau}$.*

Proof: *In the proof of Lemma 8.1, one has $\|\tilde{x}_a(t)\| < \sigma(t)$ for $\forall t \in [t_k, s_{k+1})$. Subsequently, we intend to study the dynamics of $\|\tilde{x}_a(t)\|$ over the time interval $[s_{k+1}, t_{k+1})$. With the combination of (8.3) and (8.9a), it is readily seen that, for $\forall t \in [s_{k+1}, t_{k+1})$, one has $\|\tilde{x}_a(t)\| \leq e^{\|A\|(t-s_{k+1})}\|\tilde{x}_a(s_{k+1})\| = e^{\lambda_1(t-s_{k+1})}\sigma(t) \leq e^{\lambda_1 \tau}\sigma(t)$. Therefore, it is obvious that $\|\tilde{x}_a(t)\| \leq \bar{\sigma}_0 e^{-\lambda t}$ for $\forall t \geq 0$, which shows that $\tilde{x}_a(t)$ exponentially converges to zero.*

It is obvious from $x_a(t) \equiv x_d(t)$ $\forall t \geq 0$ that $\tilde{x}_a(t) \equiv \tilde{x}_d(t)$ $\forall t \geq 0$. Thus, it follows immediately from Lemma 8.2 that the decoding error $\tilde{x}_d(t)$ is exponentially convergent under the encoding-decoding scheme (8.3)–(8.6b).

Remark 8.2: *It can be observed from Lemma 8.1 and the convergence analysis of decoding error that the value of κ is associated with two parameters, namely, the convergence rate of the triggering threshold λ (also the convergence rate of the decoding error) and the system dimension n. On one hand, a larger λ leads to a smaller κ, which means that as the convergence rate of the decoding error becomes larger, the execution of the encoding operation would be more frequent. On the other hand, the increase of the system dimension n would cause a decline of κ, which signifies that a higher dimensional system may cost more resources in order to ensure the feasibility of the coding-decoding algorithm.*

8.4 A Sufficient Condition of the Bit Rate to Guarantee the Convergence of the Decoding Error

The following theorem provides a sufficient condition of the required bit rate of the communication channel to guarantee the convergence of the decoding error.

Theorem 8.1: *Consider the system (8.1) with the event-based encoding-decoding scheme (8.3)–(8.6b). The convergence of the decoding error $\tilde{x}_d(t)$ is guaranteed (i.e. $\lim_{t \to \infty} \|\tilde{x}_d(t)\| = 0$) if the bit rate of the communication channel $\mathcal{R}$ satisfies $\mathcal{R} \geq \mathcal{R}_1 \triangleq n\lceil \frac{1}{\kappa} \rceil$.*

Proof: *It follows from Lemmas 8.1-8.2 that the convergence of the decoding error is guaranteed. In addition, it is easy to see from $s_{k+1} - s_k \geq \kappa$ in Lemma 8.1 that there are at most $\lceil \frac{1}{\kappa} \rceil$ times' packet transmissions per time unit. Since each transmission of the packet $\theta(s_k)$ occupies n bits, it suffices to use the bit rate of $n\lceil \frac{1}{\kappa} \rceil$ to guarantee that $\lim_{t\to\infty} \|\tilde{x}_d(t)\| = 0$, which completes the proof of this theorem.*

8.5 A Necessary Condition of the Bit Rate

Having obtained a sufficient condition of the bit rate under which the convergence of the decoding error is guaranteed, we are also interested in seeking a necessary condition of the bit rate below which the decoding error is divergent. It is noted that the sufficient condition in Theorem 8.1 shows that if the bit rate of the communication channel is greater than $\mathcal{R}_1$, the convergence of the decoding error $\tilde{x}_d(t)$ is guaranteed. While the necessary condition of the bit rate aims to determine a bit rate $\mathcal{R}_0$ below which the decoding error $\tilde{x}_d(t)$ diverges. In order to achieve this goal, the following analysis is first presented.

As mentioned before, since at least one of the eigenvalues of A has positive real part, it is easy to observe from (8.9a) that, for $t \in [t_k, t_{k+1})$, the evolution of $\tilde{x}_a(t)$ is divergent. Also, noting that the triggering threshold (i.e. $\sigma(t) = \sigma_0 e^{-\lambda t}$) is an exponential decay function, it is not difficult to verify that there must exist an triggering instant $s_{k+1} \in [t_k, t_{k+1})$. Consequently, in the next stage, we plan to determine an upper bound ι of the inter-event execution interval, that is $s_{k+1} - s_k \leq \iota$.

Lemma 8.3: *For the proposed event-based encoding-decoding scheme (8.3)–(8.6b), if the positive parameter λ satisfies $\lambda > \|A\|$, then an upper bound ι of the inter-event execution interval can be obtained, that is, $s_{k+1} - s_k \leq \iota$, where $\iota = \dfrac{\ln \frac{1}{\delta_0}}{\lambda_0}$ with $\lambda_0 = \lambda - \|A\|$.*

Proof: *It follows from (8.14) that, for $\forall t \in [t_k, t_{k+1})$, the dynamics of $\tilde{x}_a(t)$ can be rewritten as*

$$
\begin{aligned}
\tilde{x}_a(t) &= e^{A(t-t_k)} e^{A\tau} \left(\Sigma(s_k) - \frac{1}{2n} I_n \right) \theta(s_k) e^{\lambda \tau} e^{\lambda(t-t_k)} \sigma(t) \\
&= e^{A(t-s_k)} \left(\Sigma(s_k) - \frac{1}{2n} I_n \right) \theta(s_k) e^{\lambda(t-s_k)} \sigma(t).
\end{aligned}
\tag{8.22}
$$

In addition, noting that $e^{A(t-s_k)}$ is invertible with the inverse matrix $e^{-A(t-s_k)}$, one obtains

$$e^{-A(t-s_k)}\tilde{x}_a(t) = \left(\Sigma(s_k) - \frac{1}{2n}I_n\right)\theta(s_k)e^{\lambda(t-s_k)}\sigma(t), \tag{8.23}$$

which further derives

$$e^{\|-A\|(t-s_k)}\|\tilde{x}_a(t)\|_2 \geq \|e^{-A(t-s_k)}\tilde{x}_a(t)\|$$
$$= \left\|\left(\Sigma(s_k) - \frac{1}{2n}I_n\right)\theta(s_k)\right\|e^{\lambda(t-s_k)}\sigma(t). \tag{8.24}$$

Moreover, bearing in mind that $\left\|\left(\Sigma(s_k) - \frac{1}{2n}I_n\right)\theta(s_k)\right\| \geq \delta_0$, *one derives that, for* $t \in [t_k, t_{k+1})$, *the following relation*

$$\|\tilde{x}_a(t)\| \geq \left\|\left(\Sigma(s_k) - \frac{1}{n}I_n\right)\theta(s_k)\right\|e^{(\lambda-\|A\|)(t-s_k)}\sigma(t)$$
$$\geq \delta_0 e^{\lambda_0(t-s_k)}\sigma(t) \tag{8.25}$$

holds. Letting $\delta_0 e^{\lambda_0(t-s_k)} = 1$ *yields* $t-s_k = \frac{\ln\frac{1}{\delta_0}}{\lambda_0} \triangleq \iota$. *Considering both the encoding event generation condition (8.3) and the fact that the dynamics* $\tilde{x}_a(t)$ *satisfying* $\|\tilde{x}_a(t)\| \geq \sigma(t)$ *for* $t \in [t_k + \frac{\ln\frac{1}{\delta_0}}{\lambda_0} - \tau, t_{k+1})$, *it is easy to see that* $s_{k+1} - t_k \leq \Delta_{T_b}$, *where* $\Delta_{T_b} = \frac{\ln\frac{1}{\delta_0}}{\lambda_0} - \tau$.

Theorem 8.2: *Consider the system (8.1) with the event-based encoding-decoding scheme (8.3)–(8.6b). The decode error* $\tilde{x}_d(t)$ *diverges if the bit rate of the communication channel is less than* $\mathcal{R}_0 \triangleq n\lfloor\frac{1}{\iota}\rfloor$.

Proof: *It is inferred from (8.25) that there are at least* $\lfloor\frac{1}{\iota}\rfloor$ *times' data transmissions each time unit. Along the similar line in the proof of Theorem 8.1 that each transmission of the packet* $\theta(s_k)$ *occupies* n *bits, we can conclude that it is impossible to transmit all the generated codewords to the decoder if the bit rate of the communication channel is less than* $n\lfloor\frac{1}{\iota}\rfloor$, *which consequently results in the divergence of the decoding error* $\tilde{x}_d(t)$. *The proof of this theorem is ended.*

8.6 Analysis and Synthesis of the Addressed Event-Based Control Issue

In what follows, our effort is devoted to the analysis and synthesis of the addressed event-based control problem. Moreover, the input-delay approach is applied to derive the main results. To do this, first, defining $d_k(t) \triangleq t - t_k$ for $\forall t \in [t_k, t_{k+1})$, it is readily seen that $0 \leq d_k(t) < \iota$ and $\dot{d}_k(t) = 1$. Then, substituting the control input (8.7) into (8.1), the resulted closed-loop system is obtained as follows:

$$\begin{aligned}
\dot{x}(t) = {} & Ax(t) + BKx(t - d_k(t)) \\
& - BK\tilde{x}_d(t - d_k(t)), \ \forall \ t_k \leq t < t_{k+1}
\end{aligned} \tag{8.26}$$

where $\tilde{x}_d(t - d_k(t)) \triangleq x(t - d_k(t)) - x_d(t - d_k(t))$.

By bringing in the function $d(t)$ which is defined as $d(t) \triangleq d_k(t)$ for $t_k \leq t < t_{k+1}, k = 0, 1, \ldots, \infty$, the closed-loop system can be further rewritten as

$$\dot{x}(t) = Ax(t) + BKx(t - d(t)) - BK\tilde{x}_d(t - d(t)). \tag{8.27}$$

Now, the event-based control problem is converted into the stability analysis issue of the time-delayed system (8.27). In the following, a sufficient condition is provided to guarantee the exponential stability of the addressed closed-loop system.

Theorem 8.3: *Let the positive scalars μ, σ_0 and α be given. Under the conditions in Lemmas 8.1 and 8.3, the closed-loop system (8.27) is exponentially stable by the event-based control strategy (8.8) subject to the bit-rate condition $\mathcal{R} \geq \mathcal{R}_1$, if there exist three positive definite matrices $\bar{P} > 0$, $\bar{Q} > 0$, $\bar{R} > 0$ and a real-valued matrix X such that the condition*

$$\Pi = \begin{bmatrix} \bar{\Xi}_{11} & \bar{\Xi}_{12} & 0 & BX & \bar{\Xi}_{15} \\ * & -\frac{2}{\iota}\bar{R} & \frac{1}{\iota}\bar{R} & 0 & \bar{\Xi}_{25} \\ * & * & \bar{\Xi}_{33} & 0 & 0 \\ * & * & * & -\mu^2\bar{R} & \bar{\Xi}_{45} \\ * & * & * & * & \bar{\Xi}_{55} \end{bmatrix} < 0 \tag{8.28}$$

holds, where

$$\bar{\Xi}_{11} = \alpha\bar{P} + \bar{Q} - \frac{1}{\iota}\bar{R} + \bar{P}A^T + A\bar{P}, \ \bar{\Xi}_{12} = BX + \frac{1}{\iota}\bar{R},$$

$$\bar{\Xi}_{15} = \sqrt{\beta_1}\bar{P}A^T, \ \bar{\Xi}_{25} = \sqrt{\beta_1}X^TB^T,$$

$$\bar{\Xi}_{33} = -\frac{1}{\iota}\bar{R} - \beta_0\bar{Q}, \ \bar{\Xi}_{45} = -\sqrt{\beta_1}X^TB^T,$$

$$\bar{\Xi}_{55} = \bar{R} - 2\bar{P}, \ \beta_0 = e^{-\alpha\iota}, \ \beta_1 = \frac{e^{\alpha\iota} - 1}{\alpha}.$$

In addition, the expected controller parameter is designed as $K = X\bar{P}^{-1}$.

Proof: *First, define positive definite matrices $P \triangleq \bar{P}^{-1}$, $Q \triangleq \bar{P}^{-1}\bar{Q}\bar{P}^{-1}$ and $R \triangleq \bar{P}^{-1}\bar{R}\bar{P}^{-1}$. Then, choose the Lyapunov-Krasovskii functional candidate as*

$$V(t) = V_1(t) + V_2(t) + V_3(t) \tag{8.29}$$

where

$$V_1(t) = x^T(t)Px(t),$$

$$V_2(t) = \int_{t-\iota}^{t} e^{\alpha(s-t)} x^T(s)Qx(s)\,ds,$$

$$V_3(t) = \int_{-\iota}^{0} e^{-\alpha\theta} \int_{t+\theta}^{t} e^{\alpha(s-t)} \dot{x}^T(s)R\dot{x}(s)\,ds\,d\theta.$$

Along the trajectory of (8.27), taking the derivative of $V(t)$ with respect to the time t results in

$$\dot{V}(t) = \dot{V}_1(t) + \dot{V}_2(t) + \dot{V}_3(t) \tag{8.30}$$

where

$$\dot{V}_1(t) = 2x^T(t)P\dot{x}(t),$$

$$\dot{V}_2(t) = -\alpha V_2(t) + x^T(t)Qx(t) - \beta_0 x^T(t-\iota)Qx(t-\iota), \tag{8.31}$$

$$\dot{V}_3(t) = -\alpha V_3(t) + \beta_1 \dot{x}^T(t)R\dot{x}(t) - \int_{t-\iota}^{t} \dot{x}^T(s)R\dot{x}(s)\,ds.$$

By resorting to the well-known Jenson integral inequality, we have

$$-\int_{t-\iota}^{t} \dot{x}^T(s)R\dot{x}(s)\,ds$$

$$= -\int_{t-d(t)}^{t} \dot{x}^T(s)R\dot{x}(s)\,ds - \int_{t-\iota}^{t-d(t)} \dot{x}^T(s)R\dot{x}(s)\,ds$$

$$\leq -\frac{1}{d(t)}(x(t) - x(t-d(t)))^T R(x(t) - x(t-d(t)))$$

$$-\frac{1}{\iota - d(t)}(x(t-d(t)) - x(t-\iota))^T R \tag{8.32}$$

$$\times (x(t-d(t)) - x(t-\iota))$$

$$\leq -\frac{1}{\iota}(x(t) - x(t-d(t)))^T R(x(t) - x(t-d(t)))$$

$$-\frac{1}{\iota}(x(t-d(t)) - x(t-\iota))^T R(x(t-d(t)) - x(t-\iota)).$$

Based on (8.29)–(8.32), it can be found that

$$
\begin{aligned}
\dot{V}(t) + \alpha V(t) &\leq \alpha x^T(t)Px(t) + 2x^T(t)P\dot{x}(t) + x^T(t)Qx(t) \\
&\quad - \beta_0 x^T(t-\iota)Qx(t-\iota) + \beta_1 \dot{x}^T(t)R\dot{x}(t) \\
&\quad - \frac{1}{d}(x(t) - x(t-d(t)))^T R(x(t) - x(t-d(t))) \\
&\quad - \frac{1}{d}(x(t-d(t)) - x(t-\iota))^T R(x(t-d(t)) - x(t-\iota)) \\
&= \xi^T(t)\Xi\xi(t) + \mu^2 \|\tilde{x}_d(t-d(t))\|_R^2
\end{aligned}
\tag{8.33}
$$

where

$$
\Xi = \Xi_1 + \Xi_2^T R\Xi_2, \quad \Xi_2 = \sqrt{\beta_1}[A \quad BK \quad 0 \quad -BK],
$$

$$
\Xi_1 = \begin{bmatrix}
\Xi_{11} & PBK + \frac{1}{\iota}R & 0 & -PBK \\
* & -\frac{2}{\iota}R & \frac{1}{\iota}R & 0 \\
* & * & -\frac{1}{\iota}R - \beta_0 Q & 0 \\
* & * & * & -\mu^2 R
\end{bmatrix},
$$

$$
\xi(t) = [x^T(t) \quad x^T(t-d(t)) \quad x^T(t-\iota) \quad \tilde{x}_d^T(t-d(t))]^T,
$$

$$
\Xi_{11} = \alpha P + Q - \frac{1}{\iota}R + A^T P + PA,
$$

$$
\|\tilde{x}_d(t-d(t))\|_R = (\tilde{x}_d^T(t-d(t))R\tilde{x}_d(t-d(t)))^{\frac{1}{2}}.
$$

In order to draw the main results, we are in a position to show that $\Xi < 0$ under the condition (8.28). First, noting that $P^{-1}RP^{-T} - 2P^{-1} \geq -R^{-1}$, it follows from (8.28) that $\Pi < 0$, where $\bar{\Pi}$ has the same structure with Π and only $\bar{R} - 2\bar{P}$ is replaced by $-R^{-1}$. Then, pre-multiplying and post-multiplying Π by $\mathrm{diag}\{P,P,P,P,I\}$ and its transpose, one has

$$
\begin{bmatrix} \Xi_1 & \Xi_2^T \\ * & -R^{-1} \end{bmatrix} < 0.
\tag{8.34}
$$

Using the Schur Complement Lemma to the inequality (8.34), it immediately finds that $\Xi < 0$. Moreover, considering that both the observation error and decoding error satisfy $\|\tilde{x}_a(t)\| = \|\tilde{x}_d(t)\| \leq \sigma_0 e^{\lambda_1 \tau} e^{-\lambda t}$, it is obvious that $\mu^2 \|\tilde{x}_d(t-d(t))\|_R^2 \leq \hbar(t)$, where $\hbar(t) = \mu^2 \lambda_{\max}(R)\sigma_0^2 e^{2\lambda_1 \tau} e^{-2\lambda(t-\iota)}$. Consequently, it is inferred from (8.33) that

$$
\dot{V}(t) \leq -\alpha V(t) + \hbar(t).
\tag{8.35}
$$

Furthermore, by applying the Comparison Lemma, (8.35) implies that

$$
V(t) \leq e^{-\alpha t}V(0) + \int_0^t e^{-\alpha(t-\theta)}\hbar(\theta)d\theta.
\tag{8.36}
$$

Then, according to the fact that $V(t) \geq \lambda_{\min}(P)$ and $\hbar(t) \leq \hbar(t_0)$, one further has

$$
\begin{aligned}
\|x(t)\| &\leq \sqrt{\frac{e^{-\alpha t}V(0)}{\lambda_{\min}(P)} + \frac{\int_0^t e^{-\alpha(t-\theta)}\hbar(\theta)d\theta}{\lambda_{\min}(P)}} \\[2mm]
&\leq \sqrt{\frac{e^{-\alpha t}V(0)}{\lambda_{\min}(P)} + \frac{1-e^{-\alpha t}}{\alpha\lambda_{\min}(P)}\hbar(t_0)} \\[2mm]
&\leq \sqrt{\frac{e^{-\alpha t}V(0)}{\lambda_{\min}(P)} + \frac{1}{\alpha\lambda_{\min}(P)}\hbar(t_0)}.
\end{aligned}
\tag{8.37}
$$

In fact, since $\hbar(t)$ is a monotonically non-increasing function with $\lim_{t\to\infty} \hbar(t) = 0$, it follows from the L'Hôpital's rule that

$$
\lim_{t\longrightarrow\infty} \int_0^t e^{-\alpha(t-\theta)}\hbar(\theta)d\theta = 0.
\tag{8.38}
$$

Therefore, we can draw the conclusion that the closed-loop control system (8.27) is exponentially stable, namely, $\lim_{t\to\infty} \|x(t)\| \leq \sqrt{\frac{e^{-\alpha t}V(0)}{\lambda_{\min}(P)}} = 0$, which completes this proof.

Remark 8.3: *In this paper, we endeavor to offer a satisfactory solution to the resource-saving-oriented control problem for a class of continuous-time systems. By utilizing the event-triggered symbolic-based difference coding scheme, some bit rate conditions are derived. In addition, in terms of the input-delay approach, a unified framework has been established to deal with the proposed control issue. Compared with the existing literature, the research conducted in this paper possesses the following distinct features: 1) a sufficient condition of the bit rate is obtained to guarantee the convergence of the decoding error and the explicit relationship among the required bit rate of the control issue and the system dimension as well as the convergence rate of the decoding error is revealed; 2) a necessary condition of the bit rate is given below which the decoding error is divergent; and 3) the input-delay approach is employed to facilitate the controller analysis and design issue of the addressed systems.*

8.7 Illustrative Examples

In this section, a simulation example is presented to demonstrate the usefulness of both the proposed coding algorithm and the controller design scheme for the addressed systems.

For system (8.1), the parameter matrices are given as follows:

$$A = \begin{bmatrix} 0.01 & 0.02 \\ -0.01 & 0.01 \end{bmatrix}, \ B = \begin{bmatrix} 1.12 \\ -0.6 \end{bmatrix}.$$

It is easily seen that the eigenvalues of matrix A are $\chi_1 = 0.01 + 0.0141i$ and $\chi_2 = 0.01 - 0.0141i$, respectively, which indicates that the open-loop system (8.1) without control signal $u(t)$ is unstable. For the event generation condition (8.3), the threshold parameters are chosen as $\sigma_0 = 50$ and $\lambda = 0.23$. The encoding delay is assumed as $\tau = 0.004$, the convergence rate of the system state is set as $\alpha = 0.12$ and the parameter μ is given by $\mu = 2$. Moreover, it can be confirmed from Lemma 8.1 that the inter-event execution interval satisfies $s_{k+1} - s_k > \kappa = 0.9288$ and an upper bound of the required bit rate $\mathcal{R}$ guaranteeing the feasibility of the proposed coding strategy is obtained as $\mathcal{R} \leq \mathcal{R}_1 = \frac{n}{\kappa} = 2.1534$. Let the initial conditions be $x(0) = [0.7 \quad -0.5]^T$ and $x_a(0) = x_d(0) = [-23.7 \quad 13.5]^T$, respectively. By solving the matrix inequality (8.28) in Theorem 8.3, the controller gain is calculated as $K = [0.4360 \quad 1.2921]$.

After setting the initial condition and deriving the controller parameter, the simulation results are achieved and shown in Figures 8.2–8.6. Figure 8.2

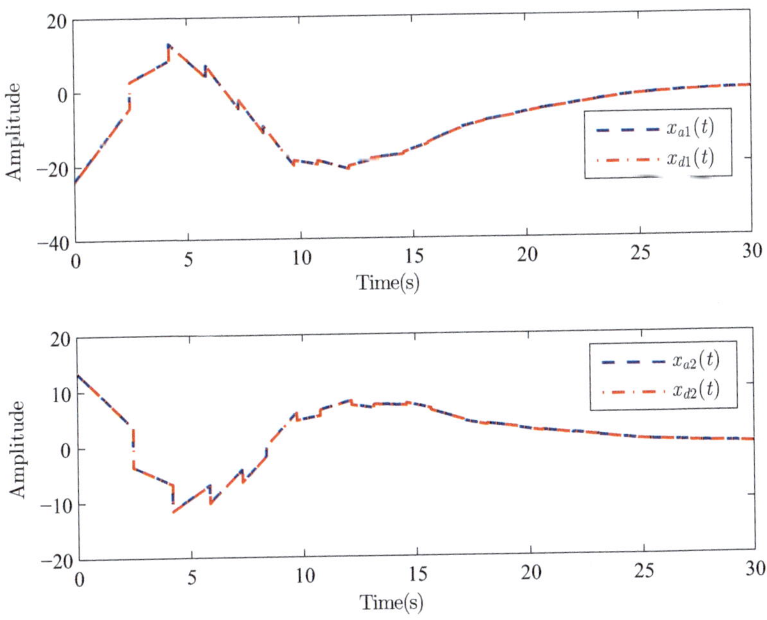

FIGURE 8.2

Dynamical evolutions of auxiliary system and decoder.

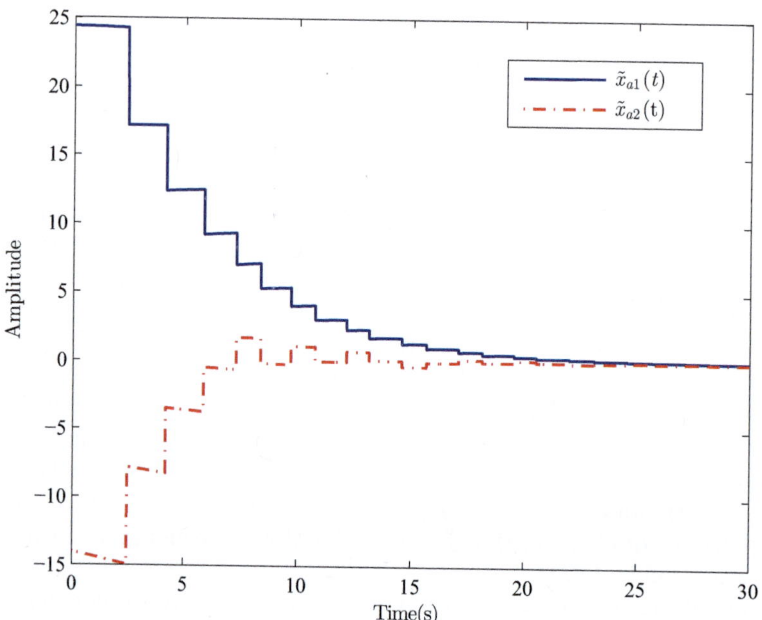

FIGURE 8.3

The decoding errors between the actual system states and their decoded values.

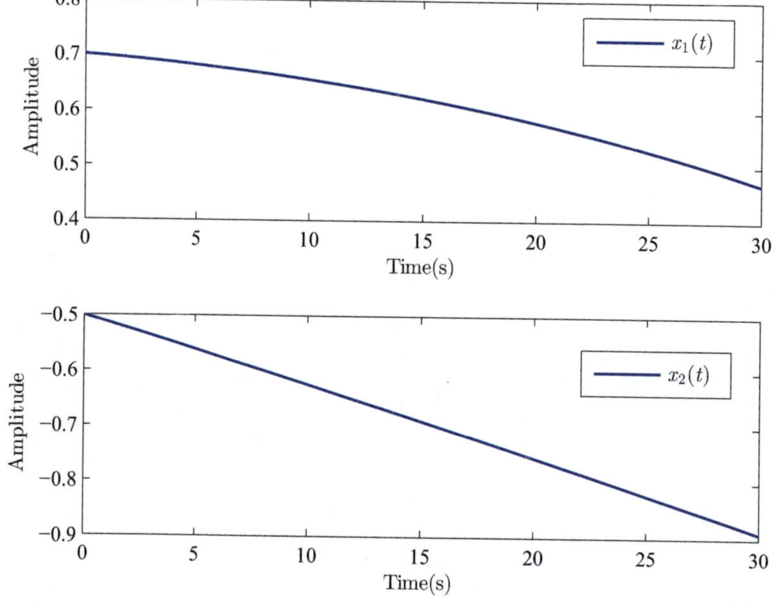

FIGURE 8.4

State response without control signal $u(t)$.

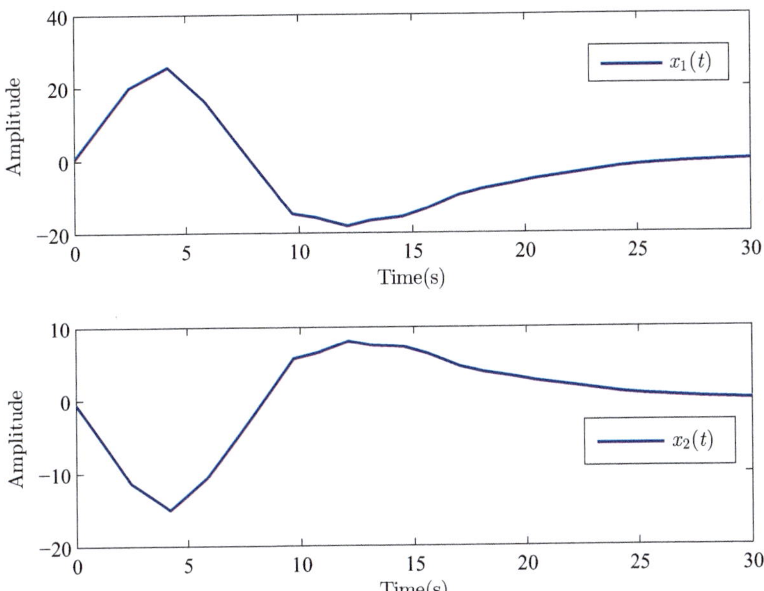

FIGURE 8.5
State response with control signal $u(t)$.

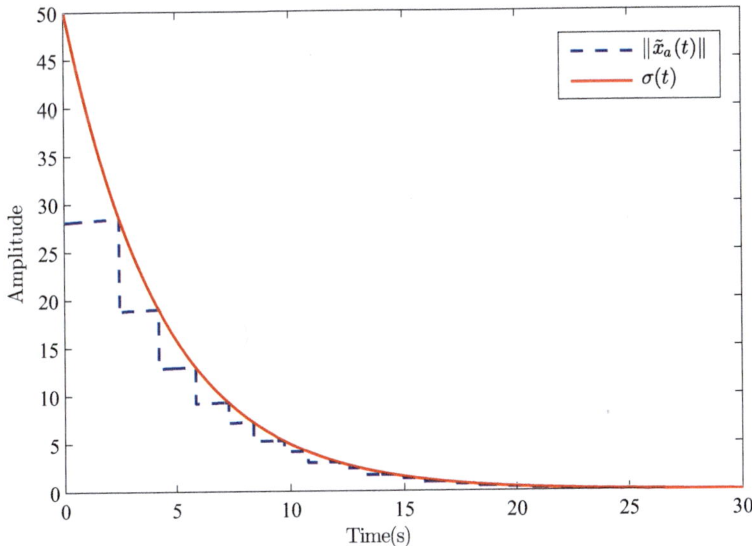

FIGURE 8.6
The threshold of the event generation function and the norm of $\tilde{x}_a(t)$.

depicts the dynamics of the auxiliary system state $x_a(t)$ and the decoder state $x_d(t)$. Figure 8.3 describes the decoding error between the system state $x(t)$ and its decoded value $x_d(t)$. Figure 8.4 plots the state trajectory of the open-loop system, while Figure 8.5 gives the state response of the closed-loop system with control input $u(t)$. Figure 8.6 illustrates the evolution of the event generation function $\sigma(t)$ and the triggering instants. It is inferred from Figures 8.2–8.3 that the decoding error converges to zero as time goes to infinity and hence the developed coding strategy works well. Figures 8.4–8.5 show that the proposed control scheme is effective for the unstable systems. From Figure 8.6, it is not difficult to see that the intersections of the two curves are the triggering instants. Consequently, all the simulation results confirm the superiority of the proposed control policy.

8.8 Summary

In this paper, the event-triggered control issue has been considered for a class of continuous-time linear systems via the digital communication channel. A novel digital data transmission mechanism has been introduced, where the communication media has been subjected to limited transmission capacity. A difference coding scheme has been developed, which has transferred the original information to one-bit symbolic data. In addition, a unified event-based encoding and sampling mechanism has been proposed in order to alleviate the communication burden as well as the equipment losses. The "Zeno" behavior has been excluded by determining a lower bound of the inter-event execution times. A sufficient condition of the required bit rate has been obtained to guarantee the convergence of the decoding error. Moreover, a necessary condition has also been derived with the bit rate below which the decoding error diverges. By obtaining the decoded data, the input-delay method has been adopted to deal with both the analysis and the design of the addressed control problem. Finally, the effectiveness of our coding/control strategies has been well verified by a simulation example.

9

Event-Based State Estimation under Constrained Bit Rate: An Encoding-Decoding Approach

In typical NSs, signal transmissions in the sensor-to-estimator and controller-to-actuator channels are realized through a *digital* communication network. Compared with traditional analogue communication, digital communication has been preferred by industry because of its distinguishing features of high reliability, strong anti-interference ability and low power consumption [6,187]. In particular, data encoding-decoding serves as a crucially important procedure in digital communication whose aim is to convert numerical data to binary codeword, thereby catering to the needs for reliable, secure and real-time communication. In recent years, encoding-decoding-based analysis/synthesis problems have drawn much attention from both academia and industry, and some representative research results can be found in and the references therein.

In the encoding-decoding procedure, the numerical data is mapped to the binary codeword with a finite number of bits through the coding operation, and the codeword is then restored to the original numerical data as accurately as possible through the decoding operation. It is worth noting that, in the design process of the encoding-decoding mechanism, the decoding accuracy is largely dependent on the *bit rate* of the network (i.e. the number of transmitted bits per second), which is recognized as an important indicator of the network bandwidth [36]. Intuitively, a larger bit rate leads to less distortion of decoded data, thereby facilitating a higher decoding accuracy. In most engineering practice, however, only limited bits are allowed to be delivered for each transmission due mainly to limited network bandwidth, and this brings great challenges to the analysis/synthesis of communication-based systems/networks. As a consequence, a great number of researchers have devoted themselves to developing efficient encoding-decoding schemes with high accuracy for NSs under constrained bit rates.

In the context of NSs with bit rate constraints, most existing results have been based on the *periodic* encoding/transmission principle under which the data are encoded/exchanged at *fixed* time instants, see [161,163]. Such

DOI: 10.1201/9781003534853-9

a periodic mechanism, though easy to be implemented, might result in unnecessarily frequent data transmissions (especially in the steady-state situation) which, in turn, give rise to the overconsumption of the scarce bit rates. For resource-saving purposes, a natural idea is to encode/transmit the data only when certain prescribed "event" occurs, and such an even-based mechanism has recently stirred much attention [35,192]. Compared with its periodic counterpart, the event-based encoding-decoding scheme would require much lower data bit rate (for achieving a favorable performance and is therefore particularly suitable in analyzing NSs with constrained bit rate, which constitutes the main focus of this paper).

Recently, there has been a trend in utilizing sign-based encoding approach (SEA) to stabilizing control systems in networked environments, where the main idea of SEA is to encode the original signal to certain sign information occupying one bit. For example, pioneering yet excellent results have been pursued in [89,90] on the SEA-based state-feedback control problem. In [89], a bit-rate condition has been obtained to stabilize a continuous-time scalar linear systems and then the results have been extended to multiple dimensional systems subject to processing delay and network delay in [90]. In this chapter, we take the initiative to deal with another fundamental research problem, namely, state estimation problem, in which the measurement signal needs to be encoded before being transmitted to the estimator. Note that the encoding approach adopted in [89,90] is highly dependent on the system dynamics, and is therefore inapplicable to the state estimation problem addressed here as the dynamics evolution of the system measurement is not readily available.

Based on the above discussion, we are motivated to develop an appropriate event-based encoding-decoding approach for solving the state estimation problem under bit rate constraints. The difficulties we are going to face in pursuit of our goal are foreseen as follows: 1) it is unclear as how to design an effective encoding-decoding scheme for the measurement outputs whose dynamical evolution is often unavailable; 2) it is nontrivial to derive a lower bound of the required bit rate that ensures the convergence of the decoding error; and 3) it is difficult to establish quantitative relationship between the system parameters and the encoding/estimation performance.

9.1 Problem Formulation

9.1.1 System Model

Consider the following linear continuous-time system:

$$\dot{x}(t) = Ax(t) + Bw(t) \tag{9.1}$$

where $x(t) \in \mathbb{R}^n$ is the state vector and $w(t) \in \mathbb{R}^p$ is the bounded disturbance input satisfying $\|w(t)\| \leq \bar{w}$ with $\bar{w}$ being a known positive scalar. A and B are known matrices with appropriate dimensions.

The measurement model is given as follows:

$$y(t) = Cx(t) + Dv(t) \tag{9.2}$$

where $y(t) \in \mathbb{R}^m$ is the measurement output and $v(t) \in \mathbb{R}^q$ is the measurement noise satisfying $\|v(t)\| \leq \bar{v}$ with $\bar{v}$ being a given positive scalar. Moreover, $v(t)$ is differentiable and its derivative satisfies $\|\dot{v}(t)\| \leq \bar{\bar{v}}$ with $\bar{\bar{v}} > 0$. C and D are known matrices with compatible dimensions and the pair (A, C) is assumed to be observable.

9.1.2 Event-Based Encoder-Decoder Design

In practical applications of NSs, it is often the case that those measurement signals of large bits fail to be directly transmitted to the receiver side due to the bit-rate restriction of the communication network. To this end, an event-based data encoding scheme is introduced to improve the efficiency of the data transmission by i) compressing the data through encoding and ii) reducing transmission frequency through event-based scheme.

The encoding scheme is characterized by the vector-valued function $\theta(\cdot)$: $\mathbb{R}^m \rightarrow \mathbb{R}^m$ with the following form:

$$\theta(\hbar) \triangleq \mathrm{vec}\{\theta_i(\hbar_i)\} \tag{9.3}$$

where $\theta_i(\hbar_i) \triangleq \mathrm{sign}(\hbar_i) \in \mathbb{R}$ denotes the scalar-valued codeword with fewer bits, $\mathrm{sign}(\cdot)$ is the signum function, and $\hbar_i$ is the ith component of the data $\hbar$ to be encoded.

Denoting the time sequence $\{s_k\}_{k\in\mathbb{Z}}$ as the transmission instants of the codeword sequence, s_k can be determined by the following rule:

$$s_{k+1} = \inf\{t | t > s_k, \|y(t) - y_a(t)\| = \sigma(t)\} \tag{9.4}$$

where $y_a(t)$ is the estimate of $y(t)$ whose dynamics shall be given later and $\sigma(t)$ is the exponential decay function defined as

$$\sigma(t) \triangleq \sigma_0 e^{-\lambda t} + \varrho.$$

Here, σ_0 and ϱ are given positive constants, and λ is a parameter to be determined later. The encoding instant and the transmission instant of the codeword are assumed to be the same one, that is, there is no time delay caused by the data encoding process.

In this paper, the differential encoding scheme is adopted, which means that the encoding operation is applied to the difference between the system measurement $y(t)$ and its estimate $y_a(t)$. Letting $\tilde{y}_a(t) \triangleq y(t) - y_a(t)$, the encoded signal is expressed as follows:

$$\theta_i\big(\tilde{y}_{ai}(s_k)\big) = \text{sign}\big(\tilde{y}_{ai}(s_k)\big) \tag{9.5}$$

where $\tilde{y}_{ai}(t)$ $(i = 1, 2, \ldots, m)$ is the i-th component of $\tilde{y}_a(t)$. In what follows, we slightly abuse the notation $\theta_i(s_k)$ to denote the encoded signal $\theta_i\big(\tilde{y}_{ai}(s_k)\big)$. Furthermore, we assume that $\theta_i(s_k)$ will not be sent to the decoder side if $\tilde{y}_{ai}(s_k) = 0$.

In the sequel, we aim to characterize the dynamics of the measurement output $y(t)$ by using the following assumption and lemmas.

Assumption 9.1: *The measurement matrix C is of full-row rank, namely, rank$\{C\} = m$.*

Lemma 9.1 (Singular value decomposition): *For a given matrix $C \in \mathbb{R}^{m \times n}$ of full-row rank, there exist two orthogonal matrices $U \in \mathbb{R}^{m \times m}$ and $V \in \mathbb{R}^{n \times n}$ such that*

$$C = UCV^T \tag{9.6}$$

where $\sigma_i > 0$ $(i = 1, 2, \ldots, m)$ are the singular values of C and $C \triangleq [C_0 \quad 0_{m \times (n-m)}]$ with $C_0 \triangleq diag\{\sigma_1, \sigma_2, \ldots, \sigma_m\}$.

Lemma 9.2: *For a given matrix $A \in \mathbb{R}^{n \times n}$ and a given matrix $C \in \mathbb{R}^{m \times n}$ of full-row rank, there exists a matrix*

$$Z \triangleq UC_0 V_1^T A V_1 (UC_0)^{-1} \in \mathbb{R}^{m \times m}$$

such that $CA = ZC$ if A satisfies $V_1^T A V_2 = 0$ where $V_1 \in \mathbb{R}^{n \times m}$ and $V_2 \in \mathbb{R}^{n \times (n-m)}$ are defined through $V \triangleq [V_1 \quad V_2]$, with orthogonal matrices $U \in \mathbb{R}^{m \times m}$ and $V \in \mathbb{R}^{n \times n}$ defined in Lemma 9.1.

Proof: *The proof of this lemma is easily accessible from Lemma 9.1 and is therefore omitted.*

Lemma 9.3: *The dynamic evolution of the measurement output $y(t)$ is governed by the differential equation*

$$\dot{y}(t) = Zy(t) + G\xi(t) \tag{9.7}$$

where

$$G \triangleq \begin{bmatrix} CB & -ZD & D \end{bmatrix}$$

$$\xi(t) \triangleq \begin{bmatrix} w^T(t) & v^T(t) & \dot{v}^T(t) \end{bmatrix}^T.$$

Proof: *It is seen from (9.1)–(9.2) that*

$$\dot{y}(t) = CAx(t) + CBw(t) + D\dot{v}(t).$$

Then, in terms of Lemma 9.2, one further calculates that

$$\dot{y}(t) = Z(y(t) - Dv(t)) + CBw(t) + D\dot{v}(t)$$
$$= Zy(t) + G\xi(t), \tag{9.8}$$

which ends the proof.

Remark 9.1: *Inspired by [89,90], in this paper, we make the first attempt to deal with the state estimation problem under the sign-information-based encoding-decoding scheme. Note that, since the measurement matrix C is generally non-invertible, there is a substantial difficulty in characterizing the dynamic evolution of y(t) by means of a differential equation. To overcome such an intrinsic difficulty, the singular value decomposition technique is applied to the measurement matrix C, which plays a vital role in deriving the dynamic evolution of y(t) under certain assumptions. In addition, the establishment of equation (9.8) facilitates, to a great extent, the subsequent encoder/decoder analysis and synthesis.*

In order to capture the dynamics of the measurement output $y(t)$ at the encoder side, we construct the following auxiliary system which is regarded as a "copy" of the decoder function

$$\begin{cases} \dot{y}_a(t) = Zy_a(t), & t \in [s_k, s_{k+1}) \\[2mm] y_a(s_k^+) = y_a(s_k) + \dfrac{1}{m}\theta(s_k)\sigma(s_k) \\[2mm] y_a(0) = 0, & t = 0. \end{cases} \tag{9.9}$$

Here, $\theta(s_k) \triangleq \begin{bmatrix} \theta_1(s_k) & \theta_2(s_k) & \cdots & \theta_m(s_k) \end{bmatrix}^T$, $\sigma(s_k) \triangleq \sigma_0 e^{-\lambda s_k} + \varrho$ and s_k^+ stands for the right limit of the triggering instant s_k, namely, $s_k^+ \triangleq \lim_{\epsilon > 0, \epsilon \to 0}\{s_k + \epsilon\}$. Similar to the results in [81–83], the auxiliary system (9.9) is a kind of hybrid systems where s_k can be viewed as the impulsive sequence determined by the event function (9.4).

According to Lemma 9.2 and (9.9), the dynamics of $\tilde{y}_a(t)$ is derived as

$$\begin{cases} \dot{\tilde{y}}_a(t) = Z\tilde{y}_a(t) + G\xi(t), & t \in [s_k, s_{k+1}) \\[2mm] \tilde{y}_a(s_k^+) = \tilde{y}_a(s_k) - \dfrac{1}{m}\theta(s_k)\sigma(s_k) \\[2mm] \tilde{y}_a(0) = y_0, & t = 0 \end{cases} \qquad (9.10)$$

where y_0 is the known initial value of the measurement output.

Following the dynamics $y_a(t)$ in (9.9), the decoder dynamics of the measurement output $y(t)$ is governed by

$$\begin{cases} \dot{y}_d(t) = Zy_d(t), & t \in [s_k, s_{k+1}) \\[2mm] y_d(s_k^+) = y_d(s_k) + \dfrac{1}{m}\theta(s_k)\sigma(s_k) \\[2mm] y_d(0) = 0, & t = 0 \end{cases} \qquad (9.11)$$

where $y_d(t)$ is the decoded value of $y(t)$.

Denoting $\tilde{y}_d(t) \triangleq y(t) - y_d(t)$, the decoding error dynamics is derived as follows:

$$\begin{cases} \dot{\tilde{y}}_d(t) = Z\tilde{y}_d(t) + G\xi(t), & t \in [s_k, s_{k+1}) \\[2mm] \tilde{y}_d(s_k^+) = \tilde{y}_d(s_k) - \dfrac{1}{m}\theta(s_k)\sigma(s_k) \\[2mm] \tilde{y}_d(0) = y_0, & t = 0. \end{cases} \qquad (9.12)$$

It is not difficult to verify from (9.9)–(9.12) that $y_a(t) \equiv y_d(t)$ and $\tilde{y}_a(t) \equiv \tilde{y}_d(t)$.

9.1.3 State Estimator Structure

After obtaining the decoded measurement $\tilde{y}_d(t)$ generated by (9.11), we construct the following Luenberger-type state estimator to observe the states of system (9.1):

$$\dot{\hat{x}}(t) = A\hat{x}(t) + K(y_d(t) - C\hat{x}(t)) \qquad (9.13)$$

where $\hat{x}(t) \in \mathbb{R}^n$ is an estimate of $x(t)$.

Denoting the estimation error as $\tilde{x}(t) \triangleq x(t) - \hat{x}(t)$, together with (9.1)–(9.2) and the definition of $\tilde{y}_d(t)$, we have

$$\dot{\tilde{x}}(t) = \bar{A}\tilde{x}(t) + \bar{B}\vartheta(t) \qquad (9.14)$$

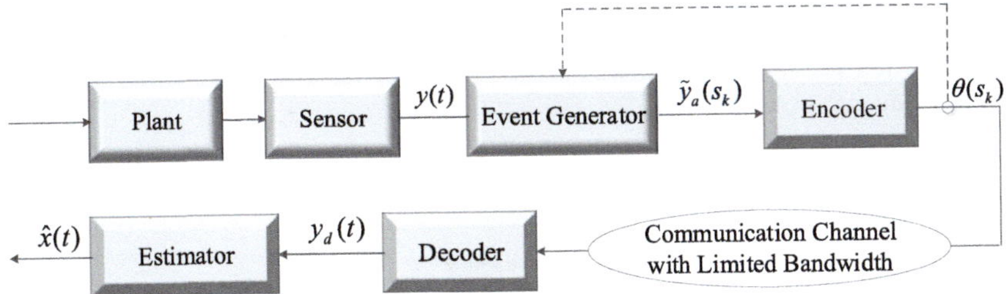

FIGURE 9.1

Schematic of state estimation problem under constrained bit rate.

where

$$\bar{A} \triangleq A - KC$$
$$\bar{B} \triangleq [B \quad -KD \quad K]$$
$$\vartheta(t) \triangleq [w^T(t) \quad v^T(t) \quad \tilde{y}_d^T(t)]^T.$$

In this paper, we investigate the state estimation problem under constrained bit rate, where the schematic structure is shown in Figure 9.1. Our main objective is to develop a state estimator under an event-based encoding-decoding scheme for the continuous-time system (9.1)–(9.2) such that the estimation error is exponentially bounded subject to the system noise $w(t)$, the measurement noise $v(t)$ and the decoding error $\tilde{y}_d(t)$.

9.2 Main Results

In this section, a positive lower bound is first derived for the event-triggered condition to exclude the Zeno phenomenon. Then, a sufficient condition of the bit rate is established to ensure that the decoding error $\tilde{y}_d(t)$ is ultimately bounded. Subsequently, a necessary condition of the bit rate is derived to provide a lower bound of the bit rate required to ensure the convergence of the decoding error. Furthermore, by utilizing the decoded measurement $y_d(t)$, a state estimator is constructed with guaranteed exponential boundedness of the estimation error.

9.2.1 Elimination of the Zeno Phenomenon

For event-triggered mechanism implemented in continuous time setting, the notorious Zeno phenomenon could damage the data sampling/sensing devices and induce the network congestion, and such a phenomenon should be strictly avoided. In what follows, we shall exclude the Zeno phenomenon from the proposed event-triggered mechanism by analyzing the dynamics of $\tilde{y}_a(t)$.

Lemma 9.4: *Under the event-triggered mechanism (9.4), if*

$$\zeta_1 + \frac{\bar{\xi}\|G\|}{\varrho} < 1,$$

then the triggering sequence $\{s_k\}_{k\in\mathbb{Z}}$ satisfies

$$s_{k+1} - s_k \geq \delta$$

where

$$\delta \triangleq \frac{-\ln\left(\zeta_1 + \frac{\bar{\xi}\|G\|}{\varrho}\right)}{\|Z\| + \bar{\lambda}}, \quad \bar{\xi} \triangleq \bar{w} + \bar{v} + \bar{\nu}$$

$$\bar{\lambda} \triangleq \max\{\lambda, 1\}, \quad \zeta_1 \triangleq \sqrt{1 - \frac{1}{m}}.$$

Proof: *Let us first focus our attention on the dynamics evolution of $\tilde{y}_a(t)$ over the time interval $[s_k, s_{k+1})$. Based on (9.10), $\tilde{y}_a(t)$ is calculated as*

$$\begin{aligned}
\tilde{y}_a(t) &= e^{Z(t-s_k)}\tilde{y}_a(s_k^+) + \int_{s_k}^{t} e^{Z(t-s)}G\xi(s)ds \\
&= e^{Z(t-s_k)}\left(\tilde{y}_a(s_k) - \frac{1}{m}\theta(s_k)\sigma(s_k)\right) \\
&\quad + \int_{s_k}^{t} e^{Z(t-s)}G\xi(s)ds.
\end{aligned} \tag{9.15}$$

Setting $\sigma_i(s_k) \triangleq |y_i(s_k) - y_{ai}(s_k)|$ $(i = 1, 2, \ldots, m)$ and $\Sigma(s_k) \triangleq \text{diag}_m\left\{\frac{\sigma_i(s_k)}{\sigma(s_k)}\right\}$, together with the definition of $\theta(s_k)$, we rewrite $\tilde{y}_a(s_k)$ as

$$\tilde{y}_a(s_k) = \Sigma(s_k)\theta(s_k)\sigma(s_k),$$

which results in

$$\begin{aligned}
\tilde{y}_a(t) &= e^{Z(t-s_k)}\left(\Sigma(s_k) - \frac{1}{m}I_m\right)\theta(s_k)\sigma(s_k) \\
&\quad + \int_{s_k}^{t} e^{Z(t-s)}G\xi(s)ds.
\end{aligned} \tag{9.16}$$

Subsequently, it is inferred from (9.16) that

$$\|\tilde{y}_a(t)\| \leq \left\| e^{Z(t-s_k)} \left(\Sigma(s_k) - \frac{1}{m} I_m \right) \theta(s_k) \sigma(s_k) \right\|$$

$$+ \left\| \int_0^{t-s_k} e^{Zs} G\xi(t-s)ds \right\|$$

$$\leq e^{\|Z\|(t-s_k)} \left\| \left(\Sigma(s_k) - \frac{1}{m} I_m \right) \theta(s_k) \right\| \sigma(s_k)$$

$$+ \left\| \int_0^{t-s_k} e^{Zs} G\xi(t-s)ds \right\|. \tag{9.17}$$

By applying some algebraic manipulations, the second term in the right-hand side of (9.16) can be further reduced to

$$\left\| \int_0^{t-s_k} e^{Zs} G\xi(t-s)ds \right\| = \left[\left(\int_0^{t-s_k} e^{Zs} G\xi(t-s)ds \right)^T \left(\int_0^{t-s_k} e^{Zs} G\xi(t-s)\, ds \right) \right]^{\frac{1}{2}}$$

$$\leq \left[(t-s_k) \int_0^{t-s_k} \left(e^{Zs} G\xi(t-s) \right)^T \left(e^{Zs} G\xi(t-s) \right) ds \right]^{\frac{1}{2}}$$

$$= \left[(t-s_k) \int_0^{t-s_k} \left\| e^{Zs} G\xi(t-s) \right\|^2 ds \right]^{\frac{1}{2}}$$

$$\leq (t-s_k)^{\frac{1}{2}} \left[(t-s_k) \left\| e^{Z(t-s_k)} \right\|^2 \|G\|^2 \right.$$

$$\left. \times \max_{s\in[0,\,t-s_k]} \left\{ \|\xi(t-s)\|^2 \right\} \right]^{\frac{1}{2}}$$

$$\leq (t-s_k)\bar{\xi} \|G\| \left\| e^{Z(t-s_k)} \right\|$$

$$\leq (t-s_k)\bar{\xi} \|G\| e^{\|Z\|(t-s_k)} \tag{9.18}$$

where the first inequality is based on the well-known Jensen inequality. Recalling that $\|w(t)\| \leq \bar{w}$, $\|v(t)\| \leq \bar{v}$, $\|\dot{v}(t)\| \leq \bar{\dot{v}}$, $\xi(t) = \begin{bmatrix} w^T(t) & v^T(t) & \dot{v}^T(t) \end{bmatrix}^T$ *and* $\bar{\xi} \triangleq \sqrt{\bar{w}^2 + \bar{v}^2 + \bar{\dot{v}}^2}$, *the third inequality follows from the fact* $\left[\max_{s\in[0,\,t-s_k]} \left\{ \|\xi(t-s)\|^2 \right\} \right]^{\frac{1}{2}} \leq \sqrt{\bar{w}^2 + \bar{v}^2 + \bar{\dot{v}}^2} = \bar{\xi}$. *Then, it is readily calculated that*

$$\left\| \left(\Sigma(s_k) - \frac{1}{m} I_m \right) \theta(s_k) \right\|$$

$$= \sqrt{ \sum_{i=1}^{m} \frac{\theta_i^2(s_k)}{\sigma^2(s_k)} \left(\sigma_i^2(s_k) - \frac{2\sigma_i(s_k)\sigma(s_k)}{m} + \frac{\sigma^2(s_k)}{m^2} \right) } \tag{9.19}$$

$$= \sqrt{ \sum_{i=1}^{m} \frac{\theta_i^2(s_k)\sigma_i^2(s_k)}{\sigma^2(s_k)} - 2 \sum_{i=1}^{m} \frac{\theta_i^2(s_k)\sigma_i(s_k)}{m\sigma(s_k)} + \sum_{i=1}^{m} \frac{\theta_i^2(s_k)}{m^2} }.$$

Furthermore, based on the facts $\sigma_i(s_k) > 0$ $(i = 1, 2, \ldots, m)$, $\sum_{i=1}^{m} \sigma_i^2(s_k) = \sigma^2(s_k)$ *and* $\theta_i^2(s_k) = 1$, *it is easy to verify that*

$$\sum_{i=1}^{m} \frac{\theta_i^2(s_k)\sigma_i^2(s_k)}{\sigma^2(s_k)} = \frac{\sum_{i=1}^{m} \sigma_i^2(s_k)}{\sigma^2(s_k)} = \frac{\sigma^2(s_k)}{\sigma^2(s_k)} = 1$$

and

$$1 \leq \frac{\sigma_1(s_k) + \sigma_2(s_k) + \cdots + \sigma_m(s_k)}{\sigma(s_k)} \leq \sqrt{m}. \tag{9.20}$$

Thus, the term $\sum_{i=1}^{m} \frac{\theta_i^2(s_k)\sigma_i(s_k)}{m\sigma(s_k)}$ *is bounded by*

$$\frac{1}{m} \leq \sum_{i=1}^{m} \frac{\theta_i^2(s_k)\sigma_i(s_k)}{m\sigma(s_k)} \leq \frac{\sqrt{m}}{m},$$

which yields

$$-\frac{2\sqrt{m}}{m} \leq -2 \sum_{i=1}^{m} \frac{\theta_i^2(s_k)\sigma_i(s_k)}{m\sigma(s_k)} \leq -\frac{2}{m}.$$

Consequently, in accordance with the above analysis, we draw the conclusion that

$$\zeta_0 \leq \left\| \left(\Sigma(s_k) - \frac{1}{m} I_m \right) \theta(s_k) \right\| \leq \zeta_1 \tag{9.21}$$

where

$$\zeta_0 \triangleq \sqrt{1 - \frac{2\sqrt{m}}{m} + \frac{1}{m^2}}, \quad \zeta_1 \triangleq \sqrt{1 - \frac{1}{m}}.$$

In addition, one derives that

$$\sigma(s_k) = \sigma_0 e^{-\lambda s_k} + \varrho$$
$$= \sigma_0 e^{\lambda(t-s_k)} e^{-\lambda t} + \varrho$$
$$\leq e^{\lambda(t-s_k)} (\sigma_0 e^{-\lambda t} + \varrho)$$
$$= e^{\lambda(t-s_k)} \sigma(t).$$

Also, it follows from (9.15)–(9.21) that

$$\|\tilde{y}_a(t)\| \leq \zeta_1 e^{(\|Z\|+\lambda)(t-s_k)} \sigma(t) + (t-s_k)\bar{\xi}\|G\|e^{\|Z\|(t-s_k)}$$
$$\triangleq \varphi(t). \tag{9.22}$$

It is obvious that $\varphi(t)$ is a monotonically increasing continuous function with regard to t over the interval $[s_k, s_{k+1})$ with $\varphi(s_k) < \sigma(s_k)$. As such, there must exist a time instant $t^* \in [s_k, s_{k+1})$ such that $\varphi(t^*) = \sigma(t^*)$, that is,

$$\sigma(t^*) = \zeta_1 e^{(\|Z\|+\lambda)(t^*-s_k)} \sigma(t^*)$$
$$+ (t^* - s_k)\bar{\xi}\|G\|e^{\|Z\|(t^*-s_k)}. \tag{9.23}$$

Noticing the form of $\sigma(t)$, it is straightforward to see that

$$\left(1 - \zeta_1 e^{(\|Z\|+\lambda)(t^*-s_k)}\right) \varrho \leq (t^* - s_k)\bar{\xi}\|G\|e^{\|Z\|(t^*-s_k)}. \tag{9.24}$$

Moreover, it can be concluded from the fact

$$t - s_k < e^{t-s_k} \ (\forall t \geq s_k)$$

and the form of $\bar{\lambda}$ that

$$1 - \zeta_1 e^{(\|Z\|+\bar{\lambda})(t^*-s_k)} < \frac{\bar{\xi}\|G\|}{\varrho} e^{(\|Z\|+\bar{\lambda})(t^*-s_k)}, \tag{9.25}$$

which means that

$$e^{(\|Z\|+\bar{\lambda})(t^*-s_k)} > \frac{1}{\left(\zeta_1 + \frac{\bar{\xi}\|G\|}{\varrho}\right)}. \tag{9.26}$$

Taking the logarithmic operation on both sides of (9.26) yields

$$t^* - s_k > \frac{\ln \frac{1}{\left(\zeta_1 + \frac{\bar{\xi}\|G\|}{\varrho}\right)}}{\|Z\| + \bar{\lambda}} \triangleq \delta, \tag{9.27}$$

which indicates that

$$s_{k+1} - s_k > \delta > 0.$$

Therefore, the Zeno behavior is eliminated and the proof of this lemma is ended.

9.2.2 The Boundedness Analysis of the Decoding Error

As we know, the decoding error has played a vital role in the estimation performance. In other words, with the increase of the decoding error, the estimation performance would accordingly get worse. Therefore, in this subsection, we are interested in establishing some criteria to analyze the boundedness of the decoding error under the bit-rate-constraint. First, a sufficient condition on the network bit rate is given under which the boundedness of the measurement decoding error $\tilde{y}_d(t)$ is ensured.

Bearing in mind that

$$\tilde{y}_d(t) \equiv \tilde{y}_a(t), \quad \forall t \geq 0,$$

we know that the boundedness of $\tilde{y}_d(t)$ is equivalent to that of $\tilde{y}_a(t)$. Then, we can infer from (9.22)–(9.23) that

$$\|\tilde{y}_a(t)\| < \sigma(t), \quad t \in [s_k, t^*].$$

In addition, it is observed from the event-triggered mechanism (9.4) that

$$\|\tilde{y}_a(t)\| \leq \sigma(t), \quad \forall t \in [t^*, s_{k+1}].$$

Hence, we have

$$\|\tilde{y}_d(t)\| = \|\tilde{y}_a(t)\| \leq \sigma(t), \quad \forall t \geq 0,$$

which implies that the decoding error $\tilde{y}_d(t)$ is ultimately bounded.

Theorem 9.1: *Under the event-based encoding-decoding scheme (9.4), (9.5) and (9.11), the decoding error $\tilde{y}_d(t)$ is ultimately bounded if the bit rate $\mathcal{R}$ of the communication channel satisfies*

$$\mathcal{R} \geq \mathcal{R}_0 \triangleq \frac{m}{\delta}.$$

Proof: *It follows from the boundedness analysis of $\tilde{y}_d(t)$ that the decoding error is ultimately bounded under the event-based encoding-decoding scheme (9.4), (9.5) and (9.11).*

Next, the encoding rule (9.5) indicates that, at each triggering instant s_k, only one bit data is transmitted for each measurement component $y_i(t)$ ($i=1$, $2,\ldots,m$), and thus m-bits data are required for $y(t)$ during each data transmission.

Now, based on Lemma 9.4, the required bit rate is $\frac{m}{\delta}$ for the sake of guaranteeing the boundedness of the decoding error. The proof is complete.

Remark 9.2: *A sufficient condition is proposed in Theorem 9.1 to guarantee that the decoding error $\tilde{y}_d(t)$ is ultimately bounded under the event-based encoding-decoding scheme (9.4). It is observed from (9.27) that the derived bit-rate condition is dependent on i) the system parameters (i.e. A which is reflected in Z), ii) the dimension of the system measurement m, iii) the bound of the disturbances (i.e. $\bar{\xi} = \bar{w} + \bar{v} + \bar{\nu}$), and iv) the threshold parameter ϱ. To be specific, a large bound of system disturbances leads to a large bit rate requirement. The increase of the dimension of the system measurement leads to the increase of bit rate. Furthermore, a larger threshold contributes to less bit rate while giving rise to a larger upper bound of the decoding error. These observations reveal the relationship between the bit rate and certain system performance, which is consistent with the practice.*

Having established a sufficient condition to ensure the boundedness of the decoding error, it is time for us to derive a corresponding necessary condition regarding the bit rate of the communication network. To this end, we shall obtain an upper bound of two arbitrarily adjacent triggering instants. Before proceeding further, the following useful lemma is provided.

Lemma 9.5: *Under the event-based encoding-decoding scheme (9.4), (9.5) and (9.11), the decoding error is divergent if the bit rate of the communication channel is less than $\mathcal{R}_1$, namely, $\mathcal{R} < \mathcal{R}_1$, where $\mathcal{R}_1 \triangleq \frac{m}{\delta_1}$, $\delta_1 \triangleq \dfrac{\ln(\zeta_0 \chi) - \ln \frac{\bar{u\xi}\|G\|}{\varrho}}{\frac{1}{d}+\|Z\|-\underline{a}}$, $\underline{a}$ is the smallest real part of the eigenvalues of Z, d is a positive scalar and other parameters are defined in Lemma 9.4.*

Proof: *The main idea of deriving the bit rate $\mathcal{R}_1$ is to seek an upper bound of the time interval between two successive triggering instants. To proceed further, we first give the following assertion.*

Assertion: For the matrix $Z \in \mathbb{R}^{m\times m}$, there exists a positive real constant $\underline{\varsigma} > 0$ and a positive-definite matrix $P \in \mathbb{R}^{m\times m}$ such that the following inequality holds

$$(e^{Z(t-s_k)})^T e^{Z(t-s_k)} \geq \chi e^{2\underline{a}(t-s_k)} I, \ \forall t \geq s_k \tag{9.28}$$

where $\chi \triangleq \lambda_{\min}\{P^T P\}\lambda_{\min}\{P^{-T}P^{-1}\}\underline{\varsigma}$ and $\underline{a}$ is the smallest real part of the eigenvalues of Z.

In what follows, we aim to prove the matrix inequality (9.28). To begin with, let us recall some properties of the matrix transformation.

Letting the matrix Z has r ($r \leq m$) different eigenvalues $\lambda_1, \lambda_2, \ldots, \lambda_r$, the characteristic polynomial of Z is written as

$$\det |\lambda I - Z| = (\lambda - \lambda_1)^{m_1} (\lambda - \lambda_2)^{m_2} \cdots (\lambda - \lambda_r)^{m_r}$$

where m_i ($i = 1, 2, \ldots, r$) is the algebraic multiplicity of λ_i and satisfies $\sum_{i=1}^{r} m_i = m$.

Let the geometric multiplicity of λ_i be l_i and denote $l \triangleq \sum_{i=1}^{r} l_i$. It is known that

$$l_i \leq m_i, \ \forall i \in \{1, 2, \ldots, r\},$$

which indicates $l \leq m$.

In terms of the theory of linear algebra, there exists a non-singular matrix $P \in \mathbb{R}^{m \times m}$ such that

$$P^{-1} Z P = J$$

where $J \triangleq \text{diag}\{J_1, J_2, \ldots, J_l\}$ with J_i ($i = 1, 2, \ldots, l$) being the Jordan blocks.

By properly arranging the order of the Jordan blocks J_i, we know that the first l_1 Jordan blocks $J_1, J_2, \ldots, J_{l_1}$ are with the eigenvalue λ_1, the eigenvalue of the Jordan blocks from J_{l_1+1} to $J_{l_1+l_2}$ is λ_2 and, by analogy, the eigenvalue of the Jordan blocks from $J_{\sum_{i=1}^{r-1} l_i + 1}$ to J_l is λ_r.

Since λ_i ($i = 1, 2, \ldots, r$) is probably a complex number, we denote $\lambda_i \triangleq a_i + b_i i$, where a_i and b_i are, respectively, the real and imaginary parts of λ_i.

Without loss of generality, suppose that the real parts of λ_i ($i = 1, \ldots, r$) satisfy $a_1 \leq a_2 \leq \cdots \leq a_r$. Then, for any Jordan block $J_i(\lambda_s)$ ($i = 1, 2, \ldots, l$), it can be rewritten as

$$J_i(\lambda_s) \triangleq \begin{bmatrix} \lambda_s & 1 & 0 & \cdots & 0 \\ 0 & \lambda_s & 1 & \cdots & 0 \\ \vdots & \vdots & \vdots & \vdots & 1 \\ 0 & 0 & 0 & \cdots & \lambda_s \end{bmatrix}, \quad s \in \{1, 2, \ldots, r\}.$$

In the sequel, for notation simplicity, J_i is slightly abused to denote $J_i(\lambda_s)$.

It follows from the property of the matrix exponent that

$$P^{-1} e^{Z(t-s_k)} P = e^{(t-s_k)P^{-1}ZP}$$

$$= e^{(t-s_k)J}$$

$$= \text{diag}\{e^{(t-s_k)J_1}, e^{(t-s_k)J_2}, \ldots, e^{(t-s_k)J_l}\},$$

which indicates

$$e^{Z(t-s_k)} = P e^{(t-s_k)P^{-1}ZP} P^{-1}$$
$$= P \, \mathrm{diag}\{e^{(t-s_k)J_1}, e^{(t-s_k)J_2}, \ldots, e^{(t-s_k)J_l}\} P^{-1}.$$

As such, one further obtains

$$e^{Z^T(t-s_k)} e^{Z(t-s_k)} = P^{-T} \mathrm{diag}\{e^{(t-s_k)J_1^T}, e^{(t-s_k)J_2^T}, \ldots, e^{(t-s_k)J_l^T}\} P^T P$$
$$\times \mathrm{diag}\{e^{(t-s_k)J_1}, e^{(t-s_k)J_2}, \ldots, e^{(t-s_k)J_l}\} P^{-1}$$
$$\geq \lambda_{\min}\{P^T P\} P^{-T} \mathrm{diag}\{e^{(t-s_k)J_1^T}, e^{(t-s_k)J_2^T}, \ldots, e^{(t-s_k)J_l^T}\}$$
$$\times \mathrm{diag}\{e^{(t-s_k)J_1}, e^{(t-s_k)J_2}, \ldots, e^{(t-s_k)J_l}\} P^{-1}. \qquad (9.29)$$

For J_i ($i \in \{1, 2, \ldots, l\}$), noticing that

$$J_i = \lambda_s I + J_i(0)$$

and

$$(\lambda_s I) J_i(0) = J_i(0) \lambda_s I,$$

e^{J_i} *can be rewritten as*

$$e^{J_i} = e^{\lambda_s I + J_i(0)} = e^{\lambda_s I} e^{J_i(0)}.$$

Hence, one obtains

$$e^{(t-s_k)J_i^T} e^{(t-s_k)J_i} = e^{(t-s_k)\bar{\lambda}_s I} e^{(t-s_k)J_i^T(0)} e^{(t-s_k)J_i(0)} e^{(t-s_k)\lambda_s I}$$

with $\bar{\lambda}_s$ being the conjugate of λ_s.

Since $e^{(t-s_k)J_i^T(0)} e^{(t-s_k)J_i(0)}$ is semi-positive definite and $e^{(t-s_k)J_i(0)}$ is invertible, one has that $e^{(t-s_k)J_i^T(0)} e^{(t-s_k)J_i(0)}$ is a positive definite matrix. Thus, there exists a positive scalar ς_i such that

$$e^{(t-s_k)J_i^T(0)} e^{(t-s_k)J_i(0)} \geq \varsigma_i I.$$

In this sense, one derives that

$$e^{(t-s_k)J_i^T} e^{(t-s_k)J_i} \geq \varsigma_i e^{2(t-s_k)a_s} I$$

where a_s is the real part of λ_s and $\bar{\lambda}_s$.

Finally, it is inferred from (9.29) and the above discussion that

$$e^{Z^T(t-s_k)}e^{Z(t-s_k)} \geq \lambda_{\min}\{P^TP\}\lambda_{\min}\{P^{-T}P^{-1}\}\underline{\varsigma}e^{2\underline{a}(t-s_k)}I \tag{9.30}$$

*where $\underline{\varsigma} \triangleq \min\{\varsigma_i\}$ and $\underline{a} \triangleq \min\{a_i\}$ ($i = 1, 2, \ldots, r$). The proof of the **Assertion** is now complete.*

Then, we focus our attention on analysis of the dynamics of the term $\|\tilde{y}_a(t)\|$. It is inferred from (9.15) that

$$\begin{aligned}
\|\tilde{y}_a(t)\| &= \left\| e^{Z(t-s_k)}\left(\Sigma(s_k) - \frac{1}{m}I_m\right)\theta(s_k)\sigma(s_k) \right. \\
&\quad \left. + \int_{s_k}^{t} e^{Z(t-s)}G\xi(s)ds \right\| \\
&\geq \left\| e^{Z(t-s_k)}\left(\Sigma(s_k) - \frac{1}{m}I_m\right)\theta(s_k)\sigma(s_k) \right\| \\
&\quad - \left\| \int_{s_k}^{t} e^{Z(t-s)}G\xi(s)ds \right\|.
\end{aligned} \tag{9.31}$$

With the help of (9.21), (9.28) and the Jensen inequality, one further calculates

$$\begin{aligned}
\|y(t) - y_a(t)\| &\geq \varsigma_0\chi e^{\underline{a}(t-s_k)}\sigma(s_k) - \left\| \int_{s_k}^{t} e^{Z(t-s)}G\xi(s)ds \right\| \\
&\geq \varsigma_0\chi e^{\underline{a}(t-s_k)}\sigma(s_k) - (t-s_k)\bar{\xi}\|G\|e^{\|Z\|(t-s_k)} \tag{9.32} \\
&\geq \left(\varsigma_0\chi e^{\underline{a}(t-s_k)} - \frac{d\bar{\xi}\|G\|}{\varrho}e^{(\frac{1}{d}+\|Z\|)(t-s_k)} \right)\sigma(t)
\end{aligned}$$

where the first inequality comes from (9.18) and the second inequality is based on the facts

$$\begin{cases} \sigma(s_k) \geq \sigma(t) \\ (t-s_k) \leq de^{\frac{t-s_k}{d}} \\ \varrho \leq \sigma(t) \end{cases} \tag{9.33}$$

with $d > 0$.

Letting

$$\left(\varsigma_0\chi e^{\underline{a}(t-s_k)} - \frac{d\bar{\xi}\|G\|}{\varrho}e^{(\frac{1}{d}+\|Z\|)(t-s_k)} \right) = 1,$$

we can obtain

$$t - s_k \leq \frac{\ln(\zeta_0 \chi) - \ln \frac{d\bar{\xi}\|G\|}{\varrho}}{\frac{1}{d} + \|Z\| - \underline{a}} \triangleq \delta_1$$

as long as

$$\zeta_0 \chi - \frac{d\bar{\xi}\|G\|}{\varrho} > 0.$$

Thus, by recalling the event-triggered mechanism (9.4), we draw a conclusion that the time interval of two successive triggering instants is bounded by δ_1, namely,

$$s_{k+1} - s_k \leq \delta_1.$$

As such, we conclude that the decoding error is divergent if

$$\mathcal{R} < \mathcal{R}_1 \triangleq \frac{m}{\delta_1}.$$

The proof of this Lemma is complete.

Remark 9.3: *Except for the sufficient condition on the required bit rate $\mathcal{R}_0$ obtained from Theorem 9.1, we further derive a bit rate condition $\mathcal{R}_1$ in Lemma 9.5, whose theoretical and practical significance lies in that 1) a necessary condition is established for the bit-rate-constrained state estimation problem, namely, below which the estimation error would be divergent; 2) a lower bound of the bandwidth requirement on the allocation of the network resource is given; and 3) a new performance index $\Delta\mathcal{R} \triangleq \mathcal{R}_0 - \mathcal{R}_1$ is provided to evaluate the conservativeness of the sufficient condition in Theorem 9.1. To be more specific, the smaller the $\Delta\mathcal{R}$, the less the conservativeness of the bit rate condition in Theorem 9.1. For the extreme case $\Delta\mathcal{R} = 0$, the sufficient condition in Theorem 1 becomes the necessary and sufficient condition. Therefore, in terms of the bit rates $\mathcal{R}_0$ and $\mathcal{R}_1$, a more accurate relationship between the estimation performance and the communication resource is established, which is of both theoretical significance and practical importance.*

9.2.3 The Design of the State Estimator

For now, we have dealt with the boundedness analysis of the decoding error dynamics $\tilde{y}_d(t)$ and the corresponding data rate condition. Next, we are going to tackle the state estimation issue by utilizing the decoded measurement $y_d(t)$.

In the following theorem, a sufficient condition is established to ensure the ultimate boundedness of the estimation error.

Theorem 9.2: *Let the bit rate of the communication network satisfy $\mathcal{R} \geq \mathcal{R}_0$. The error dynamics (9.14) of the state estimation is ultimately bounded under the event-based encoding-decoding scheme (9.4), (9.5) and (9.11) if there exist positive scalars α, γ, a positive definite matrix Q and a matrix $\bar{K}$ satisfying*

$$\Xi \triangleq \begin{bmatrix} \bar{A}^T Q + Q\bar{A} + \alpha Q & \bar{Q} \\ * & -\gamma I \end{bmatrix} < 0 \tag{9.34}$$

where

$$\bar{Q} \triangleq [QB \quad -\bar{K}D \quad \bar{K}].$$

Moreover, the estimator gain matrix is designed as

$$K = Q^{-1}\bar{K}$$

and the ultimate upper bound of the estimation error $\|\tilde{x}(t)\|$ is calculated as

$$\sqrt{\frac{\gamma \bar{\vartheta}^T \bar{\vartheta}}{\alpha \lambda_{\min}\{Q\}}}$$

with

$$\bar{\vartheta} \triangleq [\bar{w} \quad \bar{v} \quad \sigma_0 + \varrho]^T.$$

Proof: *It is known from subsection 9.2.2 that the boundedness of the decoding error $\tilde{y}_d(t)$ can be guaranteed under the bit rate condition $\mathcal{R} \geq \mathcal{R}_0$, which implies that the vector $\vartheta(t)$ is bounded.*

Construct the Lyapunov function as follows:

$$V(t) = \tilde{x}^T(t)Q\tilde{x}(t).$$

Then, the derivative of $V(t)$ along the trajectory of the estimation error dynamics (9.14) can be calculated as

$$\dot{V}(t) = 2\tilde{x}^T(t)Q\dot{\tilde{x}}(t)$$

$$= 2\tilde{x}^T(t)Q(\bar{A}\tilde{x}(t) + \bar{B}\vartheta(t)) \tag{9.35}$$

$$= \tilde{x}^T(t)(\bar{A}^T Q + Q\bar{A})\tilde{x}(t) + 2\tilde{x}^T(t)Q\bar{B}\vartheta(t),$$

which, together with $\Xi < 0$ *in (9.34), results in*

$$\dot{V}(t) + \alpha V(t) < \gamma \vartheta^T(t)\vartheta(t).$$

As such, we have

$$V(t) \leq e^{-\alpha t}V(0) + \gamma \int_0^t e^{-\alpha(t-s)}\vartheta^T(s)\vartheta(s)ds$$

$$\leq e^{-\alpha t}V(0) + \frac{\gamma}{\alpha}(1 - e^{-\alpha t})\bar{\vartheta}^T\bar{\vartheta},$$

which indicates that the estimation error satisfies

$$\|\tilde{x}(t)\| \leq \sqrt{\frac{\lambda_{\max}\{Q\}}{\lambda_{\min}\{Q\}}} e^{-\frac{\alpha t}{2}} \|\tilde{x}(0)\| + \sqrt{\frac{\gamma(1 - e^{-\alpha t})\bar{\vartheta}^T\bar{\vartheta}}{\alpha\lambda_{\min}\{Q\}}},$$

and is therefore exponentially bounded. Accordingly, the ultimate upper bound can be expressed by

$$\sqrt{\frac{\gamma\bar{\vartheta}^T\bar{\vartheta}}{\alpha\lambda_{\min}\{Q\}}}$$

and the proof of this theorem is complete.

Remark 9.4: *It follows from subsection 9.2.2 that the decoding error* $\tilde{y}_d(t)$ *asymptotically tends to* ϱ. *Based on this fact, we have*

$$\lim_{t\to\infty} \int_0^t e^{-\alpha(t-s)}\vartheta^T(s)\vartheta(s)ds \leq \lim_{t\to\infty} \frac{\vartheta^T(t)\vartheta(t)}{\alpha} \leq \frac{(\bar{w}^2 + \bar{v}^2 + \varrho)}{\alpha},$$

which means that, when the time t tends to infinity, a tighter upper bound of the estimation error is obtained as

$$\sqrt{\frac{\gamma(\bar{w}^2 + \bar{v}^2 + \varrho)}{\alpha\lambda_{\min}\{Q\}}}.$$

Remark 9.5: *Till now, a systematic study has been conducted on the encoding-decoding-based state estimation problem for a class of continuous-time systems under a bit-rate constrained network. The emphasis is placed on the research with respect to the joint design issue of encoder/decoder/estimator. To be more specific, in Lemma 9.4, the unexpected Zeno phenomenon has been excluded in order to ensure the normal operation of the designed event-based encoder. In Theorem 9.1, a sufficient condition has been first established to guarantee the boundedness of the decoding error. Then, in Lemma 9.5, a necessary condition has been given to* *illustrate that the decoding error would be divergent when the bit rate is below a given condition. Furthermore, Theorem 9.2 has provided a sufficient condition to ensure the exponential boundedness of the estimation errors on the premise of the convergence of the decoding errors.*

Remark 9.6: *Comparing with existing literature, the main novelties of our current results lie in that: 1) the research problem addressed is new that represents the first attempt to investigate the event-based encoding-decoding state estimation problem under bit rate constraints; 2) the dynamical evolution of the measurement output is proposed with the help of the singular value decomposition technique that offers a better performance for the developed sign-information-based encoding-decoding scheme; and 3) the derived bit rate condition is viewed as a bridge to establish the tight connection among the network resources and the decoding errors as well as the estimation performance.*

9.3 An Illustrative Example

Two illustrative simulation examples are presented in this section to show the effectiveness of the proposed state estimation method.

Example 9.1: *Consider a linear continuous-time system with the following parameters:*

$$A = \begin{bmatrix} 0.5 & 0.25 \\ 0.8 & 0.1333 \end{bmatrix}, \quad B = \begin{bmatrix} 0.1 \\ 0.1 \end{bmatrix},$$
$$C = [0.8 \quad 0.3], \quad D = 0.1.$$

The eigenvalues of A are 0.8 and -0.1667, which indicate that system (9.1) is unstable and it is therefore non-trivial to verify the developed estimation scheme on this system.

By applying the singular value decomposition technique to matrix C, U, C_0 and V, one can obtain

$$U = 1, \quad C_0 = 0.8544$$

$$V = \begin{bmatrix} 0.9363 & -0.3511 \\ 0.3511 & 0.9363 \end{bmatrix}$$

$$V_1 = [0.9363 \quad 0.3511]^T$$

$$V_2 = [-0.3511 \quad -0.9363]^T,$$

which ensures that $V_1^T A V_2 = 0$. *Moreover, the matrix Z is computed as* $Z = 0.8$. *The disturbances* $w(t)$ *and* $v(t)$ *are chosen as* $w(t) = v(t) = 0.1 \sin(t)$, *which imply that* $\bar{w} = \bar{v} = \bar{v} = 0.1$.

By solving the matrix inequality (9.34) *using Matlab toolbox, the estimator gain matrix K is obtained as*

$$K = [2.3726 \quad 2.1875].$$

According to Theorem 9.1, the required bit rate to achieve the ultimate boundedness of the estimation error is $\mathcal{R} = 0.6034$ *bits/second.*

The simulation results are shown in Figures 9.2-9.6. The measurement output $y(t)$ *and its decoded value are illustrated in Figure 9.2, and the corresponding decoding error is depicted in Figure 9.3. The system states* $x_i(t)$ $(i = 1, 2)$ *and their estimations are described in Figure 9.4. The estimation errors are shown in Figure 9.5. In Figure 9.6, the dynamics of the triggering function* $\|y(t) - y_a(t)\|$ *and the corresponding threshold are given, from which it is observed that the x-axis of the intersection point of the two curves is the triggering instants. All the simulation results have confirmed the validity of the proposed encoding/estimation method.*

Next, in order to make our theoretical results more convincing, a practical example (i.e. a maneuvering target tracking system) is introduced.

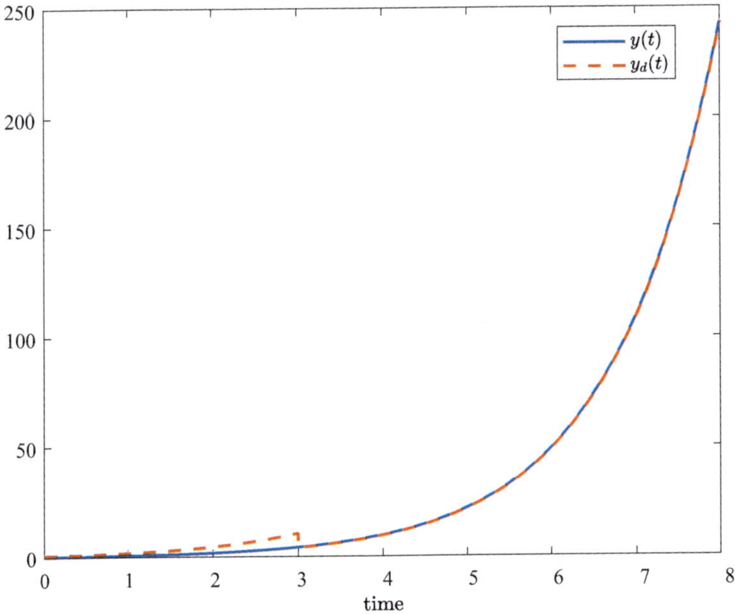

FIGURE 9.2

Measurement $y(t)$ and its decoded value $y_d(t)$.

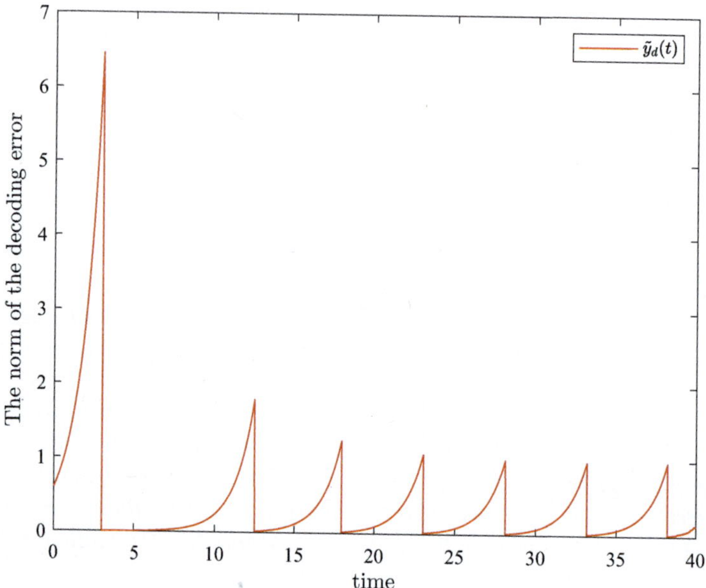

FIGURE 9.3

The norm of the decoding error $\tilde{y}_d(t)$.

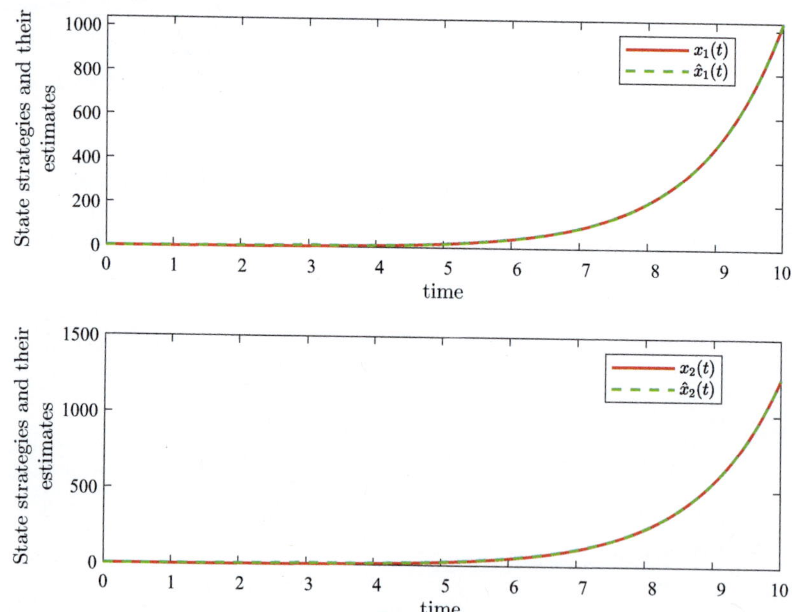

FIGURE 9.4

The state trajectories $x_i(t)$ and their estimates $\hat{x}_i(t)$ ($i = 1, 2$).

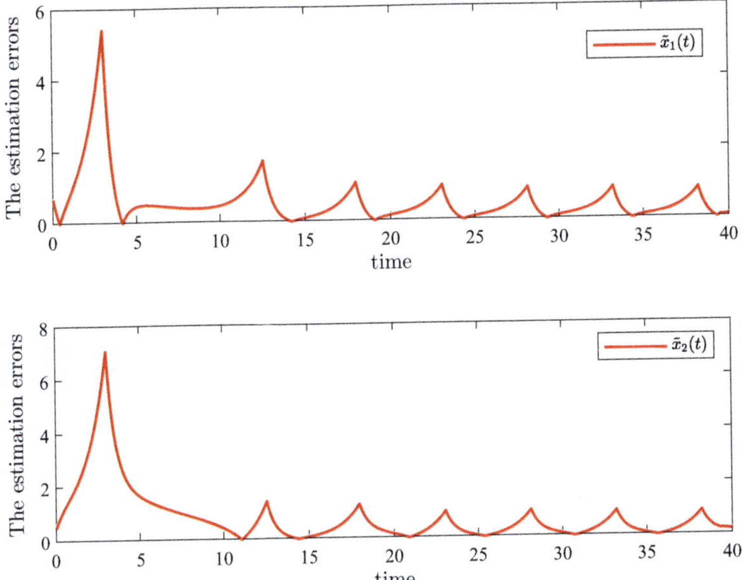

FIGURE 9.5

The estimation errors $\tilde{x}_i(t)$ $(i = 1, 2)$.

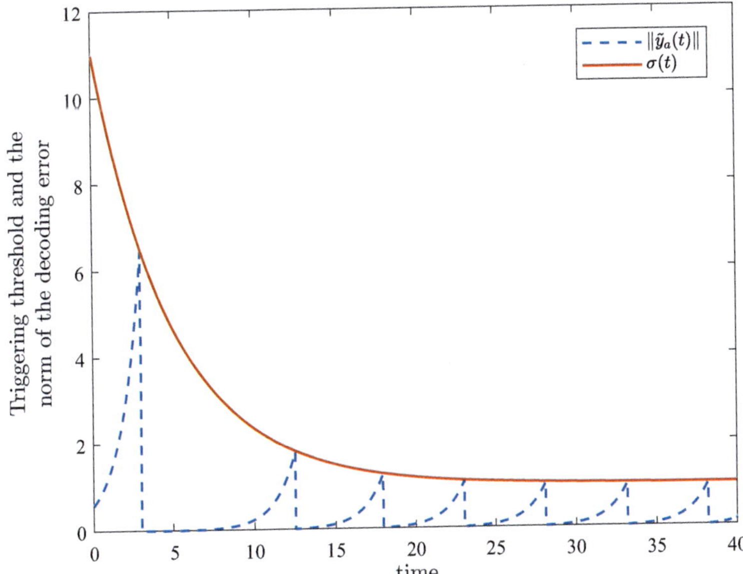

FIGURE 9.6

Triggering threshold $\sigma(t)$ and the norm of the decoding error $\|\tilde{y}_d(t)\|$.

Example 9.2: *The target plant is described as follows:*

$$\begin{cases} \dot{x}_1(t) = x_2(t) + 0.1w(t) \\ \dot{x}_2(t) = x_3(t) + 0.1w(t) \\ \dot{x}_3(t) = u(t) \end{cases}$$

where $x_1(t)$, $x_2(t)$ and $x_3(t)$ stand for, respectively, the position, the velocity and the acceleration of the target. In this case, the matrices A and B are

$$A = \begin{bmatrix} 0 & 1 & 0 \\ 0 & 0 & 1 \\ 0 & 0 & 0 \end{bmatrix}, \ B = \begin{bmatrix} 0.1 & 0.1 & 0 \end{bmatrix}.$$

The measurement matrix is given as

$$C = \begin{bmatrix} 0.01 & 0.1 & 0 \\ 0 & 0 & 0.9 \end{bmatrix}.$$

Let $u(t) = 0$ and the other parameters are set the same as the ones in Example 9.1. By solving the linear matrix inequality in Theorem 9.2, the estimator gain matrix is given by

$$K = \begin{bmatrix} 11.0827 & -0.7379 \\ 3.7083 & 1.3152 \\ 0.1407 & 0.1326 \end{bmatrix}.$$

It is calculated that $V_1^T A V_2 = [0 \ \ 0.0099]$ and thus the equality constraint in Lemma 9.2 is considered to be met. The initial conditions are set as $x(0) = [0.2 \ \ 0 \ \ -0.15]^T$, $\hat{x}(0) = [0 \ \ 0 \ \ 0]^T$, $y(0) = Cx(0)$ and $y_a(0) = y_d(0) = [0.5 \ \ 0]^T$. As expected, it is observed from Figures 9.7–9.8 that simulation results well confirm the validity of the proposed estimation scheme. Moreover, in order to examine the impact of the intensity of the disturbances $w(t)$ and $v(t)$ and the triggering threshold ϱ on the bit-rate condition, some comparisons are made as shown in Tables 9.1 and 9.2. It is can be concluded from Tables 9.1 and 9.2 that: 1) the severer the intensities of the disturbances $w(t)$ and $v(t)$ are, the larger bit rate of the communication channel is required and 2) the larger the triggering threshold ϱ is, the smaller the required bit rate would be. These comparison results are consistent with the theoretical analysis.

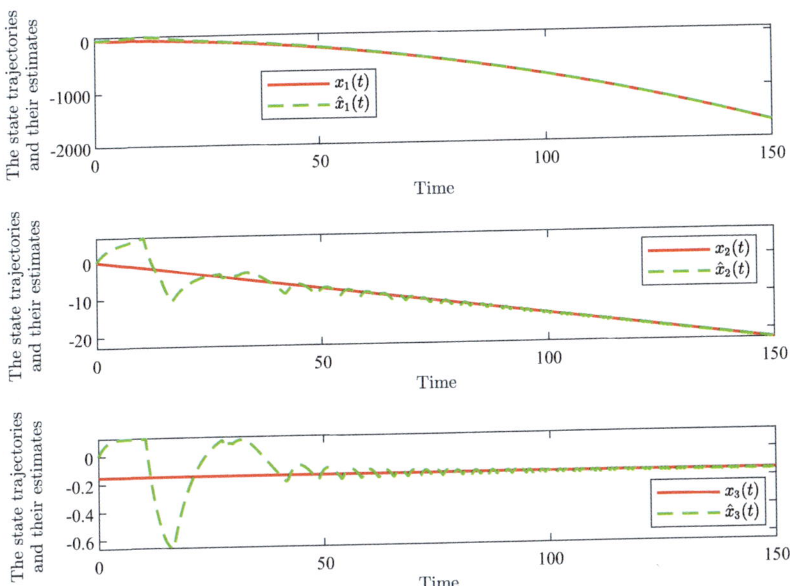

FIGURE 9.7

The state trajectories $x_i(t)$ $(i = 1, 2, 3)$ and their estimates.

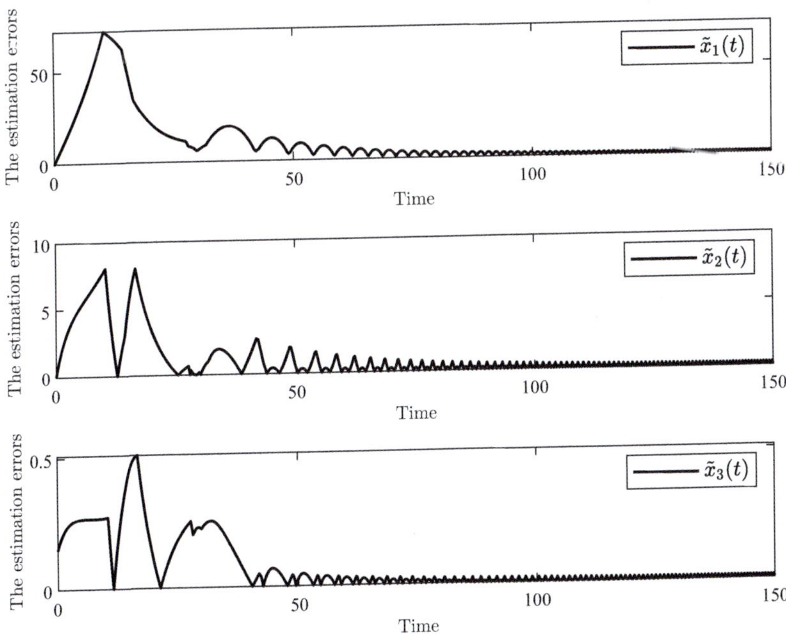

FIGURE 9.8

The estimation errors $\tilde{x}_i(t)$ $(i = 1, 2, 3)$.

TABLE 9.1

Bit Rate $\mathcal{R}_0$ Under Different Threshold Values ϱ

ϱ	0.2	0.4	0.6	0.8	1.0
$\mathcal{R}_0$	27.83	11.16	9.17	8.39	7.98

TABLE 9.2

Bit Rate $\mathcal{R}_0$ Under Different Disturbance Intensities

$\bar{w} = \bar{v}$	0.1	0.15	0.2	0.25	0.3
$\mathcal{R}_0$	7.98	8.84	9.88	11.16	12.77

9.4 Summary

In this paper, the state estimation problem has been investigated for a class of NSs under bit rate constraints. The measurement outputs have been encoded into the binary codeword with 1 bit and then transmitted to the estimator via a digital communication channel. For the resource-saving purpose, an event-triggered coding-decoding mechanism has been employed in the process of data encoding/transmission. A bit rate condition has been established to guarantee the ultimate boundedness of both the decoding error and the estimation error. Furthermore, the desired estimator gain has been acquired in terms of the solution to a matrix inequality. Finally, a numerical simulation has been proposed to demonstrate the validity of the main results.

10

Conclusions and Future Topics

In this book, the control and state estimation problems have been thoroughly investigated for networked systems under a data-rate-constrained network environment. Latest results on data-rate-constrained control and state estimation issues have been first surveyed under different types of encoding-decoding schemes and bandwidth constraints originating from real-world engineering practice. Then, in each chapter, regarding the addressed control and state estimation/filtering with data rate constraint, various encoding-decoding strategies have been developed. Meanwhile, a set of sufficient conditions have been established for designed encoding-decoding strategies to achieve certain pre-specified system performances including minimum variance specification, uniform boundedness index, consensus and synchronization requirements. Subsequently, both numerical and practical illustrative demonstrations have been presented that are capable of verifying the feasibility and efficacy of all developed data-constrained control/state estimation approaches.

This book has established a unified theoretical framework for the synthesis/analysis problems for networked systems with constrained network bandwidth. A great variety of engineering-oriented data rate constraints and corresponding encoding-decoding schemes have been taken into account in a systematical yet effective procedure. It is worth mentioning that the acquired results are still quite limited, and some concluding remarks and prospective topics are highlighted as follows:

(1) *Channel Encoding.* So far, most of the existing results focus their attention on the aspect of source encoding and decoding. In fact, the channel encoding-decoding is also very important for the reliability of the data transmission in the communication network. Therefore, it is quite meaningful to study the co-design issue of the source and channel encoding-decoding in NSs.

(2) *Combining with Other Communication Protocols.* Recently, some effective data scheduling strategies such as the Round-Robin protocol, the Try-Once-Discard protocol, the stochastic protocol, etc., have been successfully applied to the NSs in order to cope with the limited communication resources. For the purpose of making full use of the

DOI: 10.1201/9781003534853-10

network resources, the development of several encoding-decoding-based data scheduling protocols would be a trend for future research.

(3) *Data Encryption Issues.* Since NSs share the common network links for data exchange, they are prone to be attacked by adversaries, and the secure communication in NSs becomes more and more significant. Thereby, how to combine the encoding-decoding schemes with the data encryption techniques in a unified framework is another interesting topic.

(4) *Combining with Self-Triggering Mechanism.* Although there have been some results with respect to the combination of the event-triggering mechanism and the decoding-decoding communication schemes recently, the additional hardware is still needed to detect the specific "event". In this sense, by replacing the detection task according to certain admissible "prediction" [193], the self-triggering mechanism is one preferable candidate to facilitate the encoding-decoding-based control/filtering problems.

Bibliography

[1] F. Amato and M. Ariola, Finite-time control of discrete-time linear systems, *IEEE Transactions on Automatic Control*, vol. 50, no. 5, pp. 724–729, 2005.

[2] S. Azuma and T. Sugie, Dynamic quantization of nonlinear control systems, *IEEE Transactions on Automatic Control*, vol. 57, no. 4, pp. 875–888, 2012.

[3] H. Bai, M. Zhang, A. Wang, M. Liu, and Y. Zhao, Multiple description video coding using inter- and intra-description correlation at macro block level, *IEICE Transactions on Information and Systems*, vol. E97D, no. 2, pp. 384–387, 2014.

[4] Y. Barshalom, Redundancy and data compression in recursive estimation, *IEEE Transactions on Automatic Control*, vol. AC17, no. 5, pp. 684–689, 1972.

[5] Y. Berrouche and R. E. Bekka, Improved multiple description wavelet based image coding using Hadamard transform, *AEU-International Journal of Electronics and Communications*, vol. 68, no. 10, pp. 976–982, 2014.

[6] R. Caballero-Águila, A. Hermoso-Carazo, J. Linares-Perez, and Z. Wang, A new approach to distributed fusion filtering for networked systems with random parameter matrices and correlated noises, *Information Fusion*, vol. 45, pp. 324–332, 2019.

[7] R. Carli, F. Fagnani, and P. Frasca, Efficient quantized techniques for consensus algorithms, *NeCST Workshop*, pp. 1–8, Nancy, France, 2007.

[8] J. Chen and Q. Ling, Bit-rate conditions for the consensus of quantized multiagent systems with network-induced delays based on event triggering, *IEEE Transactions on Cybernetics*, vol. 51, no. 2, pp. 984–993, 2021.

[9] X. Chen, X. Liao, L. Gao, S. Yang, H. Wang, and H. Li, Event-triggered consensus for multi-agent networks with switching topology under quantized communication, *Neurocomputing*, vol. 230, pp. 294–301, 2017.

[10] T. M. Cheng, Robust output feedback stabilization of nonlinear networked systems via a finite data-rate communication channel, *Journal of Systems Science & Complexity*, vol. 24, no. 1, pp. 1–13, 2011.

[11] D. Ciuonzo, A. Aubry, and V. Carotenuto, Rician MIMO channel- and jamming-aware decision fusion, *IEEE Transactions on Signal Processing*, vol. 65, no. 15, pp. 3866–3880, 2017.

[12] G. Como, F. Fagnani, and S. Zampieri, Anytime reliable transmission of real-valued information through digital noisy channels, *SIAM Journal on Control and Optimization*, vol. 48, no. 6, pp. 3903–3924, 2010.

[13] P. Correia, P. A. Assuncao, and V. Silva, Multiple description of coded video for path diversity streaming adaptation, *IEEE Transactions on Multimedia*, vol. 14, no. 3, pp. 923–935, 2012.

[14] Y. Cui, L. Yu, Y. Liu, W. Zhang, and F.-E. Alsaadi, Dynamic event-based non-fragile state estimation for complex networks via partial nodes information, *Journal of the Franklin Institute-Engineering and Applied Mathematics*, vol. 358, no. 18, pp. 10193–10212, 2021.

[15] C. E. Curry, P. Mirchandani, and C. F. Price, State estimation with coarsely quantized, high data-rate measurements, *IEEE Transactions on Aerospace and Electronic Systems*, vol. 11, no. 4, pp. 613–621, 1975.

[16] D. F. Delchamps, Stabilizing a linear-system with quantized state feedback, *IEEE Transactions on Automatic Control*, vol. 35, no. 8, pp. 916–924, 1990.

[17] S. Dey, A. Chiuso, and L. Schenato, Event-triggered stabilization of linear systems under bounded bit rates, *IEEE Transactions on Automatic Control*, vol. 62, no. 6, pp. 3054–3061, 2017.

[18] D. Ding, Z. Wang, D. W. C. Ho, and G. Wei, Distributed recursive filtering for stochastic systems under uniform quantizations and deception attacks through sensor networks, *Automatica*, vol. 78, pp. 231–240, 2017.

[19] D. Ding, Z. Wang, B. Shen, and H. Dong, Event-triggered distributed $\mathcal{H}_\infty$ state estimation with packet dropouts through sensor networks, *IET Control Theory & Applications*, vol. 9, no. 13, pp. 1948–1955, 2015.

[20] D. Ding, G. Wei, S. Zhang, Y. Liu, and F. E. Alsaadi, On scheduling of deception attacks for discrete-time networked systems equipped with attack detectors, *Neurocomputing*, vol. 219, pp. 99–106, 2017.

[21] H. Dong, X. Bu, N. Hou, Y. Liu, F. E. Alsaadi, and T. Hayat, Event-triggered distributed state estimation for a class of time-varying systems over sensor networks with redundant channels, *Information Fusion*, vol. 36, pp. 243–250, 2017.

[22] H. Dong, Z. Wang, J. Lam, and H. Gao, Distributed filtering in sensor networks with randomly occurring saturations and successive packet dropouts, *International Journal of Robust and Nonlinear Control*, vol. 24, pp. 1743–1759, 2014.

[23] A. A. Elgamal and T. M. Cover, Achievable rates for multiple descriptions, *IEEE Transactions on Information Theory*, vol. 28, no. 6, pp. 851–857, 1982.

[24] J. Fang and Q.-Q. Liu, Quantized feedback control over packet dropout communication channels, *Information Technology and Control*, vol. 41, no. 3, pp. 229–238, 2012.

[25] J. Fang and H. Li, Distributed adaptive quantization for wireless sensor networks: From delta modulation to maximum likelihood, *IEEE Transactions on Signal Processing*, vol. 56, no. 10, pp. 5246–5257, 2008.

[26] A. Farhadi, Feedback channel in linear noiseless dynamic systems controlled over the packet erasure network, *International Journal of Control*, vol. 88, no. 8, pp. 1490–1503, 2015.

[27] A. Farhadi and N. U. Ahmed, Tracking nonlinear noisy dynamic systems over noisy communication channels, *IEEE Transactions on Communications*, vol. 59, no. 4, pp. 955–961, 2011.

[28] A. Farhadi and C. D. Charalambous, Robust coding for a class of sources: Applications in control and reliable communication over limited capacity channels, *Systems & Control Letters*, vol. 57, no. 12, pp. 1005–1012, 2008.

[29] A. Farhadi and C. D. Charalambous, Stability and reliable data reconstruction of uncertain dynamic systems over finite capacity channels, *Automatica*, vol. 46, no. 5, pp. 889–896, 2010.

[30] Y. W. Feng and G. Ge, Optimal coding-decoding for systems controlled via a communication channel, *International Journal of Systems Science*, vol. 44, no. 12, pp. 2190–2198, 2013.

[31] A. L. Fradkov, B. Andrievsky, and M. Ananyevskiy, State estimation and synchronization of pendula systems over digital communication channels, *European Physical Journal-Special Topics*, vol. 223, no. 4, pp. 773–793, 2014.

[32] A. L. Fradkov, B. Andrievsky, and R. J. Evans, Chaotic observer-based synchronization under information constraints, *Physical Review E*, vol. 73, no. 6, Art. No. 066209, 2006.

[33] A. L. Fradkov, B. Andrievsky, and R. J. Evans, Synchronization of passifiable Lurie systems via limited-capacity communication channel, *IEEE Transactions on Circuits And Systems I-Regular Papers*, vol. 56, no. 2, pp. 430–439, 2009.

[34] L. Gao, X. Liao, H. Li, and G. Chen, Event-triggered control for multi-agent network with limited digital communication, *Nonlinear Dynamics*, vol. 82, no. 4, pp. 1659–1669, 2015.

[35] X. Ge, Q.-L. Han, X.-M. Zhang, L. Ding, and F. Yang, Distributed event-triggered estimation over sensor networks: A survey, *IEEE Transactions on Cybernetics*, vol. 50, no. 3, pp. 1306–1320, 2020.

[36] H. Geng, H. Liu, L. Ma, and X. Yi, Multi-sensor filtering fusion meets censored measurements under a constrained network environment: Advances, challenges and prospects, *International Journal of Systems Science*, vol. 52, no. 16, pp. 3410–3436, 2021.

[37] V. K. Goyal, Multiple description coding: Compression meets the network, *IEEE Transactions on Signal Processing Magazine*, vol. 18, no. 5, pp. 74–93, 2001.

[38] S. Hattori, K. Kobayashi, H. Okada, and M. Katayama, On-off error control coding scheme for minimizing tracking error in wireless feedback, *IEEE Transactions on Industrial Informatics*, vol. 11, no. 6, pp. 1411–1421, 2015.

[39] F. Han, G. Wei, D. Ding, and Y. Song, Finite-horizon $\mathcal{H}_\infty$-consensus control for multi-agent systems with random parameters: The local condition case, *Journal of the Franklin Institute-Engineering and Applied Mathematics*, vol. 354, no. 14, pp. 6078–6097, 2017.

[40] F. Han, G. Wei, D. Ding, and Y. Song, Local condition based consensus filtering with stochastic nonlinearities and multiple missing measurements, *IEEE Transactions on Automatic Control*, vol. 62, no. 9, pp. 4784–4790, 2017.

[41] L. He, D. Han, X. Wang, and L. Shi, State estimation over a lossy network using linear temporal coding, in *Proceedings of the 51st IEEE Conference on Decision and Control*, pp. 2002–2007, Maui, HI, USA, 2012.

[42] W. He, G. Chen, Q.-L. Han, and F. Qian, Network-based leader-following consensus of nonlinear multi-agent systems via distributed impulsive control, *Information Sciences*, vol. 380, pp. 145–158, 2017.

[43] X. He, Z. Wang, X. Wang, and D. H. Zhou, Networked strong tracking filtering with multiple packet dropouts: Algorithms and applications, *IEEE Transactions on Industrial Electronics*, vol. 61, no. 3, pp. 1454–1463, 2014.

[44] J. Hespanha, A. Ortega, and L. Vasudevan, Towards the control of linear systems with minimum bit-rate, in *Proceedings of the 15th International Symposium on Mathematical Theory of Networks and Systems*, pp. 1–15, Notre Dame, IN, USA, 2002.

[45] M. Hernandez-Gonzalez, M. Basin, and O. Stepanov, Discrete-time state estimation for stochastic polynomial systems over polynomial observations, *International Journal of General Systems*, vol. 47, no. 5, pp. 512–528, 2018.

[46] N. Hou, H. Dong, Z. Wang, W. Ren, and F. E. Alsaadi, Non-fragile state estimation for discrete Markovian jumping neural networks, *Neurocomputing*, vol. 179, pp. 238–245, 2016.

[47] N. Hou, H. Dong, Z. Wang, W. Ren, and F. E. Alsaadi, $\mathcal{H}_\infty$ state estimation for discrete-time neural networks with distributed delays and randomly occurring uncertainties through fading channels, *Neural Networks*, vol. 89, pp. 61–73, 2017.

[48] C. Hu and H. Jiang, Pinning synchronization for directed networks with node balance via adaptive intermittent control, *Nonlinear Dynamics*, vol. 80, nos. 1–2, pp. 295–307, 2015.

[49] J. Hu, Z. Wang, and H. Gao, Recursive filtering with random parameter matrices, multiple fading measurements and correlated noises, *Automatica*, vol. 49, no. 11, pp. 3440–3448, 2013.

[50] L. Hu, Z. Wang, Q.-L. Han, and X. Liu, State estimation under false data injection attacks: Security analysis and system protection, *Automatica*, vol. 219, pp. 176–183, 2018.

[51] L. Hu, Z. Wang, Q.-L. Han, and X. Liu, Event-based input and state estimation for linear discrete time-varying systems, *International Journal of Control*, vol. 91, no. 1, pp. 101–113, 2018.

[52] J. Hu, Z. Wang, S. Liu, and H. Gao, A variance-constrained approach to recursive state estimation for time-varying complex networks with missing measurements, *Automatica*, vol. 64, pp. 155–162, 2016.

[53] H. Ishii and B. A. Francis, Quadratic stabilization of sampled-data systems with quantization, *Automatica*, vol. 39, no. 10, pp. 1793–1800, 2013.

[54] H. Ishii and S. Hara, A subband coding approach to control under limited data rates and message losses, *Automatica*, vol. 44, no. 4, pp. 1141–1148, 2008.

[55] A. Isidori, L. Marconi, and C. de Persis, Remote tracking via encoded information for nonlinear systems, *Systems & Control Letters*, vol. 55, no. 10, pp. 809–818, 2006.

[56] Y. Isidori, K. Takaba, and D. E. Quevedo, Stability analysis of networked control systems subject to packet-dropouts and finite-level quantization, *Systems & Control Letters*, vol. 60, no. 5, pp. 325–332, 2011.

[57] X.-W. Jiang, B. Hu, Z.-H. Guan, X.-H. Zhang, and L. Yu, Best achievable tracking performance for networked control systems with encoder-decoder, *Information Sciences*, vol. 305, pp. 184–195, 2015.

[58] Z. Jiang and Y. Wang, Input-to-state stability for discrete-time nonlinear systems, *Automatica*, vol. 37, no. 6, pp. 857–869, 2001.

[59] Z. Jin, V. Gupta, and R. M. Murray, State estimation over packet dropping networks using multiple description coding, *Automatica*, vol. 42, no. 9, pp. 1141–1152, 2006.

[60] D. Kapetanovic, S. Chatzinotas, and B. Ottersten, Index assignment for multiple description repair in distributed storage systems, in *Proceedings of the 2014 IEEE International Conference on Communications*, pp. 3896–3901, Sydney, NSW, Australia, Jun. 10–14, 2014.

[61] M. J. Khojasteh, P. Tallapragada, J. Cortes, and M. Franceschetti, The value of timing information in event-triggered control: The scalar case, in *Proceedings of the 54th Annual Allerton Conference on Communication, Control, and Computing*, pp. 1165–1172, Monticello, IL, USA, Sept. 27–30, 2016.

[62] M. J. Khojasteh, P. Tallapragada, J. Cortes, and M. Franceschetti, Time-triggering versus event-triggering control over communication channels, in *Proceedings of the 2017 IEEE 56th Annual Conference on Decision and Control*, Melbourne, VIC, Australia, 2017.

[63] E. Kofman and J. H. Braslavsky, Level crossing sampling in feedback stabilization under data-rate constraints, in *Proceedings of the 45th IEEE Conference on Decision and Control*, pp. 4423–4428, San Diego, CA, USA, 2006.

[64] D. Lehmann and J. Lunze, Event-based control using quantized state information, in *Proceedings of the 2nd IFAC Workshop on Distributed Estimation and Control in Networked Systems*, pp. 1989–1994, Annecy, France, 2010.

[65] A. S. Leong, D. Subhrakanti, and G. N. Nair, Quantized filtering schemes for multi-sensor linear state estimation: Stability and performance under high rate quantization, *IEEE Transactions on Signal Processing*, vol. 61, no. 15, pp. 3852–3865, 2013.

[66] H. Li, G. Chen, T. Huang, Z. Dong, and L. Gao, Event-triggered distributed average consensus over directed digital networks with limited communication bandwidth, *IEEE Transactions on Cybernetics*, vol. 46, no. 12, pp. 3098–3110, 2016.

[67] J. Li, H. Dong, F. Han, N. Hou, and X. Li, Filter design, fault estimation and reliable control for networked time-varying systems: A survey, *Systems Science & Control Engineering: An Open Access Journal*, vol. 5, pp. 331–341, 2017.

[68] J.-Y. Li, Z. Wang, R. Lu, and Y. Xu, Partial-nodes-based state estimation for complex networks with constrained bit rate, *IEEE Transactions on Network Science and Engineering*, vol. 8, pp. 2, pp. 1887–1899, 2021.

[69] T. Li, M. Fu, L. Xie, and J. Zhang, Distributed consensus with limited communication data rate, *IEEE Transactions on Automatic Control*, vol. 56, no. 2, pp. 279–292, 2011.

[70] H. Li, C. Huang, G. Chen, X. Liao, and T. Huang, Distributed consensus optimization in multiagent networks with time-varying directed topologies and quantized communication, *IEEE Transactions on Cybernetics*, vol. 47, no. 8, pp. 2044–2057, 2017.

[71] L. Li, M. Lemmon, and X. Wang, Stabilizing bit-rates in quantized event triggered control systems, in *Proceedings of The 15th ACM International Conference on Hybrid Systems: Computation and Control*, pp. 245–254, Beijing, P. R. China, 2012.

[72] H. Li, S. Liu, Y. C. Soh, and L. Xie, Event-triggered communication and data rate constraint for distributed optimization of multiagent systems, *IEEE Transactions on Systems, Man, and Cybernetics: Systems*, vol. 48, no. 11, pp. 1908–1919, 2018.

[73] L. Li, X. Wang, and M. D. Lemmon, Efficiently attentive event-triggered systems with limited bandwidth, *IEEE Transactions on Automatic Control*, vol. 62, no. 3, pp. 1491–1497, 2017.

[74] Q. Li, B. Shen, Z. Wang, and F. E. Alsaadi, A sampled-data approach to distributed $\mathcal{H}_\infty$ resilient state estimation for a class of nonlinear time-delay systems over sensor networks, *Journal of the Franklin Institute*, vol. 354, pp. 7139–7157, 2017.

[75] W. Li, Y. Jia, and J. Du, Recursive state estimation for complex networks with random coupling strength, *Neurocomputing*, vol. 219, pp. 1–8, 2017.

[76] W. Li, G. Wei, D. Ding, Y. Liu, and F. E. Alsaadi, A new look at boundedness of error covariance of Kalman filtering, *IEEE Transactions on Systems, Man, and Cybernetics: Systems*, vol. 48, no. 2, pp. 309–314, 2018.

[77] W. Li, G. Wei, F. Han, and Y. Liu, Weighted average consensus-based unscented Kalman filtering, *IEEE Transactions on Cybernetics*, vol. 46, no. 2, pp. 558–567, 2016.

[78] T. Li and L. Xie, Distributed consensus over digital networks with limited bandwidth and time-varying topologies, *Automatica*, vol. 47, no. 9, pp. 2006–2015, 2011.

[79] T. Li and L. Xie, Distributed coordination of multi-agent systems with quantized-observer based encoding-decoding, *IEEE Transactions on Automatic Control*, vol. 57, no. 12, pp. 3023–3037, 2012.

[80] X. Li, C. Li, and M. Z. Q. Chen, Stabilisation of non-linear DISS systems with uncertainty via encoded feedback, *IET Control Theory & Applications*, vol. 11, no. 5, pp. 732–739, 2017.

[81] X. Li, Y. Yu, and L. Guo, Stability of time-delay systems with impulsive control involving stabilizing delays, *Automatica*, vol. 124, Art. No. 109336, 2021.

[82] X. Li, D. Peng, and J. Cao, Lyapunov stability for impulsive systems via event-triggered impulsive control, *IEEE Transactions on Automatic Control*, vol. 65, no. 11, pp. 4908–4913, 2020.

[83] X. Li, T. Zhang, and J. Wang, Input-to-state stability of impulsive systems via event-triggered impulsive control, *IEEE Transactions on Cybernetics*, vol. 52, no. 7, pp. 7187–7195, 2022.

[84] Z. Li, Z. Duan, and G. Chen, Consensus of discrete-time linear multi-agent systems with observer-type protocols, *Discrete and Continuous Dynamical Systems-Series B*, vol. 16, no. 2, pp. 489–505, 2011.

[85] D. Liberzon, On stabilization of linear systems with limited information, *IEEE Transactions on Automatic Control*, vol. 48, no. 2, pp. 304–307, 2003.

[86] D. Liberzon, Hybrid feedback stabilization of systems with quantized signals, *Automatica*, vol. 39, no. 3, pp. 1543–1554, 2003.

[87] D. Liberzon and J. P. Hespanha, Stabilization of nonlinear systems with limited information feedback, *IEEE Transactions on Automatic Control*, vol. 50, no. 6, pp. 910–915, 2005.

[88] N. Lin and Q. Ling, Bit-rate conditions for the consensus of quantized multiagent systems based on event triggering, *IEEE Transactions on Cybernetics*, vol. 52, no. 1, pp. 116–127, 2022.

[89] Q. Ling, Bit rate conditions to stabilize a continuous-time scalar linear system based on event triggering, *IEEE Transactions on Automatic Control*, vol. 62, no. 8, pp. 4093–4100, 2017.

[90] Q. Ling, Bit-rate conditions to stabilize a continuous-time linear system with feedback dropouts, *IEEE Transactions on Automatic Control*, vol. 63, no. 7, pp. 2176–2183, 2018.

[91] G. Liu, H. Liu, H. Chen, C. Zhou, and L. Shu, Position-based adaptive quantization for target location estimation in wireless sensor networks using one-bit data, *Wireless Communications & Mobile Computing*, vol. 16, no. 8, pp. 929–941, 2016.

[92] Q. Liu and J. Fan, State estimation for networked control systems using fixed data rates, *International Journal of Systems Science*, vol. 48, no. 9, pp. 1818–1828, 2017.

[93] S. Liu, T. Li, and L. Xie, Distributed consensus for multiagent systems with communication delays and limited data rate, *SIAM Journal on Control and Optimization*, vol. 49, no. 6, pp. 2239–2262, 2011.

[94] S. Liu, T. Li, L. Xie, M. Fu, and J. Zhang, Continuous-time and sampled-data-based average consensus with logarithmic quantizers, *Automatica*, vol. 49, no. 11, pp. 3329–3336, Nov. 2013.

[95] S. Liu, Y. Song, D. Ding, and Y. Liu, Event-triggered dynamic output feedback RMPC for polytopic systems with redundant channels: Input-to-state stability, *Journal of the Franklin Institute-Engineering and Applied Mathematics*, vol. 354, no. 7, pp. 2871–2892, 2017.

[96] S. Liu, Y. Song, G. Wei, and X. Huang, RMPC-based security problem for polytopic uncertain system subject to deception attacks and persistent disturbances, *IET Control Theory & Applications*, vol. 11, no. 10, pp. 1611–1618, 2017.

[97] S. Liu, Z. Wang, Y. Chen, and G. Wei, Protocol-based unscented Kalman filtering in the presence of stochastic uncertainties, *IEEE Transactions on Automatic Control*, vol. 65, no. 3, pp. 1303–1309, 2020.

[98] S. Liu, Z. Wang, B. Shen, and G. Wei, Partial-neurons-based state estimation for delayed neural networks with state-dependent noises under redundant channels, *Information Sciences*, vol. 547, pp. 931–944, 2021.

[99] S. Liu, Z. Wang, G. Wei, and M. Li, Distributed set-membership filtering for multirate systems under the Round-Robin scheduling over sensor networks, *IEEE Transactions on Cybernetics*, vol. 50, no. 5, pp. 1910–1920, 2020.

[100] S. Liu, G. Wei, Y. Song, and Y. Liu, Extended Kalman filtering for stochastic nonlinear systems with randomly occurring cyber attacks, *Neurocomputing*, vol. 207, pp. 708–716, 2016.

[101] W. Liu, Z. Wang, and N. Mingkang, Controlled synchronization for chaotic systems via limited information with data packet dropout, *Automatica*, vol. 49, no. 11, pp. 3329–3336, 2013.

[102] H. Liu, Z. Wang, B. Shen, and X. Liu, Event-triggered $\mathcal{H}_\infty$ state estimation for delayed stochastic memristive neural networks with missing measurements: The discrete time case, *IEEE Transactions on Neural Networks and Learning Systems*, vol. 29, no. 8, pp. 3726–3737, 2018.

[103] H. Liu, Z. Wang, B. Shen, and F. E. Alsaadi, $\mathcal{H}_\infty$ state estimation for discrete-time memristive recurrent neural networks with stochastic time-delays, *International Journal of General Systems*, vol. 45, no. 5, pp. 633–647, 2016.

[104] M. B. Loiola, R. R. Lopes, and J. M. T. Romano, Modified Kalman filters for channel estimation in orthogonal space-time coded systems, *IEEE Transactions on Automatic Control*, vol. 60, no. 1, pp. 533–538, 2012.

[105] J. Lu, D. W. C. Ho, and Z. Wang, Pinning stabilization of linearly coupled stochastic neural networks via minimum number of controllers, *IEEE Transactions on Neural Networks*, vol. 20, no. 10, pp. 1617–1629, 2009.

[106] Y. Luo, Z. Wang, G. Wei, and F. E. Alsaadi, Robust $\mathcal{H}_\infty$ filtering for a class of two-dimensional uncertain fuzzy systems with randomly occurring mixed delays, *IEEE Transactions on Fuzzy Systems*, vol. 25, no. 1, pp. 70–83, 2017.

[107] Y. Luo, Z. Wang, G. Wei, F. E. Alsaadi, and T. Hayat, State estimation for a class of artificial neural networks with stochastically corrupted measurements under Round-Robin protocol, *Neural Networks*, vol. 77, pp. 70–79, 2016.

[108] L. Ma, Z. Wang, Q.-L. Han, and H. K. Lam, Variance-constrained distributed filtering for time-varying systems with multiplicative noises and deception attacks over sensor networks, *IEEE Sensors Journal*, vol. 17, no. 7, pp. 2279–2288, 2017.

[109] L. Ma, Z. Wang, and H. K. Lam, Event-triggered mean-square consensus control for time-varying stochastic multi-agent system with sensor saturations, *IEEE Transactions on Automatic Control*, vol. 62, no. 7, pp. 3524–3531, 2017.

[110] V. Malyavej and A. V. Savkin, Set-valued robust Kalman state estimation via digital communication channels with bit-rate constraints, in *Proceedings of the American Control Conference*, Denver, CO, USA, pp. 1896–1901, 2003.

[111] A. S. Matveev, State estimation via limited capacity noisy communication channels, *Mathematics of Control Signals and Systems*, vol. 20, no. 1, pp. 1–35, 2008.

[112] A. S. Matveev and A. V. Savkin, Shannon zero error capacity in the problems of state estimation and stabilization via noisy communication channels, *International Journal of Control*, vol. 80, no. 2, pp. 241–255, 2007.

[113] S. Movaghati and M. Ardakani, Optimum bit-sensor assignment for distributed estimation in inhomogeneous sensor networks, *IEEE Communications Letters*, vol. 18, no. 4, pp. 668–671, 2014.

[114] G. N. Nair and R. J. Evans, Stabilization with data-rate-limited feedback: Tightest attainable bounds, *Systems & Control Letters*, vol. 41, no. 1, pp. 49–56, 2000.

[115] G. N. Nair and R. J. Evans, Exponential stabilisability of finite-dimensional linear systems with limited data rates, *Automatica*, vol. 39, no. 4, pp. 585–593, 2003.

[116] G. N. Nair and R. J. Evans, Stabilizability of stochastic linear systems with finite feedback data rates, *SIAM Journal on Control and Optimization* vol. 43, no. 2, pp. 413–436, 2004.

[117] G. N. Nair, A nonstochastic information theory for communication and state estimation, *IEEE Transactions on Automatic Control*, vol. 58, no. 6, pp. 1479–1510, 2013.

[118] G. Nagamani, B. Adhira, and G. Soundararajan, Non-fragile extended dissipative state estimation for delayed discrete-time neural networks: Application to quadruple tank process model, *Nonlinear Dynamics*, vol. 104, no. 1, pp. 451–466, 2023.

[119] D. Nešić and D. S. Laila, A note on input-to-state stabilization for nonlinear sampled-data systems, *IEEE Transactions on Automatic Control*, vol. 47, no. 7, pp. 1153–1158, 2002.

[120] J. Ostergaard, D. E. Quevedo, and A. Ahlen, Predictive power control and multiple-description coding for wireless sensor networks, in *Proceedings of the IEEE International Conference on Acoustics, Speech, and Signal Processing*, Taipei, Taiwan, Apr. 19–24, 2009.

[121] J. Ostergaard and D. E. Quevedo, Multiple descriptions for packetized predictive control, *Eurasip Journal on Advances in Signal Processing*, vol. 2016, Art. No. 45, 2016.

[122] C. de Persis and A. Isidori, Stabilizability by state feedback implies stabilizability by encoded state feedback, *Systems & Control Letters*, vol. 53, nos. 3–4, pp. 249–258, 2004.

[123] C. de Persis, n-Bit stabilization of n-dimensional nonlinear systems in feedforward form, *IEEE Transactions on Automatic Control*, vol. 50, no. 3, pp. 299–311, 2005.

[124] C. de Persis, Nonlinear stabilizability via encoded feedback: The case of integral ISS systems, *Automatica*, vol. 42, no. 10, pp. 1813–1816, 2006.

[125] C. de Persis and D. Nešić, Practical encoders for controlling nonlinear systems under communication constraints, *Systems & Control Letters*, vol. 57, no. 8, pp. 654–662, 2008.

[126] C. de Persis, Minimal data rate stabilization of nonlinear systems over networks with large delays, *International Journal of Robust and Nonlinear Control*, vol. 20, no. 10, pp. 1097–1111, 2010.

[127] Z. Qiu, L. Xie, and Y. Hong, Quantized leaderless and leader-following consensus of high-order multi-agent systems with limited data rate, *IEEE Transactions on Automatic Control*, vol. 61, no. 9, pp. 2432–2447, 2016.

[128] F. Qu, E. Tian, and X. Zhao, Chance-constrained H_∞ state estimation for recursive neural networks under deception attacks and energy constraints: The finite-horizon case, *IEEE Transactions on Neural Networks and Learning Systems*, vol. 34, no. 9, pp. 6492–6503. DOI: 10.1109/TNNLS.2021.3137426.

[129] D. E. Quevedo, A. Ahlen, and J. Ostergaard, Energy efficient state estimation with wireless sensors through the use of predictive power control and coding, *IEEE Transactions on Signal Processing*, vol. 58, no. 9, pp. 4811–4823, 2010.

[130] D. E. Quevedo, J. Ostergaard, and A. Ahlen, Power control and coding formulation for state estimation with wireless sensors, *IEEE Transactions on Control Systems Technology*, vol. 22, no. 2, pp. 413–427, 2014.

[131] M. Rabi, C. Ramesh, and K. H. Johansson, Separated design of encoder and controller for networked linear quadratic optimal control, *SIAM Journal on Control and Optimization*, vol. 54, no. 2, pp. 662–689, 2016.

[132] M. M. Rana, L. Li, and S. Su, Distributed state estimation using RSC coded smart grid communications, *IEEE Access*, vol. 3, pp. 1340–1349, 2015.

[133] M. A. Rana, F. Ndiaye, S. A. Chowdhury, and N. Mansoor, Multiple description image transmission for diversity systems over unreliable communication networks, in *Proceedings of 10th International Conference on Computer and Information Technology*, pp. 240–244, Dhanmondi, Bangladesh, Dec. 27–29, 2007.

[134] T. S. Rappaport, *Wireless Communications: Principles and Practice*, Prentice Hall, 2001.

[135] W. Ren and R. W. Beard, Consensus seeking in multiagent systems under dynamically changing interaction topologies, *IEEE Transactions on Automatic Control*, vol. 50, no. 5, pp. 655–661, 2005.

[136] A. Ribeiro and B. Giannakis, Bandwidth-constrained distributed estimation for wireless sensor networks-Part I: Gaussian case, *IEEE Transactions on Signal Processing*, vol. 54, no. 3, pp. 1131–1143, 2006.

[137] A. Ribeiro, B. Giannakis, and S. I. Roumeliotis, SOI-KF: Distributed Kalman filtering with low-cost communications using the sign of innovations, *IEEE Transactions on Signal Processing*, vol. 54, no. 12, pp. 4782–4795, 2006.

[138] A. J. Rojas, J. S. Freudenberg, J. H. Braslavsky, and R. H. Middleton, Optimal signal to noise ratio in feedback over communication channels with memory, in *Proceedings of the 45th IEEE Conference on Decision & Control*, pp. 240–244, San Diego, CA, USA, Dec. 13–15, 2006

[139] A. V. Savkin, Analysis and synthesis of networked control systems: Topological entropy, observability, robustness and optimal control, *Automatica*, vol. 42, no. 1, pp. 51–62, 2006.

[140] A. V. Savkin and T. M. Cheng, Detectability and output feedback stabilizability of nonlinear networked control systems, *IEEE Transactions on Automatic Control*, vol. 52, no. 4, pp. 730–735, 2007.

[141] A. V. Savkin and I. R. Petersen, Set-valued state estimation via a limited capacity communication channel, *IEEE Transactions on Automatic Control*, vol. 48, no. 4, pp. 676–680, 2003.

[142] C. E. Shannon, A mathematical theory of communication, *Bell System Technical Journal*, vol. 27, no. 3, pp. 379–423, 1948.

[143] C. E. Shannon, A mathematical theory of communication, *Bell System Technical Journal*, vol. 27, no. 4, pp. 623–666, 1948.

[144] A. Shirazinia, A. A. Zaidi, L. Bao, and M. Skoglund, Dynamic source-channel coding for estimation and control over binary symmetric channels, *IET Control Theory & Applications*, vol. 9, no. 9, pp. 1444–1454, 2015.

[145] B. Shen, Z. Wang, D. Ding, and H. Shu, $\mathcal{H}_\infty$ state estimation for complex networks with uncertain inner coupling and incomplete measurements, *IEEE Transactions on Neural Networks and Learning Systems*, vol. 24, no. 12, pp. 2027–2037, 2013.

[146] B. Shen, Z. Wang, and X. Liu, Sampled-data synchronization control of dynamical networks with stochastic sampling, *IEEE Transactions on Automatic Control*, vol. 57, no. 10, pp. 2644–2650, 2012.

[147] B. Shen, Z. Wang, and H. Qiao, Event-triggered state estimation for discrete-time multidelayed neural networks with stochastic parameters and incomplete measurements, *IEEE Transactions on Neural Networks and Learning Systems*, vol. 28, no. 5, pp. 1152–1163, 2017.

[148] B. Shen, Z. Wang, D. Wang, and Q. Li, State-saturated recursive filter design for stochastic time-varying nonlinear complex networks under deception attacks, *IEEE Transactions on Neural Networks and Learning Systems*, vol. 31, no. 10, pp. 3788–3800, 2020.

[149] B. Shen, Z. Wang, D. Wang, and H. Liu, Distributed state-saturated recursive filtering over sensor networks under Round-Robin protocol, *IEEE Transactions on Cybernetics*, vol. 50, no. 8, pp. 3605–3615, 2020.

[150] T. Sui, K. You, and M. Fu, Kalman filtering with intermittent observations using measurements coding, in *10th IEEE International Conference on Control and Automation (ICCA' 2013)*, pp. 1127–1132, 2013.

[151] T. Sui, K. You, M. Fu, and D. Marelli, Stability of MMSE state estimators over lossy networks using linear coding, *Automatica*, vol. 51, pp. 167–174, 2015.

[152] J. Suo, N. Li, and Q. Li, Event-triggered H_∞ state estimation for discrete-time delayed switched stochastic neural networks with persistent dwell-time switching regularities and sensor saturations, *Neurocomputing*, vol. 455, pp. 297–307, 2021.

[153] P. Tallapragada and J. Cortes, Event-triggered stabilization of linear systems under bounded bit rates, *IEEE Transactions on Automatic Control*, vol. 61, no. 6, pp. 1575–1589, 2016.

[154] Y. Tang, H. Gao, and J. Kurths, Robust H_∞ self-triggered control of networked systems under packet dropouts, *IEEE Transactions on Cybernetics*, vol. 46, no. 12, pp. 3294–3305, 2016.

[155] S. Tatikonda and S. Mitter, Control-under communication constraints, *IEEE Transactions on Automatic Control*, vol. 49, no. 7, pp. 1056–1068, 2004.

[156] V. A. Vaishampayan, Design of multiple description scalar quantizers, *IEEE Transactions on Information Theory*, vol. 39, no. 3, pp. 8214–834, 1993.

[157] V. A. Vaishampayan and J. C. Batllo, Asymptotic analysis of multiple description quantizers, *IEEE Transactions on Information Theory*, vol. 44, no. 1, pp. 278–284, 1998.

[158] J. Wang and H. Li, Stabilization of a continuous linear system over channel with network-induced delay and communication constraints, *European Journal of Control*, vol. 31, pp. 72–78, 2016.

[159] F. Wang, J. Liang, Z. Wang, and X. Liu, A variance-constrained approach to recursive filtering for nonlinear two-dimensional systems with measurement degradations, *IEEE Transactions on Cybernetics*, vol. 48, no. 6, pp. 1877–1887, 2018.

[160] L. Wang, Z. Wang, Q.-L. Han, and G. Wei, Event-based variance-constrained $\mathcal{H}_\infty$ filtering for stochastic parameter systems over sensor networks with successive missing measurements, *IEEE Transactions on Cybernetics*, vol. 48, no. 3, pp. 1007–1017, 2018.

[161] L. Wang, Z. Wang, Q.-L. Han, and G. Wei, Synchronization control for a class of discrete-time dynamical networks with packet dropouts: A coding-decoding-based approach, *IEEE Transactions on Cybernetics*, vol. 48, no. 8, pp. 2437–2448, 2018.

[162] L. Wang, Z. Wang, T. Huang, and G. Wei, An event-triggered approach to state estimation for a class of complex networks with mixed time delays and nonlinearities, *IEEE Transactions on Cybernetics*, vol. 46, no. 11, pp. 2497–2508, 2016.

[163] L. Wang, Z. Wang, B. Shen, and G. Wei, Recursive filtering with measurement fading: A multiple description coding scheme, *IEEE Transactions on Automatic Control*, vol. 66, no. 11, pp. 5144–5159, 2021.

[164] L. Wang, Z. Wang, G. Wei, and F. E. Alsaadi, Finite-time state estimation for recurrent delayed neural networks with component-based event-triggering protocol, *IEEE Transactions on Neural Networks and Learning Systems*, vol. 29, no. 4, pp. 1046–1057, 2018.

[165] J. Wang and Z. Yan, Coding scheme based on spherical polar coordinate for control over packet erasure channel, *International Journal of Robust and Nonlinear Control*, vol. 51, pp. 1159–1176, 2014.

[166] X. Wang, E. E. Yaz, and J. Long, Robust and resilient state-dependent control of discrete-time nonlinear systems with general performance criteria, *Systems Science & Control Engineering: An Open Access Journal*, vol. 2, pp. 48–54, 2014.

[167] G. Wei, S. Liu, Y. Song, and Y. Liu, Probability-guaranteed set-membership filtering for systems with incomplete measurements, *Automatica*, vol. 60, pp. 12–16, 2015.

[168] G. Wei, S. Liu, L. Wang, and Y. Wang, Event-based distributed set-membership filtering for a class of time-varying non-linear systems over sensor networks with saturation effects, *International Journal of General Systems*, vol. 45, no. 5, pp. 532–547, 2016.

[169] N. Wiener, *Cybernetics: Or Control and Communication in the Animal and the Machine*, Hermann & Cie and MIT Press, 1948.

[170] T. Wen, L. Zou, J. Liang, and C. Roberts, Recursive filtering for communication-based train control systems with packet dropouts, *Neurocomputing*, vol. 275, pp. 948–957, 2018.

[171] W. S. Wong and R. W. Brockett, Systems with finite communication bandwidth constraints-II: Stabilization with limited information feedback, *IEEE Transactions on Automatic Control*, vol. 44, no. 5, pp. 1049–1053, 1999.

[172] J. Wu, H. Li, and X. Chen, Leader-following consensus of nonlinear discrete-time multi-agent systems with limited communication channel capacity, *Journal of the Franklin Institute-Engineering and Applied Mathematics*, vol. 354, no. 10, pp. 4179–4195, 2017.

[173] X. Xiao, L. Zhou, and G. Lu, Detection of singular systems via a limited communication channel with missing measurements capacity, *Information Sciences*, vol. 228, pp. 192–202, 2013.

[174] K. Xiong, C. Wei, and L. Liu, Robust extended Kalman filtering for nonlinear systems with stochastic uncertainties, *IEEE Transactions on Systems Man and Cybernetics Part A-Systems and Humans*, vol. 40, no. 2, pp. 399–405, 2010.

[175] T. Xu, S. Liu, and G. Wei, Event-based finite-time H_∞ synchronization control for switched complex networks with average dwell time, *International Journal of Robust and Nonlinear Control*, vol. 36, no. 4, pp. 3923–3943, 2022.

[176] Y. Xu, C. Zhu, and L. Yu, Multipath routing of multiple description coded images in wireless networks, *Journal of Computer Science and Technology*, vol. 29, no. 4, pp. 576–588, 2014.

[177] K. You, Control over communication networks: A personal perspective, in *Proceedings of the 11th World Congress on Intelligent Control and Automation (WCICA'2014)*, pp. 2002–2007, Shenyang, P. R. China, 2012.

[178] K. You, Recursive algorithms for parameter estimation with adaptive quantizer, *Automatica*, vol. 52, pp. 192–201, 2015.

[179] K. You and L. Xie, Network topology and communication data rate for consensusability of discrete-time multi-agent systems, *IEEE Transactions on Automatic Control*, vol. 56, no. 10, pp. 2262–2275, 2011.

[180] M. Yu, C. Yan, and D. Xie, Event-triggered control for couple-group multi-agent systems with logarithmic quantizers and communication delays, *Asian Journal of Control*, vol. 19, no. 2, pp. 681–691, 2017.

[181] W. Yu, G. Chen, and M. Cao, Some necessary and sufficient conditions for second-order consensus in multi-agent dynamical systems, *Automatica*, vol. 46, no. 6, pp. 1089–1095, 2010.

[182] W. Yu, W. X. Zheng, G. Chen, W. Ren, and J. Cao, Second-order consensus in multi-agent dynamical systems with sampled position data, *Automatica*, vol. 47, no. 7, pp. 1496–1503, 2011.

[183] Y. Yuan, P. Zhang, L. Guo, and H. Yang, Towards quantifying the impact of randomly occurred attacks on a class of networked control systems, *Journal of the Franklin Institute-Engineering and Applied Mathematics*, vol. 354, no. 12, pp. 4966–4988, 2017.

[184] S. Zhang, D. Ding, G. Wei, J. Mao, Y. Liu, and F. E. Alsaadi, Design and analysis of $\mathcal{H}_\infty$ filter for a class of T-S fuzzy system with redundant channels and multiplicative noises, *Neurocomputing*, vol. 260, pp. 257–264, 2017.

[185] S. Zhang, D. Ding, G. Wei, Y. Liu, and F. E. Alsaadi, $\mathcal{H}_\infty$ state estimation for artificial neural networks over redundant channels, *Neurocomputing*, vol. 226, pp. 117–125, 2017.

[186] W. Zhang, S. Vedantam, and U. Mitra, Joint transmission and state estimation: A constrained channel coding approach, *IEEE Transactions on Information Theory*, vol. 57, no. 10, pp. 7084–7095, 2011.

[187] X.-M. Zhang, Q.-L. Han, X. Ge, D. Ding, L. Ding, D. Yue, and C. Peng, Networked control systems: A survey of trends and techniques, *IEEE/CAA Journal of Automatica Sinica*, vol. 7, no. 1, pp. 1–17, 2020.

[188] G. Zherlitsyn and A. S. Matveev, Min-max optimal data encoding and fusion in sensor networks, *Automatica*, vol. 46, no. 9, pp. 1546–1552, 2010.

[189] D. Zhou, X. He, Z. Wang, G. P. Liu, and Y. Ji, Leakage fault diagnosis for an internet-based three-tank system: An experimental study, *IEEE Transactions on Control Systems Technology*, vol. 20, no. 4, pp. 857–870, 2012.

[190] L. Zhou and G. Lu, Detection and stabilization for discrete-time descriptor systems via a limited capacity communication channel, *Automatica*, vol. 45, no. 10, pp. 2272–2277, 2009.

[191] L. Zou, Z. Wang, and H. Gao, Set-membership filtering for time-varying systems with mixed time-delays under Round-Robin and Weighted Try-Once-Discard protocols, *Automatica*, vol. 74, pp. 341–348, 2016.

[192] L. Zou, Z. Wang, J. Hu, Y. Liu, and X. Liu, Communication-protocol-based analysis and synthesis of networked systems: Progress, prospects and challenges, *International Journal of Systems Science*, vol. 52, no. 14, pp. 3013–3034, 2021.

[193] L. Zou, Z. Wang, and D. H. Zhou, Event-based control and filtering of networked systems: A survey, *International Journal of Automation and Computing*, vol. 14, no. 3, pp. 239–253, 2017.

Index

C

channel fading, 27, 110–112, 121–122, 134–135
consensus control, 21–23, 26, 88, 91, 107, 109

D

dynamical multiple description coding, 27, 141, 151, 164
dynamical networks, 67–68, 70, 72, 85, 88

E

encoding-decoding mechanism, 4, 6, 12, 35, 40, 42, 49, 64, 185
event-based encoding-decoding scheme, 174, 176, 186, 196–197, 204

G

gain-scheduled state estimation, 25, 29–30, 35, 42, 45

M

multi-agent systems, 2, 21, 26, 88
multiple description coding, 11, 26–27, 110–111, 113, 116, 120, 140–142, 151, 164

N

network-induced phenomena, 9, 21, 31, 141

O

one-bit symbolic data, 184

P

partial-neurons-based state estimation, 46, 64

R

recursive filtering, 110, 122

S

synchronization control, 26, 28, 65–66, 68, 70, 80–81, 83–84, 86–87

Z

Zeno behavior, 27, 171, 173, 184, 196